T0138453

Science Periodicals in
Nineteenth-Century Britain

Science Periodicals in Nineteenth-Century Britain

CONSTRUCTING SCIENTIFIC COMMUNITIES

Edited by
Gowan Dawson, Bernard Lightman,
Sally Shuttleworth, and Jonathan R. Topham

The University of Chicago Press CHICAGO & LONDON

The University of Chicago Press, Chicago 60637
The University of Chicago Press, Ltd., London
© 2020 by The University of Chicago
All rights reserved. No part of this book may be used or reproduced in any manner
whatsoever without written permission, except in the case of brief quotations in critical
articles and reviews. For more information, contact the University of Chicago Press,
1427 E. 60th St., Chicago, IL 60637.
Published 2020
Printed in the United States of America

29 28 27 26 25 24 23 22 21 20   1 2 3 4 5

ISBN-13: 978-0-226-67651-7 (cloth)
ISBN-13: 978-0-226-68346-1 (e-book)
DOI: https://doi.org/10.7208/chicago/9780226683461.001.0001

Library of Congress Cataloging-in-Publication Data

Names: Dawson, Gowan, editor, author. | Lightman, Bernard V., 1950– editor, author. |
Shuttleworth, Sally, 1952– editor, author. | Topham, Jonathan R., editor, author.
Title: Science periodicals in nineteenth-century Britain : constructing scientific
communities / edited by Gowan Dawson, Bernard Lightman, Sally Shuttleworth and
Jonathan R. Topham. Other titles: Science periodicals in 19th century Britain
Description: Chicago : University of Chicago Press, 2020. | Includes bibliographical
references and index.
Identifiers: LCCN 2019035334 | ISBN 9780226676517 (cloth) | ISBN 9780226683461 (ebook)
Subjects: LCSH: Science—Great Britain—Periodicals—History—19th century. | Medicine—
Great Britain—Periodicals—History—19th century. | Science journalism—Great Britain—
History—19th century.
Classification: LCC PN5124.S35 S353 2020 | DDC 505—dc23
LC record available at https://lccn.loc.gov/2019035334

♾ This paper meets the requirements of ANSI/NISO Z39.48-1992 (Permanence of Paper).

For Gillian Beer

CONTENTS

Constructing Scientific Communities

Gowan Dawson and Jonathan R. Topham

In March 1828 the Scottish landscape gardener and author John Claudius Loudon outlined the vision for his new *Magazine of Natural History*. First, with individuals the world over being occupied in "discovering new objects, or in explaining the nature of those already known," the periodical would assist active "students of nature" in keeping up "their state of knowledge with the progress of science." Secondly, it would "extend a taste for this . . . knowledge among general readers and observers, and especially among gardeners, farmers, and young persons resident in the country" by

> subjecting every part of the science to discussion, in a language in which all technicalities are explained as they occur; by inviting every reader to communicate every circumstance, even the most trivial, respecting the native habits and habitations of plants, the localities of minerals and strata, and peculiar or striking states of the atmosphere; by encouraging all who are desirous of information to propose questions, to state their doubts, the kind of information they desire, or their particular opinion, on any part of the subject.[1]

Loudon's eager hope was that his periodical would not only become a tool to make new scientific observers, but would also draw those observers together with others who were already proficient into an enlarged community of interconnected practitioners. Observations that might be thought "trivial" would be rendered "truly valuable when viewed in reference to general conclusions." Thus might "persons wholly unacquainted with Natural History as

FIG. 0.1. "Six o'Clock p.m.: The Newspaper Window at the General Post-Office," from George Augustus Sala, *Twice Round the Clock; or, The Hours of the Day and Night in London* (London: Houlston and Wright, [1859]), 233. Reproduced by permission of the Victorian Studies Centre, University of Leicester.

a science" learn to become valuable scientific observers. "In this way," Loudon concluded, "we hope to call forth a new and numerous class of naturalists."

Loudon's editorial vision was characteristic of many of the scientific and medical periodicals that, as part of a vast proliferation of print more generally, increasingly deluged the reading public in nineteenth-century Britain, threatening at times to overwhelm the postal system by which they were often distributed (fig. 0.1). Whether motivated by the desire to maintain a sense of connectedness and common purpose among members of a learned society or professional body, by the intention to recruit and discipline new observers, or by the commercial imperative to maximize a market, periodicals played a key role in defining and developing communities of scientific practitioners and the new participants whom Loudon sought to encourage. While, of course, these ambitions were often not fully realized, as our fifth chapter shows was the case for Loudon's *Magazine of Natural History*, they nevertheless acted as powerful determinants of the role and form of scientific periodicals. This volume sets itself to examine these distinctive qualities in relation to the wholesale transformations in the scientific enterprise that occurred during the nineteenth century. In the course of doing so, it sheds an instructive sidelight on the current lively debate concerning the purpose, practices, and price of scientific journals in the twenty-first century.

For much of the last century, scientific periodicals have been thought of primarily as vehicles of certified knowledge. Underpinned by the process of peer review, scientific papers have been seen as the embodiment of scientific discovery, and as the basis of scientific authority and reputation. So pervasive have such perspectives been that historians have sometimes struggled to see how relatively recently these functions have developed, and to grasp how different scientific periodicals were in the past. In recent years, a concerted effort has been made to eschew this presentism and "denaturalize the scientific paper as the dominant genre of scientific life."[2] Yet, as scholars have striven to show that the characteristics of scientific periodicals taken for granted in the early twenty-first century only emerged by degrees over the course of the nineteenth and early twentieth centuries, a further question has arisen. If the scientific periodicals of the nineteenth century are not to be thought of in essentialist terms as the favored genre of legitimated knowledge, what is the historian to make of them? What, in other words, was their precise function and role in the making of nineteenth-century science?

This book focuses on one key aspect of the larger history that is thus unlocked. Rather than thinking of scientific periodicals primarily as the favored locus for authorized scientific discovery, our approach here is to examine the role that they played in the development and functioning of more or less coherent collectives within the sciences. One of the enduring problematics in interpreting science in nineteenth-century Britain has been to find a means of characterizing the diversity, complexity, and often disharmony of its unfamiliar preprofessional communities of practice. An early suggestion in the 1970s that scholars might use prosopography (a kind of collective biography designed to establish the common attributes of members of a group) to investigate membership of these communities fell on deaf ears. A generation of scholarship, however, has since taken us far beyond the unhelpfully anachronistic dichotomy of professionals and amateurs to appreciate the social complexity of nineteenth-century science. Artisan botanists and mechanical technicians, we now recognize, were integral to the diversified communities of practice engaged in scientific work.[3] More recently, historians have transcended another unhelpful dichotomy—that between elite and popular—in recognizing that the communication of knowledge to a diversity of audiences plays a crucial role in its very construction.[4] Understanding how science worked in nineteenth-century Britain entails improving our grasp on its social topography and the ways that communities formed and operated.

In this context, the scientific and medical periodicals whose numbers grew dramatically in nineteenth-century Britain (see fig. 1.1) offer an especially valuable and barely touched resource. As with the proliferating forms of scientific print more generally, those who produced these periodicals sought to garner more or less sizeable groups of purchasers and readers. However, periodicals were distinctive, albeit not unique, in their serial character. This had several important consequences. To begin with, the expected continuance of the periodical required the drawing together of a group of readers with sufficient commonality of interest to sustain ongoing publication. Periodicals were thus often founded by pre-existing groups, such as societies and clubs. In any case, they were produced with the ambition of developing a relatively stable and reliable group of consumers, although the prodigiously high rate of rapid failure demonstrates that that ambition was not easy to fulfill. In addition, the serial form of the periodical—extending through time—offered many more opportunities for readers to engage with those producing it, and with other readers, than was the case with a standard book. As issue followed issue, the periodical and the group of readers it fostered were shaped by the comments and responses of readers themselves, in contributed articles, letters pages, and other interactive formats which, appealing both to readers eager to exchange information and to proprietors with a vested interest in encouraging purchasers to come back for each issue, were one of the characteristic features of commercial science journals. Moreover, relationships often developed between readers, whether through learning about each others' activities in the pages of a periodical, or through the private communications that the periodical spawned.

Reexamining the scientific and medical periodicals of nineteenth-century Britain from this perspective promises radically to enhance our vision of the shifting communities and practices of science in the period. Much of the best work in this area currently focuses either on societies or on individual interactions. Martin Rudwick's classic study of Darwin in 1830s London, for instance, focused on the "social and cognitive topography of geology" very largely within the oral context of private and society discussion, especially among what Rudwick termed "gentlemanly specialists." Similarly, Anne Secord's important study of the interactions of artisanal and gentlemanly botanists offered a sophisticated analysis of correspondence as the "primary form of social interaction" within their shared community.[5] While such studies are indispensable, periodicals afford a large additional body of detailed evidence about the workings of communities of scientific practice. In particular, periodicals themselves played a crucial role in constructing, negotiating, and disrupting such communities, and all the more so as the century went on. As chapters in this volume show, sciences such as geology and botany came more and more to depend on periodicals for their community-building, practice, and identity, with societies—and even scientific fields—sometimes only coming into existence once a journal had established the existence of a group of like-minded individuals.

This volume begins to address the neglect of scientific periodicals in nineteenth-century Britain from this broader perspective. Given the large numbers of such periodicals—a conservative estimate suggests that something well in excess of a thousand titles were published, some of which lasted for the entire century—it is beyond the scope of this book to offer systematic, let alone comprehensive treatment. While this introduction outlines a new historiographical orientation to the history of science periodicals, the next chapter offers something of a chart of the historical terrain. The chapters that follow provide a series of representative examples from across the range of scientific subjects and time periods. These are not meant to add up to a tight and cohesive account, but rather should be taken in the spirit of "samplings and soundings" that exemplify the larger issues.[6] While subject coverage is wide, ranging across natural history, the physical sciences, and medicine, inevitably some areas (for example, chemistry) do not receive treatment. The emphasis on Britain is also, in part, pragmatic, allowing for a more focused exploration. The highly entrepreneurial character of British publishing, the population's high degree of literacy, and the notably voluntarist character of British science all make the British case an especially suggestive one.[7] Never have scientific periodicals been more important in constructing scientific communities than they were in nineteenth-century Britain.

This having been said, the focus on scientific communities developed in this volume has much wider relevance—both for other countries, and also for other time periods. Of course, much will be gained by further developing the analysis within a larger international perspective. While broadly similar concerns relating to the construction of scientific communities recur in different national and regional contexts, the manifold variations exhibited splendidly point up the national and regional differences in the organization of both scientific communities and scientific work. In the German lands of the late Enlightenment, for instance, the distinctively rapid expansion of specialist science periodicals reflected an interest in consolidating particular communities of knowledge in the universities of separate and to some extent competing states.[8] In the early American republic, by contrast, it was a single periodical,

Benjamin Silliman's *American Journal of Science and Arts* (1818–), published from New Haven but distributed across the Eastern seaboard, that enabled the rival scientific groups in the urban centers of Philadelphia, New York, and Boston to transcend their local enmities and come together for the first time as a unified national community, a process subsequently consolidated by the formation of the American Association for the Advancement of Science in 1848.[9] Moreover, periodicals have played a key role not only in enabling readers to imagine and engage with national communities of practice, but also in enabling them to imagine transcending such limitations in the construction of transnational communities.[10] The approach developed in this volume thus invites further development in relation to other national and supranational settings.

The historical perspectives offered here also cast important new light on other time periods, including the early twenty-first century debates concerning the future of the scientific journal. As scientists, universities, and funders become ever more radical in questioning the dominating role of modern scientific periodicals as vehicles of authorized knowledge that authenticate the expertise of scientists, so the alternative vision developed here of their role in constructing scientific communities becomes useful in offering conceptual tools to assist in that reevaluation. For instance, just as Loudon sought to use his new magazine to expand the community of scientific practitioners, so modern scientists are questioning what forms of communication can facilitate the involvement of "citizen scientists" and others beyond the academy. Likewise, as in the nineteenth century, changes in communication technology and the economics of publishing are now seen as opportunities to expand the communities of practitioners. Moreover, such changes are again being understood in relation to concerns about securing accountability for scientific knowledge among a wider public, comprising increasingly well educated citizens. As several participants observed at a recent Royal Society symposium on the future of the scientific journal, the "centrality of communities" is a key strand linking discussions of journals in the nineteenth and twenty-first centuries.[11] Thus, while this book is entirely historical in focus and approach, the themes it explores are highly topical, and its findings have bearings for contemporary science.

The remainder of this introduction offers a historical and historiographical overview of science periodicals in nineteenth-century Britain that serves to situate the more detailed studies in the individual chapters and elaborates on the larger analytical framework. We begin with a review of the existing literature, situating our approach in relation to the "communication turn" in the history of science, and to the recent growth of interest in the distinctiveness of nineteenth-century science periodicals. The following section then offers a more developed account of this volume's approach to such science periodicals, outlining some of the key themes. Finally, we provide an overview of the book's contents, showing how the various chapters exemplify and explore those themes in further depth.

British periodicals of the nineteenth century first became the subject of sustained scholarly attention from the mid-1960s, with the inception of the *Wellesley Index to Victorian Periodicals* (1966–89), the founding of the Research Society for Victorian Periodicals (1968), and the commencement of the *Waterloo Directory of Victorian Periodicals* (1970).[12] However, it was not until the late 1970s that scientific and medical periodicals became a distinct focus of scholarship.[13] A number of this first generation of studies by historians of science and medicine were naturally designed to chart and survey the historical output of periodicals, characterizing and enumerating different types and the patterns of their production.[14] Especially useful, though, were studies that pushed beyond this to consider the role of editors and publishers in producing and managing scientific periodicals. Encouraged by Roy MacLeod's historical centenary supplement to *Nature* in 1969, Jack Meadows and William Brock were prominent in opening up the practical, commercial, and strategic world of scientific periodical production. Meadows had written a biography of *Nature*'s founding editor, Norman Lockyer, while Brock became interested in the importance of commercial science journals in nineteenth-century Britain, which accounted for almost two-thirds of the 535 scientific titles listed in the first phase of the *Waterloo Directory*. In 1984 the two completed a study of the leading Victorian printers and publishers of scientific periodicals, Taylor and Francis, which still serves as the most useful overview of scientific periodical publishing in the period.[15]

Brock and Meadows' study of Taylor and Francis drew attention to many of the practical and commercial issues involved in scientific periodical publishing, but proved especially informative regarding the role of publishers and printers, who acted "as midwives in the creative process of bringing forth periodicals," making "decisions about which forms of scientific literature could survive in the market place," as well as the work of scientific editors, including Richard Taylor and William Francis themselves.[16] In placing such emphasis on the practical aspects of publishing, Brock and Meadows drew on the work of Susan Sheets-Pyenson, who in the early 1980s completed two groundbreaking studies of the publication of natural history periodicals in the 1830s, showing what could be learned from careful attention to publishing records.[17] These works also underpinned a valuable and distinctive study by David Allen, offering an account of how natural history periodicals became economically viable during the first half of the nineteenth century.[18] Meanwhile, historians of medicine began to attend to periodicals more seriously, not only charting the output and sampling the content, but also seeking to expose the important role of editors in shaping and producing them.[19]

While this first cluster of studies began to delineate some of the terrain and highlight the methodological issues involved in getting to grips with the great efflorescence of scientific and medical periodicals in nineteenth-century Britain, it did little to offer a larger historiographical vision. It is striking, indeed, that the introduction to W. F. Bynum, Stephen Lock, and Roy Porter's collection, *Medical Journals and Medical Knowledge* (1992), ran to just five pages, with the editors noting that it was "remarkable," in view of their "vast importance," how little "the history of scientific, technical, and medical journals" had been studied.[20] What larger historiographical visions there were, notably the one offered by Robert M. Young, actually cast scientific periodicals as deleterious agents of the "fragmentation of the common intellectual context," with Young arguing that it was the "popularity of *Nature* among increasingly professional scientists," along with the emergence of other similarly specialized journals, that shattered the "rich interdisciplinary culture" established by general periodicals earlier in the nineteenth century.[21] Another large-scale perspective came from those interested in exploring the rhetorical history of the scientific "article," but, as historians have shown more recently, such work has sometimes lacked historical nuance and needs to be reevaluated in the light of the growing awareness of the historical variability in scientific periodicals.[22] From the perspective of this volume, however, the most suggestive of these earlier studies was Sheets-Pyenson's examination of "popular science periodicals in Paris and London," with its innovative analysis of how such periodicals fostered the development of distinctive communities of scientific practice.

Focusing on popular periodicals in early-nineteenth-century London, Sheets-Pyenson came face to face with the ways in which such publications encouraged as one of their primary goals the active involvement of readers in the scientific enterprise, with workers being invited to contribute their own findings in the context of an open-ended vision of inductive science. Moreover, she found in many of these periodicals a divergent vision of the "canons of scientific investigation, criticism, and explanation," terming this scientific activity "low" science, in contradistinction to the "high" science of the "scientific establishment." For Sheets-Pyenson, these "low" periodicals served an important part of the "pyramid" of British scientific readers who expected to be able to contribute to the work of science.[23] She also explored the importance of such contributions to science in her innovative study of Darwin's reading of natural history journals. That such an important philosophical naturalist should be engrossed by the observations and comments of readers in the *Magazine of Natural History* helps to situate such individuals within the larger scientific domain, in which observations by "An Admirer of Nature, Ipswich" and "Miss Kent" sat cheek by jowl with those by such learned naturalists as William Kirby and William Swainson (fig. 0.2). Indeed, what Darwin most valued in the *Magazine of Natural History* was its willingness to publish what he termed "discussions & observations on what the world would call trifling points in Natural History," and, as Sheets-Pyenson showed, these ostensibly trifling points, which were a distinctive facet of many commercial scientific journals eager for cheap copy, were a vital source of descriptive information and detailed data in the period when Darwin was initially formulating the theory of natural selection.[24] While Sheets-Pyenson suggested that the open ethos of her early journals was later supplanted by a professionalizing ethos in the 1860s, chapters in this volume show that the story was not as linear as that suggests.[25]

Sheets-Pyenson's interest in periodical readers, communities of practice, and the social topography of science has been developed further in more recent studies. One major impetus for this has been the publications of the Science in the Nineteenth-Century Periodical (SciPer) Project at the Universities of Leeds and Sheffield. This project, which ran from 1999 to 2004, was devoted to examining the representation of science, medicine, and technology, and the interpenetration of scientific and literary discourse, in the *general* periodicals of nineteenth-century Britain, but the project brought the insights of periodical studies to bear on the history of science more generally. In particular, it highlighted the importance of even general periodicals in scientific communication and debate, drawing out their role in developing and defining audiences, and in facilitating discussion and interaction over time. Periodicals were, SciPer suggested, publications in which new communities of readers could be developed, sharing certain interests in and attitudes toward the sciences, negotiated in an ongoing conversation between editors, publishers, writers, and readers. In such publications, the participants in a new community of enquiry could learn to recognize each other as sharing a common enterprise, shaping their ideas into a scientific discipline, as was the case, for instance, with so-called "baby science" between the 1860s and the 1890s.[26]

This work on science in general periodicals reflected a more general turn

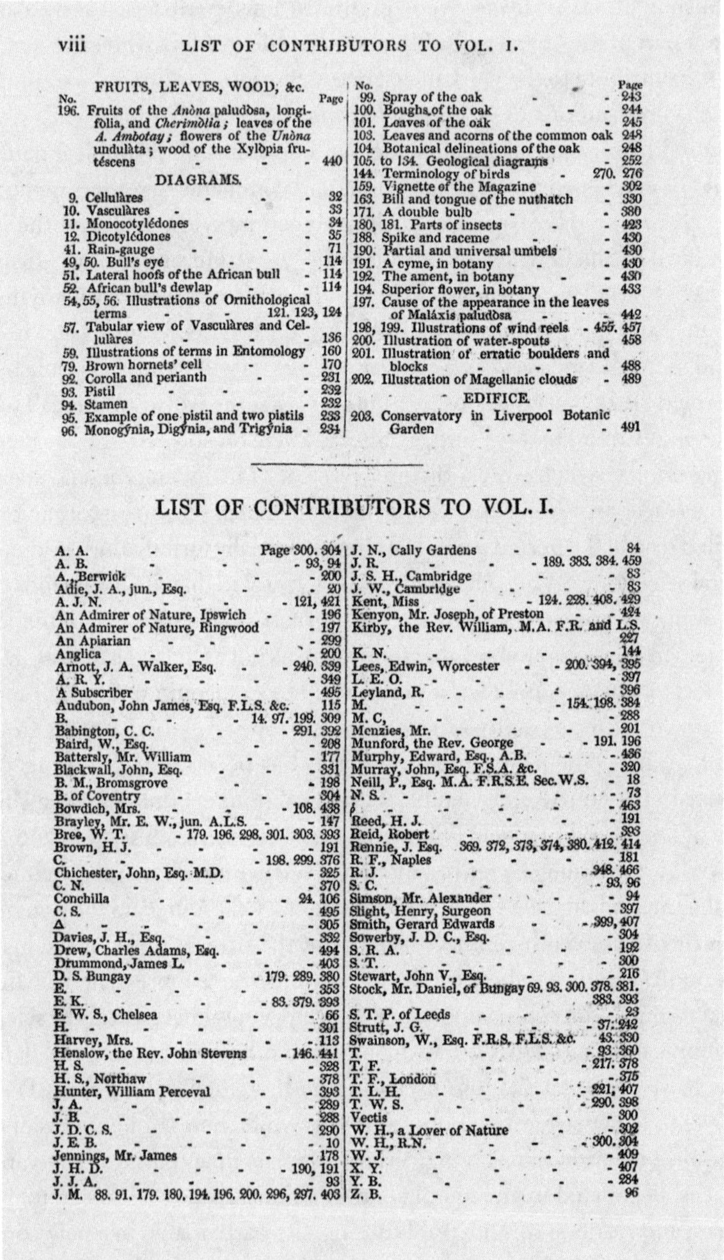

viii LIST OF CONTRIBUTORS TO VOL. I.

FRUITS, LEAVES, WOOD, &c.

No.		Page
196.	Fruits of the *Anòna* paludòsa, longifòlia, and *Cherimòlia*; leaves of the *A. Ambotay*; flowers of the *Unòna* undulàta; wood of the Xylòpia frutéscens	440

DIAGRAMS.

No.		Page
9.	Cellulàres	32
10.	Vasculàres	33
11.	Monocotylédones	34
12.	Dicotylédones	35
41.	Rain-gauge	71
49, 50.	Bull's eye	114
51.	Lateral hoofs of the African bull	114
52.	African bull's dewlap	114
54, 55, 56.	Illustrations of Ornithological terms	121. 123, 124
57.	Tabular view of Vasculàres and Cellulàres	136
59.	Illustrations of terms in Entomology	160
79.	Brown hornets' cell	170
92.	Corolla and perianth	231
93.	Pistil	232
94.	Stamen	232
95.	Example of one pistil and two pistils	233
96.	Monogýnia, Digýnia, and Trigýnia	234
99.	Spray of the oak	243
100.	Boughs of the oak	244
101.	Leaves of the oak	245
103.	Leaves and acorns of the common oak	248
104.	Botanical delineations of the oak	248
105. to 134.	Geological diagrams	252
144.	Terminology of birds	270. 276
159.	Vignette of the Magazine	302
163.	Bill and tongue of the nuthatch	330
171.	A double bulb	380
180, 181.	Parts of insects	423
188.	Spike and raceme	430
190.	Partial and universal umbels	430
191.	A cyme, in botany	430
192.	The ament, in botany	430
194.	Superior flower, in botany	433
197.	Cause of the appearance in the leaves of Maláxis paludòsa	442
198, 199.	Illustrations of wind reels	455. 457
200.	Illustration of water-spouts	458
201.	Illustration of erratic boulders and blocks	488
202.	Illustration of Magellanic clouds	489

EDIFICE.

No.		Page
203.	Conservatory in Liverpool Botanic Garden	491

LIST OF CONTRIBUTORS TO VOL. I.

A. A.	Page 300. 304
A. B.	93, 94
A., Berwick	200
Adie, J. A., jun., Esq.	20
A. J. N.	121, 421
An Admirer of Nature, Ipswich	196
An Admirer of Nature, Ringwood	197
An Apiarian	299
Anglica	200
Arnott, J. A. Walker, Esq.	240. 339
A. R. Y.	349
A Subscriber	495
Audubon, John James, Esq. F.L.S. &c.	115
B.	14. 97. 199. 309
Babington, C. C.	291. 392
Baird, W., Esq.	208
Battersly, Mr. William	177
Blackwall, John, Esq.	331
B. M., Bromsgrove	198
B. of Coventry	304
Bowdich, Mrs.	108. 438
Brayley, Mr. E. W., jun. A.L.S.	147
Bree, W. T.	179. 196. 298. 301. 303. 393
Brown, H. J.	191
C.	198. 299. 376
Chichester, John, Esq. M.D.	325
C. N.	370
Conchilla	24. 106
C. S.	495
Δ	305
Davies, J. H., Esq.	332
Drew, Charles Adams, Esq.	494
Drummond, James L.	403
D. S. Bungay	179. 289. 380
E.	353
E. K.	83. 379. 393
E. W. S., Chelsea	66
H.	301
Harvey, Mrs.	113
Henslow, the Rev. John Stevens	146. 441
H. S.	328
H. S., Northaw	378
Hunter, William Perceval	393
J. A.	289
J. B.	288
J. D. C. S.	290
J. E. B.	10
Jennings, Mr. James	178
J. H. D.	190, 191
J. J. A.	93
J. M.	88. 91. 179. 180. 194. 196. 200. 296, 297. 403
J. N., Cally Gardens	84
J. R.	189. 383. 384. 459
J. S. H., Cambridge	83
J. W., Cambridge	83
Kent, Miss	124. 228. 408. 429
Kenyon, Mr. Joseph, of Preston	424
Kirby, the Rev. William, M.A. F.R. and L.S.	227
K. N.	144
Lees, Edwin, Worcester	200. 394, 395
L. E. O.	397
Leyland, R.	396
M.	154. 198. 384
M. C,	288
Menzies, Mr.	201
Munford, the Rev. George	191. 196
Murphy, Edward, Esq., A.B.	436
Murray, John, Esq. F.S.A. &c.	320
Neill, P., Esq. M.A. F.R.S.E. Sec.W.S.	18
N. S.	73
ρ	463
Reed, H. J.	191
Reid, Robert	393
Rennie, J. Esq.	369. 372, 373, 374, 380. 412. 414
R. F., Naples	181
R. J.	348. 466
S. C.	93. 96
Simson, Mr. Alexander	94
Slight, Henry, Surgeon	397
Smith, Gerard Edwards	399, 407
Sowerby, J. D. C., Esq.	304
S. R. A.	192
S. T.	300
Stewart, John V., Esq.	216
Stock, Mr. Daniel, of Bungay	69. 93. 300. 378. 381. 383. 393
S. T. P. of Leeds	23
Strutt, J. G.	37. 242
Swainson, W., Esq. F.R.S. F.L.S. &c.	43. 330
T.	93. 360
T. F.	217. 378
T. F., London	375
T. L. H.	221, 407
T. W. S.	290. 398
Vectis	300
W. H., a Lover of Nature	302
W. H., R.N.	300. 304
W. J.	409
X. Y.	407
Y. B.	284
Z. B.	96

FIG. 0.2. "List of Contributors," *Magazine of Natural History* 1 (1829): viii. Image from the Biodiversity Heritage Library, www.biodiversitylibrary.org. Contributed by the Natural History Museum Library, London.

among historians of science toward the role of communication processes in the making of scientific knowledge.[27] James Secord's seminal 2004 article "Knowledge in Transit" reviewed the growing interest in the "movement, translation, and transmission" of scientific knowledge, suggesting that the underpinning historiography needed to be more systematic and ambitious. His inspirational vision of a systematic focus on communication as constitutive of scientific knowledge placed a new emphasis on moving beyond the "what" of scientific communication to the "how," "where," "when," and "for whom." Moreover, as Secord and others have suggested, such a reorientation offers a helpful way out of the sterile and often anachronistic distinction between "popular" science and science proper. Rather than thinking of certain forms of communication as merely popular, Secord's framework renders the full range of scientific communication of relevance in the making of science.[28] From the perspective of this volume, such a view is clearly valuable in suggesting to the historian that the communication that took place through the whole range of scientific periodicals, from the costly transactions of prestigious specialist societies to the cheapest of commercial journals, should be taken seriously as part of the work of science.

Perhaps partly in consequence of the new scholarly focus on communication, the last decade has witnessed a great resurgence of interest in scientific periodicals in nineteenth-century Britain. A common feature of this work has been a growing sense of the extent to which such periodicals were markedly different from the early-twenty-first century conception of the scientific journal as the prime location for making accredited contributions to the sciences and building reputation and careers. As Secord pointed out in an important overview of scientific print in the nineteenth century, "What it meant for something to be a scientific periodical, and the role of periodical publication regimes within the sciences, was radically uncertain right through the middle years of the nineteenth century."[29] This point has been developed in a number of more detailed studies. Jonathan Topham and Iain Watts have explored the distinctiveness of the first British commercial science journals, the purposes, uses, and readerships of which were notably different from modern journals.[30] At the other end of the century, Melinda Baldwin's important history of the journal *Nature* shows that it was founded with a strikingly unfamiliar vision of what a scientific periodical should be, and describes the long process of transformation by which it came to have its more familiar character. Indeed, the journal that is now the international benchmark for modern science publishing was initially intended to appeal to both scientific practitioners and the general public; only later, with the increasing specialization of science and the

journal's growing emphasis on news of the latest research, did it come to be
directed more narrowly to specialist practitioners. A similar series of changes
and transformations is currently being brought to light by the longue durée
history of the *Philosophical Transactions of the Royal Society* undertaken by
Aileen Fyfe and her collaborators.[31] Equally significant is the recent work of
Alex Csiszar, which offers an overarching view of the vigorous process of experimentation with scientific periodicals in nineteenth-century London and
Paris, and the developments that resulted in many of the familiar characteristics of the modern scientific journal.[32]

These studies have put the history of scientific periodicals in the nineteenth century on a new footing. By more consciously historicizing the form
of the scientific journal, they have drawn attention to new questions about the
changing purposes of such periodicals over this key transitional period. The
introduction to Baldwin's study of *Nature* argues that the journal "came to
define a scientific community" whose boundaries were "constantly shifting,
constantly being renegotiated and redefined." The *Philosophical Transactions*
project team also emphasize the role of periodicals in enabling "geographically dispersed scholars to communicate, and sometimes to coordinate, their
research," and in helping to "establish and police knowledge communities."[33]
In a related way, Csiszar has laid great emphasis on changing conceptions of
the publics for science in a post-Enlightenment world, drawing on Thomas
Broman's important study of periodicals in eighteenth-century German medicine. As Csiszar points out, the discursive category of the public became key in
the Enlightenment in securing legitimacy, and the new periodicals of the nineteenth century wrestled with how to use the notion to this end.[34] Nonetheless,
the distinctive role of scientific periodicals in the creation and management of
scientific and medical communities remains little explored. Fyfe points out
in her recent survey of the field that, "We know far too little about the distribution, circulation and readership of scientific journals."[35] More than that,
historians have much to gain by learning about how those involved—from editors and publishers to readers and contributors—used scientific periodicals
to shape communities of practice. It is this research agenda that the current
volume sets out to address.

RETHINKING THE ROLE OF THE
SCIENTIFIC PERIODICAL

As the historical review in the following chapter shows, scientific, medical,
and technical periodicals were not only extremely numerous in nineteenth-

century Britain, but also extremely diverse. Few resembled the peer-reviewed scientific journals familiar in the early twenty-first century, manifesting instead a wide variety of characteristics, purposes, and practices. This returns us to the key historiographical question with which we began. What framework can the historian bring to bear on the scientific periodicals of nineteenth-century Britain that provides a means of navigating this complicated terrain? The central claim of this volume—that scientific periodicals had a significant role to play in the development and operation of communities of scientific practice—certainly has much to offer in this context. The individual chapters that follow explore a number of the implications of that perspective in more detail, but in this introduction we now turn to consider some of the core historiographical themes in a more focused way. What does it mean to say that periodicals were involved in the development of scientific communities? How was the form of the periodical exploited and developed to facilitate such processes, and what constraints came into play? And, finally, how did these processes relate to the vexed question of who could be involved in the practice of the sciences, and who had the right to exert control over them? We focus on each of these questions in turn.

Scientific Communities

We have suggested in this introduction that periodicals have much to offer the historian in understanding "communities of scientific practice." However, as David Cahan has observed, the topic of "communities" is one that historians have been slow to pursue systematically or to examine conceptually. Cahan offers a working characterization of scientific communities as "populations of individuals who share similar cognitive interests and values that serve to provide them with a collective social identity and to advance individual scientific careers and group needs." Vital to such populations functioning as communities, Cahan suggests, is their engaging in concerted action over time and sharing a distinctive sense of social cohesion. But Cahan also observes that the notion of community can operate at a more abstract level in science, generating a sense of common identity and belonging at disciplinary, national, or international scale, independently of personal interaction.[36] In this regard, scientific communities are like the "imagined communities," sustained especially by newspapers, that Benedict Anderson suggests underpin the rise of nationalism in his landmark study of the subject. However, while for Anderson the abstract sense of being part of a nation that is cultivated by newspapers is the product of what he terms "print capitalism," in the case of science periodicals, such imagined communities were as much the creation of readers as they were of editors and publishers.[37] While the latter often had, among other motives, a vested commercial interest in generating a sense of community that ensured a sufficient number of recurrent purchasers, for readers a feeling of authentic connection with other readers could be forged through the altruistic exchange of letters and information in interactive formats such as the notes and queries column.

Cahan's characterization of scientific communities is extremely helpful from the perspective of this volume, and it is easy to see that scientific periodicals played a key role in their formation and functioning. Sometimes, scientific periodicals were established with a clear intention to cultivate a working assemblage of readers who shared a strongly cohesive sense of common purpose and identity, and from whom they could obtain scientific observations and contributions. Such might be the case with a periodical established by a club or society, the key object of which was to continue and develop a conversation begun within the confines of a meeting room, often expanding to a national or even international scale and drawing individuals who rarely or never met into a cohesive imagined community. This was also the case with more speculative ventures, such as the groundbreaking cheap weekly the *Mechanics' Magazine* (1823–72), where the editorial imagining of a geographically dispersed community of artisans sharing common interests in mechanical knowledge was only partially mapped onto the emerging social infrastructure of artisanal life in the clubs and political gatherings of 1820s Britain. The producers of some other periodicals never expected their readers to form so cohesive a group, and the commercial failure and brief duration of so many science periodicals suggests that, even when there was such an expectation, the cohesion of a community of readers could never be taken for granted, and required careful cultivation. To achieve a viable market for their product, editors and publishers accepted the need to combine multiple groups of readers, who might typically share certain cognitive interests—say, botany or astronomy—while having very different cognitive values and sharing little in the way of social identity. When, for instance, Loudon established his *Magazine of Natural History*, he was quite clear that active "students of nature" and "general readers and observers" were both to be interested by the new product, despite their very different senses of themselves and their rather different needs.

Of course, both types of readers would themselves have simultaneously belonged to other, often overlapping communities, and those whom Loudon designated "students of nature" would, depending on their standing and social class, have been members of institutions such as the prestigious Royal Society or the peripatetic British Association for the Advancement of Science. They might also have belonged to smaller, more exclusive groups predicated on sociability, such as the Red Lion Club, made up of younger members of the British Association who held raucous tavern dinners where they could indulge in boisterous behavior not acceptable at the stuffy formal banquets of the parent organization, or the X Club, an informal dining group whose strategic collaboration and institutional maneuvers helped them to become the spokesmen for science to much of the nineteenth-century public.[38] In addition, however, they would almost certainly have participated in communities beyond the strictly scientific—for instance, reading and contributing to general or literary periodicals alongside members of other intellectual and occupational groups. Prominent men of science such as Thomas Henry Huxley, as Paul White has argued, even helped to "create and sustain a single community of diverse, but complementary, élites" in mid-Victorian Britain, whose varied members, whether Anglicans or agnostics, novelists or naturalists, were brought together by their adherence to liberal views of culture and reform.[39] The communities forged by science periodicals must be viewed in relation to this broader context of community formation in the nineteenth century, but there is still considerable value in emphasizing, as this volume does, the specific ways that specialist journals fostered a sense of collective identity among their contributors and readers.

One of the benefits of approaching scientific periodicals from this perspective is precisely that it brings into focus the diversity of communities and their interaction in a way that is easily lost when focusing on particular institutions or on such inchoate social categories as "amateur," "philosophical," "practical," "popular," and "professional." Periodicals addressed to multiple imagined audiences allow the historian to gain a clearer sense of how contemporaries understood what Rudwick talks about as the different zones in the social topography of science. For Rudwick, the main distinction between the zones is in terms of levels of "ascribed competence," and this was certainly a consideration for journal editors. The Scottish natural philosopher David Brewster's vision of the *Edinburgh Philosophical Journal* (1819–64), for instance, encompassed "men of Genius" and "General Readers," with a view to maximizing the market for a new kind of authoritative but fashionable scientific monthly. But the question was not merely one of "ascribed competence." Periodical editors often perceived that a population of purchasers could be generated by addressing individuals with different cognitive values. For example, journalist Alexander Tilloch's *Philosophical Magazine* (1798–) was aimed at both "philosophers" and "mechanics," and was thus meant to advance both theoretical knowledge and practical improvements, often entailing a common subject focus, albeit with distinct objects in scientific and technical innovation. Of course, the readers envisaged by Brewster and Tilloch, whether "men of Genius" or merely "mechanics," were almost exclusively male, and science periodicals that aimed to include women in their audiences were generally confined to certain fields such as botany and horticulture, with the opening number of Loudon's *Gardener's Magazine* (1826–44) urging that its subject was "agreeable . . . especially to the female sex."[40] Only later in the nineteenth century did female contributors begin to make their mark in science journalism, with, for instance, Phebe Lankester, who often wrote under the pseudonym "Penelope," contributing regular articles on both botany and public health to the *Popular Science Review* (1862–81).[41] As the chapters by Sally Shuttleworth and Sally Frampton in this volume show, public health and medicine continued to be topics in which female contributors and readers played an important, if frequently contested, role.

At one end of the spectrum, therefore, periodicals offer insights into the overlapping nature of different communities of practice that had related but distinct interests. Indeed, publishers, editors, writers, and readers were often obliged to set out explicitly how they saw the differences and similarities among the diverse groups of readers. This sometimes occurred in forthright and controversial exchanges, but it also occurred in more programmatic statements about how editors and others viewed the scientific division of labor. At the other end of the spectrum, however, periodicals played a key role in developing tightly bound communities of practice with shared epistemic goals and a strong corporate identity. Such journals might emerge out of preexisting face-to-face interactions in clubs, societies, professional bodies, or academic institutions, though they might as easily lead to such interactions. It was, for instance, the activities of the *Mechanics' Magazine* that led to the formation of the London Mechanics' Institution, rather than the reverse. Having founded their new journal, the editors used its pages to invite the mechanics of London to associate for the purpose of establishing their own institution, and to organize and report on the subsequent meetings (fig. 0.3).[42] Periodicals could offer an opportunity for an individual or group to explore the viability of an "imagined community." Did the editor's vision of a new scientific discipline, medical specialty, or professional identity find an answering call from a coherent body of readers that might lead to a growing social consolidation? While many editors missed their mark, many others had a key role to play in developing the social topography of nineteenth-century science.

Mechanic's Magazine,
Museum, Register, Journal, & Gazette.

Bright as the pillar rose at Heaven's command,
When Israel march'd along the desert land,
Blazed through the night on lonely wilds afar,
And told the path—a never-setting star. *Campbell.*

No. 12.]　　SATURDAY, NOVEMBER 15, 1823.　　[Price 3d.

Public Meeting,
FOR THE ESTABLISHMENT OF
THE LONDON MECHANICS' INSTITUTE.

THE paramount importance to the whole Mechanics of the British empire of the proceedings which we are about to detail, will, we are assured, make all our readers well pleased to see the place usually occupied by an engraving, as well as every other corner of this week's Mechanic's Magazine, devoted to giving a full and accurate report of these proceedings.

The Public Meeting for taking into consideration the propriety of establishing a London Mechanics' Institution was held, agreeably to the invitation in our last and preceding Numbers, on the evening of Tuesday, the 11th inst. The large room of the Crown and Anchor Tavern, one of the very largest in the metropolis, was engaged for the occasion, and at the time appointed for taking the chair, it was completely filled. It is said to hold 2,500 persons; certainly more than 2,000 were present. We were glad to perceive that they consisted chiefly of that class for whose good the institution is intended, namely, *working mechanics;* and that they showed, by their conduct and demeanor, that they comprehended fully the serious magnitude of the object for which they were assembled, and came to the consideration of it with minds warmed apparently to enthusiasm in its support; yet keenly intent on examining and scrutinizing well the means by which they were to be invited to realize the promised good. It was a meeting of men resolved both *to think and act*

for themselves. During the whole evening there was nothing but expressions of applause, and yet we did not observe one single instance of applause bestowed on a sentiment or proposition which did not deserve it. We should, perhaps, except the applause given to an offer from a professional lecturer to deliver a course of lectures on some branch of science *gratis,* though perhaps the cheers in this case were a token more of gratitude to the individual for his generous intentions, than of approbation of a condition so much at variance with what they had before declared should be the fundamental principle of the London Mechanics' Institution —*namely, that the mechanics should pay as well as they can for whatever instruction they are to receive.* The earnest, discriminating, and orderly attention with which they listened to the whole of the proceedings, exceeded any thing we had ever before witnessed in so numerous an assembly. Some two or three unhappy individuals, we are told, were indiscreet enough to intrude themselves in a state which utterly disqualified them from taking a part in the deliberations of rational men; and could they have been *lifted* forward, and exhibited on the table, they might, like the slaves of Sparta, who were made drunk for the purpose, have given *point* to many an excellent moral that fell from the speakers on the effects of a debasing intemperance; but no sooner were they observed by the

N

Periodical Formats and Finances

Given the important role that periodicals played in the growth and management of scientific communities, it is important to examine further the relevance of their distinctive characteristics in that process. Of course, the serial character of the periodical is at the core in this regard. The regular, date-stamped, and open-ended character of periodical publication allows editors, writers, and readers alike to imagine an ongoing relationship based around scientific practice, the desire to communicate, and a sense of the unfolding work of science. These are all aspects of seriality recently emphasized by Nick Hopwood, Simon Schaffer, and James Secord. They also reflect that, "more than anything else, the experience of sequential reading of printed paper tied groups together," noting that "it was often said that a political group or religious sect did not really exist until it issued a periodical or newspaper."[43] As this makes clear, however, scientific periodicals were not unique in having a role in building communities. Indeed, the natural sciences were by no means the only field of knowledge considered to have a progressive character suited to serial publication, and the intimate connection between progress and periodicals had long been a source of anxiety and satire among conservative commentators.

This raises the question whether the producers of scientific periodicals exploited the form in distinctive ways, adapted to scientific purposes.[44] To what extent were scientific periodicals the same as or different from other periodicals, and did that change over time? Of course, the variety is such that no very general answer can be offered, and science journals were published in myriad periodical formats including, but not restricted to, society transactions and proceedings, weekly and monthly magazines, reviews, annuals, and digests of abstracts. It is clear that on many occasions, the publishers and editors of journals looked to existing periodicals, nonscientific as well as scientific, in conceiving the format of their new productions. Yet, with the passage of time, aspects of format were adapted to what were seen to be the distinctive demands of scientific work. For instance, the manner in which Loudon encouraged "persons wholly unacquainted with Natural History as a science" to offer observations in the pages of the *Magazine of Natural History* reflected a sense of the distinctive character of natural history as a field science, dependent upon the observations of a large body. Moreover, this active encouragement of small-scale observations by readers became a notable feature of a large array of scientific periodicals in the nineteenth century, including, as is seen in the chapters in this volume, those focused on natural history, geology, and astronomy, but also extending to public health. Often such initiatives involved making space in the pages of the periodical for readers to contribute queries, requests for information, and offers of help with specimens and observations—activities which were likely to enhance readers' sense of being part of a community of practice. Similarly, in the late nineteenth century the new periodical format of the digest of abstracts was closely adapted to more recent developments in scientific practice, particularly the massive growth in the number of professional scientific workers, based in institutional laboratories and universities, who, in order to validate and advance their own experimental studies, required rapid access to relevant information in different specialist disciplines. The particular types of scientific work undertaken by these two very different communities—casual observers in the field and professional "scientists" (a designation, of course, that only came into widespread use at the close of the nineteenth century)—were both actively shaped by the formats of the journals they read, and to which they contributed.

Of course, such facets as readers' contributions also appeared in some nonscientific periodicals, which in itself offers the historian an opportunity to explore the extent to which scientific communities were different from those with other interests. Yet for historians of science increasingly interested in the quotidian practice of science, understanding the choices made by those producing periodicals in relation to the organization of scientific work promises valuable insights. This indeed is the main focus of a recent volume on "scholarly journals in early modern Europe," where the editors' core question is, "Has the creation and development of a periodical form changed the nature not only of scientific communication but also of scientific and indeed of scholarly practice?"[45] Here, recent work on the use of "paper technologies" in the sciences offers a fertile approach.[46] The editor's management of the format and contents of the periodical can usefully be thought of as the deployment of such a technology in organizing scientific work, including doing so through organizing a division of scientific labor. Thus, for instance, just as interpersonal correspondence networks are attracting increasing scholarly attention as "scientific tools," Matthew Wale's chapter in this volume shows that the mediated correspondence networks of periodicals can also be understood in these terms.[47]

It is important to appreciate, however, that these decisions about format were never entirely motivated by editorial ambitions in regard to the practice of science. Anyone familiar with the history of periodicals will recognize that concerns about finances were rarely far from the minds of editors, and that was undoubtedly the case with scientific periodicals, too. It is interesting to speculate what Robert Jameson and David Brewster's *Edinburgh Philosophi-*

cal Journal or Norman Lockyer's *Nature* might have looked like if the editors and publishers had felt convinced that they might publish a periodical without concerning themselves with a "popular" audience beyond the active "philosophers" or "men of science" like themselves. As several of the chapters in this volume show, moreover, even the ostensibly noncommercial transactions and proceedings produced by scientific societies, which underwrote the costs of publication through the patronage of their members, generally found it necessary to shift to more commercial business models after mid-century, even if they found the requirements of the marketplace distasteful and endeavored to retain the prestige associated with being privately subsidized. The point here, of course, is that the financial considerations that troubled most producers of periodicals had substantial consequences for the character of the scientific communities that were produced and the scientific work that was undertaken through their pages. In seeking to maximize a market by selling his *Philosophical Magazine* both to manufacturers and to learned philosophers, for instance, Alexander Tilloch committed himself to a distinctive mélange in terms both of form and contents. Of course, readers did not see themselves merely as consumers, and instead the sense of community created by journals often relied on a sentiment of shared work and participation that was inspired by values of mutuality, or more old-fashioned paternalism, rather than mercantile competition. The story of scientific periodicals in nineteenth-century Britain is nevertheless in part a story of the changing economics of periodical publication, and the continuing necessity of turning a profit or at least breaking even, which endlessly redounded on editorial visions and practices.

Legitimacy and Control

While they were often made in response to financial considerations, decisions about the intended readership of scientific periodicals were often also highly political. Attempts to broaden and open out the communities of scientific practice, whether financially or ideologically motivated, were frequently perceived to be a challenge to established bodies. In particular, the growing multitude of learned societies were often alarmed, wary, or dictatorial about periodicals that seemed likely to undermine their authority by appealing to larger and more diverse constituencies of readers. Iain Watts's recent study of William Nicholson's *Journal of Natural Philosophy* (1797–1813), for instance, draws attention to the manner in which the Royal Society perceived Nicholson's editorial activities as an encroachment on their management of the community of scientific practitioners.[48] And, as Alex Csiszar shows in chapter 3 of this volume, it was in this context that scientific societies sought to claim back their control by addressing an imagined general public through "proceedings" that would serve to secure their public authority. Moreover, as Csiszar has shown elsewhere, the development of processes of refereeing and the establishment of the scientific paper as a distinctive form of publication of record also need to be read in relation to such concerns. These can be viewed as some of the "ways in which received boundaries between experts and non-experts—and the values and standards that come with them—were erected in the first place."[49]

The perception of elite institutions—especially the metropolitan learned societies—that certain scientific periodicals represented a threat in their deliberate attempts to widen participation in scientific practice was not, of course, mere paranoia. Many periodicals were founded with precisely such ends in view, as is clear in the case of the *Magazine of Natural History* with which we began, and in many other cases explored in this volume. As James Secord has perceptively observed, the imagined "futures of science" were always multiple. Moreover, it was not simply the case that the learned societies did not have a monopoly on such imaginings; they also disagreed within their own ranks about how science should operate.[50] The growing range of scientific periodicals in nineteenth-century Britain can thus be read as a contest regarding the shape of the communities of science, and concerning who had the authority to adjudicate what went on within them. As Geoffrey Belknap shows in chapter 5 of this volume, the *Magazine of Natural History* prompted a range of new editors to enter the field with periodicals seeking to define and control alternative visions of the natural history community. Competition between commercial journals was often not merely a matter of financial success or failure: it represented a battle for control within the communities of scientific practice. This continued to be the case throughout the nineteenth century, even as the machinery of scientific expertise described by Csiszar developed, with new kinds of alternative communities emerging, notably including new professional and trade groups such as architects and telegraph engineers.

This reference to professional groups leads us naturally to the medical press, where issues concerning legitimacy and control were arguably most strongly felt. As the following chapter shows, it is striking that it was in medicine, where a clearly defined market for print was known to exist, that the earliest of the new specialized commercial periodicals of the eighteenth century were produced. Unlike either the *Philosophical Transactions* (1665–) or the *Gentleman's Magazine* (1731–), such publications addressed clearly defined imagined constituencies, encompassing various combinations of physicians, surgeons, and apothecaries. In the early nineteenth century, however, periodicals were commenced with more contentious alignments, whether addressing a wide public audience alongside medical practitioners, as did the *Monthly Gazette of Health* (1816–32), a diverse constituency of supporters of an unorthodox medical doctrine, like the *Phrenological Journal* (1823–47), or indeed a politically charged section of the medical profession, as did the *Lancet* (1823–). Such diversification of medical periodicals and the communities they supported and represented only increased as the century progressed. As Sally Shuttleworth shows in chapter 10 of this volume, periodicals helped to forge communities interested in public health that went far beyond physicians and surgeons without too much controversy. Yet, as Sally Frampton demonstrates in chapter 9, other periodicals prompted the ire of medical authorities when they targeted readers who were not doctors, including nurses and medical administrators, as well as others not employed in medicine. As medical communities themselves became more specialized toward the end of the century, the desire of leading doctors to control and delimit those engaged with medical knowledge was met with a continuing defiance from dissenting editors and readers.

Altogether, therefore, we have to think of scientific and medical periodicals as key sites in which, and between which, the power structures of science and medicine were developed and negotiated. As Csiszar shows in chapter 3, the history of such periodicals takes us to the heart of "an enduring problem of science and democracy." Just as in other aspects of social and cultural life, so in the sciences the expansion of print media and the diversification of readers and contributors left members of elites who sought to establish privileged claims to authority needing to do extra work to gain public legitimacy. In a world in which a coal heaver might have an opinion on the proceedings of scientific societies, as caricatured by Robert Seymour in a famous series of lithographed sketches (fig. 0.4), periodicals not only were used to build boundaries to demarcate and legitimate privileged knowledge communities, but also frequently functioned to challenge such communities. Moreover, the situation was, if anything, all the more keenly felt in medicine, where livelihoods were very evidently at stake in the contests concerning the divergent ways that periodicals defined communities of medical knowledge and practice.

The historiographical reflections offered here are intended to flesh out the vision of a new way of exploring the importance of scientific periodicals in nineteenth-century science. For too long, the study of scientific periodicals has been limited by a misconception of their diverse history, rooted in the perspective of the present, when they have become primarily understood as certifying

F I G . 0 . 4 . "Every Day Scenes, Scene 3 (Coalheavers in the Byron Coffee House)," *Seymour's Sketches, Illustrated in Prose & Verse* (London: H. Wallis, [1838?]), plate 3. The caption reads: "[Coalheaver 1:] 'You shall have the paper directly, Sir; but really the debates are so very interesting.' [Coalheaver 2:] 'Oh, pray don't hurry, Sir; it's only the scientific notices I care about.'" Reproduced with the permission of Special Collections, Leeds University Library.

vehicles of authoritative scientific knowledge. Taking seriously the role of scientific periodicals in the development and operation of scientific communities in the nineteenth century takes us to the heart of key questions in the history of nineteenth-century science, concerning the changing and interlocking character of communities of scientific practice, the organization of scientific work, and the struggle for authority and control. Such an approach puts scientific periodicals where they belong: at the heart of the story of nineteenth-century science.

By the end of the century, science was increasingly associated with academically employed professionals in a way that, in the new century, came increasingly to supplant nineteenth-century ambitions for more inclusive scientific communities. However, recent developments make it easier to appreciate the ongoing importance of the perspectives brought to bear in this volume. For instance, "citizen science" initiatives, supported and encouraged by the flexibility of digital communication technologies, have again placed the question of the character, limits, and management of scientific communities at center stage. Similarly, the inception of the Internet and the spiraling costs of journal publication have led to a vigorous debate concerning how the format and finances of scientific periodicals affect scientific work practices and the management of working communities. At the same time, principled concerns about public access to the results of publicly funded research in an era in which public engagement has become a policy watchword link these debates to issues of public accountability and legitimacy. Underlying current discussions of the future of the scientific journal lies the question, so central to nineteenth-century debates, of how knowledge claims can be rendered authoritative, through "defining boundaries and validating membership" of knowledge communities, without making those boundaries impermeable. While twenty-first-century science periodicals are significantly different from those of the nineteenth century, they continue to "define and support communities," with many of the same issues at stake.[51]

SAMPLINGS AND SOUNDINGS

This book is divided into three parts that examine different aspects of how scientific and medical periodicals created and negotiated a variety of communities of practice across the nineteenth century. In part 1, "New Formats for New Readers," the chapters examine some of the constraints and new possibilities surrounding how scientific communities could be conceived, especially during the earlier years of the century. Chapter 1 offers an overview of the new kinds of science periodicals that were produced in nineteenth-century Britain, charting some of the most significant patterns and trends and providing a somewhat tentative map of the terrain. Chapter 2, Jonathan R. Topham's "Redrawing the Image of Science: Technologies of Illustration and the Audiences for Scientific Periodicals in Britain, 1790–1840," examines how the changing technologies of scientific illustration had important consequences for the readership of early nineteenth-century journals, and thus for the sense of how science should be configured. Chapter 3, Alex Csiszar's "Proceedings and the Public: How a Commercial Genre Transformed Science," examines the emergence of "proceedings" as a new format of periodical publication, with learned societies responding to the early nineteenth-century emergence of commercial journals and their demands for public accountability. Recognizing "proceedings" as a distinct periodical genre demonstrates the intimate connection between commercial journalism and the consolidation of specialized scientific publishing, which is a theme that runs through many of the chapters in the next part of this book.

Part 2, "Defining the Communities of Science," examines how developments in periodicals in five scientific fields—geology, natural history, entomology, physics, and astronomy—served to shape, but also responded to, changing communities. The chapters explore the changing notions of who was properly involved in scientific practice, as well as tensions and contests over who should exert control. The picture that emerges is one in which fellows of learned societies and professionalizing academics were continually engaging with larger public and professional constituencies. Chapter 4, Gowan Dawson's "'An Independent Publication for Geologists': The Geological Society, Commercial Journals, and the Remaking of Nineteenth-Century Geology," shows how the Geological Society found itself constantly responding to a commercial press that often subverted both its authority and its hierarchical conception of the earth sciences, and instead harnessed the field observations of an army of enthusiasts. It was in natural history that specialist commercial journals first found a sizeable community of readers; but here, as Geoffrey Belknap argues in chapter 5, "Natural History Periodicals and Changing Conceptions of the Naturalist Community, 1828–65," there were competing visions of how such a community should be managed. Like geology, entomology was another of the subfields of natural history that, by mid-century, increasingly had its own specialist commercial journals, and, like the earth sciences discussed in Dawson's chapter 4, it too attracted practitioners from across the social spectrum. Chapter 6, Matthew Wale's "'The Sympathy of a Crowd': Imagining Scientific Communities in Mid-Nineteenth-Century Entomology Periodicals," explores how the application of print publication to older forms of scientific correspondence had important implications for the social and geographic makeup of the communities of entomological practice.

The last two chapters in part 2 focus on the physical sciences—specifically, physics and astronomy—in the final decades of the nineteenth century, and there are both continuities with and differences from how the journals in the life sciences considered in the first three chapters of this part forged and reacted to new forms of scientific community. Chapter 7, Graeme Gooday's "Periodical Physics in Britain: Institutional and Industrial Contexts, 1870–1900," examines how in physics, perhaps surprisingly, the number of interested constituencies grew rapidly in the late nineteenth century with the rise of industry, technical professions, and school science so that, as in Csiszar's and Dawson's chapters, learned societies found themselves responding to commercial journals with different conceptions of what the science should look like. Without the same industrial and technical applications, astronomy was much closer to natural history in attracting amateur practitioners from across the social spectrum, albeit that their exclusion from the increasingly mathematical professional forms of the science gave a distinctive role to the new astronomical journals established toward the end of the nineteenth century. In chapter 8, "Late Victorian Astronomical Society Journals: Creating Scientific Communities on Paper," Bernard Lightman surveys a series of society-based astronomical periodicals appealing to amateurs that in the 1880s and 1890s helped to motivate and manage a large and various community of observers.

The rich sense of the diverse visions and practices of communities engaged in the sciences that comes from the first two parts is echoed in a distinctive way in part 3 of this volume, "Managing the Boundaries of Medicine," which focuses on medical periodicals. Unlike the natural sciences, medicine had a distinct and increasingly consolidated professional community in nineteenth-century Britain, and, as was discussed earlier, previous histories have often emphasized the role of journals in the emergence of intraprofessional specialisms. At the same time, however, new periodicals became pivotal in developing and managing larger communities of practice, as the two chapters in this part of the book both show. In chapter 9, "'A Borderland in Ethics': Medical Journals, the Public, and the Medical Profession in Nineteenth-Century Britain," Sally Frampton examines how a range of new journals in the 1880s enabled different groups of people other than medical professionals to participate in debates over health care. Chapter 10, Sally Shuttleworth's "'National Health is National Wealth': Publics, Professions, and the Rise of the Public Health Journal," shows how, in the increasingly important field of sanitary science, periodicals were central to the organization of groups involved in campaigning and in gathering information about public health, communicating sanitary science to their audiences and encouraging citizens to self-manage their health and the health of their communities.

As Shuttleworth suggests in her chapter, the communities of active and committed participants cultivated by public health periodicals in the late nineteenth century helped lay the foundations for the environmental campaigning of the present day. Such modern campaigning is often facilitated by the Internet and associated digital technologies, which, like the nineteenth-century information revolution inaugurated by steam-powered machine printing and new methods of stereotyping and distribution, have helped create distributed communities interested in and engaged with the sciences extending far beyond the confines of the academy. In their analysis of the ways in which print and its associated technologies fostered the development and operation of communities of scientific practice, the chapters in this book thus shed important new light on a theme of enduring significance. For, as we have seen above, the relations between print, technology, and community are as significant for science in the digital age as they were two centuries ago.

NOTES

1. [John Claudius Loudon], [*Prospectus for the "Gardener's Magazine"*] ([London: Longman, Hurst, Rees, Orme, Brown and Green, 1826]), 2–3. There is a copy of the prospectus, annotated with the March date, in the John Johnson Collection, Bodleian Library, Oxford.

2. Alex Csiszar, "How Lives Became Lists and Scientific Papers Became Data: Cataloguing Authorship during the Nineteenth Century," *British Journal for the History of Science* 50 (2017): 23–60, on p. 23. See also, e.g., James A. Secord, "Science, Technology, and Mathematics," in *The History of the Book in Britain*, vol. 6, 1830–1914, ed. David McKitterick (Cambridge: Cambridge University Press, 2009), 443–74; Jonathan R. Topham, "Anthologizing the Book of Nature: The Circulation of Knowledge and the Origins of the Scientific Journal in Late Georgian Britain," in *The Circulation of Knowledge between Britain, India and China: The Early-Modern World to the Twentieth Century*, ed. Bernard Lightman, Gordon McOuat, and Larry Stewart (Leiden and Boston: Brill, 2013), 119–52; Melinda Baldwin, *Making "Nature": The History of a Scientific Journal* (Chicago: University of Chicago Press, 2015); Aileen Fyfe, Julie McDougall-Waters, and Noah Moxham, "350 Years of Scientific Periodicals," *Notes and Records of the Royal Society* 69 (2015): 227–39; and Alex Csiszar, *The Scientific Journal: Authorship and the Politics of Knowledge in the Nineteenth Century* (Chicago: University of Chicago Press, 2018).

3. Steven Shapin and Arnold Thackray, "Prosopography as a Research Tool in History of Science: The British Scientific Community 1700–1900," *History of Science* 12 (1974): 1–28; Anne Secord, "Science in the Pub: Artisan Botanists in Early Nineteenth-Century Lancashire," *History of Science* 32 (1994): 269–315; Iwan Rhys Morus, *Frankenstein's Children: Electricity, Exhibition, and Experiment in Early-Nineteenth-Century London* (Princeton, NJ: Princeton University Press, 1998); and Alison Winter, "The Construction of Orthodoxies and Heterodoxies in the Early Victorian Life Sciences," in *Victorian Science in Context*, ed. Bernard Lightman (Chicago: University of Chicago Press, 1997), 24–50.

4. See, e.g., James A. Secord, "Knowledge in Transit," *Isis* 95 (2004): 654–72; Jonathan R.

Topham, "Rethinking the History of Science Popularization / Popular Science," in *Popularizing Science and Technology in the European Periphery, 1800–2000*, ed. Faidra Papanelopoulou, Agusti Nieto-Galan, and Enrique Perdiguero (Aldershot, UK: Ashgate, 2009), 1–10; and Gowan Dawson, *Show Me the Bone: Reconstructing Prehistoric Monsters in Nineteenth-Century Britain and America* (Chicago: University of Chicago Press, 2016).

    5. See, e.g., Martin Rudwick, "Charles Darwin in London: The Integration of Public and Private Science," *Isis* 73 (1982): 186–206, on p. 191; and Anne Secord, "Corresponding Interests: Artisans and Gentlemen in Nineteenth-Century Natural History," *British Journal for the History of Science* 27 (1994): 383–408.

    6. Joanne Shattock and Michael Wolff, eds., *The Victorian Periodical Press: Samplings and Soundings* ([Leicester, UK]: Leicester University Press; Toronto: University of Toronto Press, 1982).

    7. For interesting discussions on the book in international perspective, see Michael Suarez and H. R. Woudhuysen, eds., *The Book: A Global History* (Oxford: Oxford University Press, 2013).

    8. See Martin Gierl, "The 'Gelehrte Zeitung': The Presentation of Knowledge, the Representation of Göttingen University, and the Praxis of Self-Reviews in the 'Göttingische Gelehrte Anzeigen,'" *Archives internationales d'histoire des sciences* 63 (2013): 321–41.

    9. See Simon Baatz, "'Squinting at Silliman': Scientific Periodicals in the Early American Republic, 1810–1833," *Isis* 82 (1991): 223–44.

    10. See, for instance, Csiszar, *Scientific Journal*.

    11. Cameron Neylon, "Communities Need Journals," *Notes and Records of the Royal Society* 70 (2016): 383–85. See also Jason Potts, John Hartley, Lucy Montgomery, Cameron Neylon, and Ellie Rennie, "A Journal Is a Club: A New Economic Model for Scholarly Publishing," *Prometheus* 35 (2017): 75–92. Neylon's paper appears in a special issue of *Notes and Records* entitled "Science Periodicals in the Nineteenth and Twenty-First Centuries," and based on papers delivered at the symposium "The End of the Scientific Journal? Transformations in Publishing," which took place at the Royal Society on 27 November 2015. See especially the editorial, Sally Shuttleworth and Berris Charnley, "Science Periodicals in the Nineteenth and Twenty-First Centuries," *Notes and Records of the Royal Society* 70 (2016): 297–304.

    12. Walter Houghton et al., eds., *The Wellesley Index to Victorian Periodicals, 1824–1900* (Toronto: University of Toronto Press, 1966–85); Michael Wolff, John S. North, and Dorothy Deering, *The Waterloo Directory of Victorian Periodicals, 1824–1900*, Phase I (Waterloo, ON: University of Waterloo, [1970]); John S. North, ed., *The Waterloo Directory of Irish Newspapers and Periodicals, 1800–1900* (Waterloo, ON: North Waterloo Academic Press, 1986); idem., *The Waterloo Directory of Scottish Newspapers and Periodicals, 1800–1900*, 2 vols. (Waterloo, ON: North Waterloo Academic Press, 1989); and idem., *The Waterloo Directory of English Newspapers and Periodicals, 1800–1900*, series 1 and 2, 20 vols. (1997–). For the early literature in the field, see William S. Ward, *British Periodicals and Newspapers, 1789–1832: A Bibliography of Secondary Sources* (Lexington: University Press of Kentucky, 1973); Lionel Madden, *The Nineteenth-Century Periodical Press in Britain: A Bibliography of Modern Studies 1901–1971* (New York and London: Garland, 1976); Larry K. Uffelman, *The Nineteenth-Century Periodical Press in Britain: A Bibliography of Modern Studies, 1972–1987* (Edwardsville: Southern

Illinois University, 1992); J. Don Vann and Rosemary T. VanArsdel, eds., *Victorian Periodicals: A Guide to Research*, 2 vols. (New York: Modern Language Association of America, 1978–89); and J. Don Vann and Rosemary T. VanArsdel, eds., *Victorian Periodicals and Victorian Society* (Toronto: University of Toronto Press, 1994). More recent surveys include Laurel Brake and Marysa Demoor, eds., *Dictionary of Nineteenth-Century Journalism in Great Britain and Ireland* (Gent, Belgium: Academia Press, and London: British Library, 2009), and Andrew King, Alexis Easley, and John Morton, eds., *The Routledge Handbook to Nineteenth-Century British Periodicals and Newspapers* (London and New York: Routledge, 2016).

13. Various attempts had been made to list and enumerate nineteenth-century scientific and medical journals. See, for instance, Samuel H. Scudder, *Catalogue of Scientific Serials of All Countries Including the Transactions of Learned Societies in the Natural, Physical and Mathematical Sciences, 1633–1876*, reprint (New York: Kraus Reprint Corp., 1965 [1879]); Henry Carrington Bolton, *A Catalogue of Scientific and Technical Periodicals, 1665–1895, Together with Chronological Tables and a Library Check-List*, 2nd edition (Washington: Smithsonian Institution, 1897); and William R. LeFanu, *British Periodicals of Medicine: A Chronological List, 1640–1899*, revised edition, ed. Jean Loudon (Oxford: Wellcome Unit for the History of Medicine, 1984). An important early application of bibliometrics to scientific periodicals is found in Derek J. de Solla Price, *Little Science, Big Science* (New York and London: Columbia University Press, 1963). A number of in-house histories had been produced by journals and societies. See, for example, Robert J. Rowlette, *The Medical Press and Circular, 1839–1939: A Hundred Years in the Life of a Medical Journal* (London: [The Medical Press and Circular], 1939); R. Fish, "The Library and Scientific Publications of the Zoological Society of London," in *The Zoological Society of London, 1826–1976 and Beyond*, ed. Solly Zuckermann (London: Academic Press, 1976), 233–52; and *"The Economist," 1843–1943: A Centenary Volume* (London: Oxford University Press, 1943). One notable early study of the functioning of a scientific periodical was S. Lilley, "'Nicholson's Journal,' 1797–1813," *Annals of Science* 6 (1948–50): 78–101.

14. See, for instance, the essays by Manten, Meadows, Shaw, and Katzen in A. J. Meadows, ed., *The Development of Science Publishing in Europe* (Amsterdam: Elsevier, 1980); R. M. Gascoigne, *A Historical Catalogue of Scientific Periodicals, 1665–1900: With a Survey of Their Development* (New York: Garland, 1985); and Bruce M. Manzer, *The Abstract Journal, 1792–1920: Origin, Development and Diffusion* (Metuchen, NJ: Scarecrow Press, 1977).

15. Roy MacLeod, et al., "Centenary Supplement," *Nature* 224 (1969): 417–76; A. J. Meadows, *Science and Controversy: A Biography of Sir Norman Lockyer, Founder of "Nature"* (London and Basingstoke: Macmillan, 1972), and *Communication in Science* (London: Butterworths, 1974), ch. 3; W. H. Brock, "The Development of Commercial Science Journals in Victorian Britain," in Meadows, *Development of Science Publishing*, 95–122; W. H. Brock and A. J. Meadows, ed., *The Lamp of Learning: Taylor & Francis and the Development of Science Publishing*, 2nd edition (London: Taylor and Francis, 1998 [1984]); and W. H. Brock, "Brewster as a Scientific Journalist," in *Martyr of Science: Sir David Brewster, 1781–1868*, ed. A. D Morrison-Low and J. R. R. Christie (Edinburgh: Royal Scottish Museum, 1984), 37–42.

16. Brock and Meadows, *Lamp of Learning*, xii, quoted from Susan Sheets-Pyenson, "Popular Science Periodicals in Paris and London: The Emergence of a Low Scientific Culture, 1820–1875," *Annals of Science* 42 (1985): 549–72, on p. 549. See the commentary on Brock and

Meadows' contribution in Jonathan R. Topham, "Technicians of Print and the Making of Natural Knowledge," *Studies in History and Philosophy of Science* 35 (2004): 391–400.

17. Susan Sheets-Pyenson, "A Measure of Success: The Publication of Natural History Journals in Early Victorian Britain," *Publishing History* 9 (1981): 21–36; and "From the North to Red Lion Court: The Creation and Early Years of the *Annals of Natural History*," *Archives of Natural History* 10 (1981): 221–49.

18. David E. Allen, "The Struggle for Specialist Journals: Natural History in the British Periodicals Market in the First Half of the Nineteenth Century," *Archives of Natural History* 23 (1996): 107–23. See also Ray Desmond, "Loudon and Nineteenth-Century Horticultural Journalism," in *John Claudius Loudon and the Early Nineteenth Century in Great Britain*, ed. Elisabeth B. MacDougall (Washington: Dumbarton Oaks Trustees for Harvard University, 1980), 77–97; and Sarah Dewis, *The Loudons and the Gardening Press: A Victorian Cultural Industry* (Farnham, UK: Ashgate, 2014).

19. W. F. Bynum, S. Lock, and R. Porter, eds., *Medical Journals and Medical Knowledge: Historical Essays*, (London and New York: Routledge, 1992); and P. W. J. Bartrip, *Mirror of Medicine: A History of the "British Medical Journal"* (Oxford: British Medical Journal and Clarendon Press, 1990).

20. Bynum, Lock, and Porter, eds., *Medical Journals*, 2.

21. Robert M. Young, "Natural Theology, Victorian Periodicals, and the Fragmentation of a Common Context," in *Darwin's Metaphor: Nature's Place in Victorian Culture* (Cambridge: Cambridge University Press, 1985), 126–63, on pp. 155–56. For a critique of Young's claims, see Geoffrey Cantor et al., *Science in the Nineteenth-Century Periodical: Reading the Magazine of Nature* (Cambridge: Cambridge University Press, 2004), introduction.

22. Charles Bazerman, *Shaping Written Knowledge: The Genre and Activity of the Experimental Article in Science* (Madison: University of Wisconsin Press, 1988); Dwight Atkinson, *Scientific Discourse in Sociohistorical Context: The "Philosophical Transactions of the Royal Society of London," 1675–1975* (London: Routledge, 1998); and Alan G. Gross, Joseph E. Harmon, and Michael S. Reidy, *Communicating Science: The Scientific Article from the Seventeenth Century to the Present* (New York: Oxford University Press, 2002). For critical perspectives on the genre of the scientific article, see, e.g., Csiszar, *Scientific Journal*; and Thomas Broman, "J. C. Reil and the 'Journalization' of Physiology," in *The Literary Structure of Scientific Argument: Historical Studies*, ed. Peter Dear (Philadelphia: University of Pennsylvania Press, 1991), 13–42.

23. Sheets-Pyenson, "Popular Science Periodicals," 551, 554.

24. Susan Sheets-Pyenson, "Darwin's Data: His Reading of Natural History Journals, 1837–1842," *Journal of the History of Biology* 14 (1981): 231–48.

25. Sheets-Pyenson, "Popular Science Periodicals," 555; see also Ruth Barton, "Just Before *Nature*: The Purposes of Science and the Purposes of Popularization in Some English Popular Science Journals of the 1860s," *Annals of Science* 55 (1998): 1–33.

26. Cantor et al., *Science in the Nineteenth-Century Periodical*, esp. introduction and ch. 8; Louise Henson et al., eds., *Culture and Science in the Nineteenth-Century Media* (Aldershot, UK: Ashgate, 2004); Geoffrey Cantor and Sally Shuttleworth, eds., *Science Serialized: Representations of the Sciences in Nineteenth-Century Periodicals* (Cambridge, MA, and London: MIT Press, 2004).

27. See, for instance, Jonathan R. Topham, "Scientific Publishing and the Reading of Science in Nineteenth-Century Britain: A Historiographical Survey and Guide to Sources," *Studies in History and Philosophy of Science* 31A (2000): 559–612; and "Scientific Readers: A View from the Industrial Age," *Isis* 95 (2004): 431–42.

28. Secord, "Knowledge in Transit"; Topham, "Rethinking the History"; Bernard Lightman, *Victorian Popularizers of Science: Designing Nature for New Audiences* (Chicago: University of Chicago Press, 2007).

29. Secord, "Science, Technology, and Mathematics," 451. For an earlier historical reevaluation of peer review, see John C. Burnham, "The Evolution of Editorial Peer Review," *Journal of the American Medical Association* 263 (1990): 1323–29.

30. Topham, "Anthologizing the Book of Nature"; Iain Watts, "'We Want No Authors': William Nicholson and the Contested Role of the Scientific Journal in Britain, 1797–1813," *British Journal for the History of Science* 47 (2014): 397–419; Jonathan R. Topham, "The Scientific, the Literary and the Popular: Commerce and the Reimagining of the Scientific Journal in Britain, 1813–1825," *Notes and Records of the Royal Society* 70 (2016): 305–24.

31. Fyfe, McDougall-Waters, and Moxam, "350 Years"; Aileen Fyfe, "Journals, Learned Societies and Money: *Philosophical Transactions* ca. 1750–1900," *Notes and Records of the Royal Society* 69 (2015): 277–99; Aileen Fyfe and Noah Moxham, "Making Public Ahead of Print: Meetings and Publications at the Royal Society, 1752–1892," *Notes and Records of the Royal Society* 70 (2016): 361–79; and Noah Moxham and Aileen Fyfe, "The Royal Society and the Prehistory of Peer Review, 1665–1965," *Historical Journal* 61 (2017): 863–89.

32. Csiszar, *Scientific Journal*.

33. Fyfe, McDougall-Waters and Moxam, "350 Years," 227–28.

34. Csiszar, *Scientific Journal*; Thomas Broman, "Periodical Literature," in *Books and the Sciences in History*, ed. Marina Frasca-Spada and Nick Jardine (Cambridge: Cambridge University Press, 2000), 225–38.

35. Aileen Fyfe, "Journals and Periodicals," in *A Companion to the History of Science*, ed. Bernard Lightman (Chichester, UK: Wiley Blackwell, 2016), 387–99, on pp. 395–96.

36. David Cahan, "Institutions and Communities," in *From Natural Philosophy to the Sciences: Writing the History of Nineteenth-Century Science*, ed. David Cahan (Chicago: University of Chicago Press, 2003), 291–328, on p. 293.

37. Benedict R. O. Anderson, *Imagined Communities: Reflections on the Origin and Spread of Nationalism* (London: Verso, 1983), 36 and passim.

38. On the Red Lion Club, see Daniel Brown, *The Poetry of Victorian Scientists: Style, Science and Nonsense* (Cambridge: Cambridge University Press, 2013), 89–109; on the X Club, see Ruth Barton, *The X Club: Power and Authority in Victorian Science* (Chicago: University of Chicago Press, 2018). On the importance of sociability to nineteenth-century scientific communities more generally, see Hannah and John W. Gay, "Brothers in Science: Science and Fraternal Culture in Nineteenth-Century Britain," *History of Science* 35 (1997): 425–53.

39. Paul White, "Ministers of Culture: Arnold, Huxley and Liberal Anglican Reform of Learning," *History of Science* 43 (2005): 115–38, on p. 127.

40. [John Claudius Loudon], "Introduction," *Gardener's Magazine* 1 (1826), 1–9, on p. 1. On female readers of botanical publications, see Dewis, *Loudons and the Gardening Press*, 178–86.

41. See Claire Brock, "Lankester, Phebe (1825–1900)," in Brake and Demoor, eds., *Dictionary of Nineteenth-Century Journalism*, 347.

42. For the magazine's own account of its involvement, see anonymous, "Popular Political Economy," *Mechanics' Magazine*, 16 June 1827, pp. 378–84, esp. pp. 381–84.

43. Nick Hopwood, Simon Schaffer, and James A. Secord, "Seriality and Scientific Objects in the Nineteenth Century," *History of Science* 48 (2010): 251–85, on pp. 271 and 278.

44. On the adaption of serial publication to scientific purposes, see Gowan Dawson, "Paleontology in Parts: Richard Owen, William John Broderip, and the Serialization of Science in Early Victorian Britain," *Isis* 103 (2012): 637–67.

45. Jeanne Pieffer, Maria Conforti, and Patrizia Delpiano, "Introduction [to Special Issue on 'Scholarly Journals in Early Modern Europe: Communication and the Construction of Knowledge']," *Archives internationales d'histoire des sciences* 63 (2013): 5–24, on p. 14.

46. See, for instance, Ursula Klein, *Experiments, Models, Paper Tools: Cultures of Organic Chemistry in the Nineteenth Century* (Stanford, CA: Stanford University Press, 2003); and Anke te Heesen, "The Notebook: A Paper-Technology," in *Making Things Public: Atmospheres of Democracy*, ed. Bruno Latour and Peter Weibel (Cambridge, MA: MIT Press, 2005), 582–89.

47. Janet Browne, "Corresponding Naturalists," in *The Age of Scientific Naturalism: Tyndall and His Contemporaries*, ed. Bernard Lightman and Michael S. Reidy (London: Pickering & Chatto, 2014), 157–69, on p. 158; and Anne Secord, "Corresponding Interests: Artisans and Gentlemen in Nineteenth-Century Natural History," *British Journal for the History of Science* 27 (1994): 383–408. On views concerning the division of labor in nineteenth-century science, see Timothy Alborn, "The Business of Induction: Industry and Genius in the Language of British Scientific Reform, 1820–1840," *History of Science* 34 (1996): 191–221.

48. Watts, "We Want No Authors."

49. Csiszar, "How Lives Became Lists," 58. See also Csiszar, *Scientific Journal*; and Moxham and Fyfe, "The Royal Society."

50. The last part of James A. Secord, *Victorian Sensation: The Extraordinary Publication, Reception, and Secret Authorship of "Vestiges of the Natural History of Creation"* (Chicago: University of Chicago Press, 2000), addresses the divergent imagined "futures of science" in 1840s Britain. See also James A. Secord, *Visions of Science: Books and Readers at the Dawn of the Victorian Age* (Oxford, UK: Oxford University Press, 2014).

51. Neylon, "Communities Need Journals," 385.

* 1 *

New Formats for New Readers

Scientific, Medical, and Technical Periodicals in Nineteenth-Century Britain: New Formats for New Readers

Gowan Dawson and Jonathan R. Topham

In the *Monthly Review* for July 1799, the Cambridge mathematician Robert Woodhouse reflected on the effect on the "intellectual character of society" of what he presented as the first of a new kind of scientific journal—William Nicholson's *Journal of Natural Philosophy*. It was, he considered, part of a recent and larger development:

> When presses multiplied, and restraints were removed from them,—when writing became a trade, and the love of gain operated with the love of fame as motives to authorship,—the number of literary productions increased, and their nature was changed: the serious and unremitted devotion of twenty or thirty years, to the study of a particular science, was no longer considered as a necessary preparation for a work; and when a person imagined that he had some information to communicate, the means were ready.

Journals such as Nicholson's, he claimed, like the "Epitomes, Abstracts, Synopses, [and] Abridgements" that had become commonplace of late, had the regrettable effect of removing incentives for profound learning. Importantly, however, their positive achievement was that they answered the desire for knowledge that had spread itself "through all ranks." They acted as an intellectual manure that produced "more uniform utility over the whole soil" than could be produced by lumpen works of individual scholarship.[1] For Woodhouse, Nicholson's innovation thus amounted to a democratizing impulse: the new scientific journal invited the involvement of a far larger proportion of the populace in the project of natural enquiry.

As the introduction to this volume has demonstrated, the history of scientific, medical, and technical periodicals in nineteenth-century Britain is not one that can be read through the character or functions of modern scientific journals. On the contrary, the increasing range of periodicals that came into existence took a wide range of forms and functioned in a similarly wide range of ways. Only by degrees—and especially toward the end of the century—did a proportion of these periodicals come to resemble modern academic journals to a significant extent. Throughout, the history was punctuated by the production of new types of periodicals aimed at new audiences. As Woodhouse had noted on the eve of the new century, changes in the culture of print opened up new possibilities for configuring knowledge communities, and editors, publishers, and societies were quick to seize those opportunities in shaping new kinds of scientific periodicals. Charting the resultant developments in the forms and audiences of scientific periodicals is an indispensable first step in understanding their changing role in constructing scientific communities. In this chapter, therefore, we offer a broad and inevitably tentative overview of the development of scientific, medical, and technical periodicals over the course of the nineteenth century, seeking to map some of the distinctive features of that history.

Getting to grips with the scale and diversity of scientific, medical, and technical periodicals in nineteenth-century Britain is, however, a major challenge. Despite the existence of several catalogues, no satisfactorily comprehensive listing exists. The most systematic is the bibliography of British and colonial medical periodicals prepared in 1937 by W. R. LeFanu, the librarian of the Royal College of Surgeons of England, and expanded in 1984 by Jean Loudon, although even this is far from complete. It does, however, enable the historian to gain something of a sense of the rate of growth across the period, from nine British medical titles in 1800 to more than 150 in 1900.[2] Gaining a comparable sense for scientific titles is less straightforward. Robert Gascoigne's *Historical Catalogue of Scientific Periodicals, 1665–1900* (1985) lists 124 British titles, selected on the basis of the degree of their use by "scientists of the time," as measured by their citation in leading catalogues of scientific papers, rising from 11 in 1800 to 110 in 1900. The *Catalogue of Scientific Serials, 1633–1876*, compiled in 1879 by the American naturalist Samuel Scudder, is more extensive, listing a total of more than 550 titles published in Britain and Ireland. More extensive still (although excluding society publications) is the *Catalogue of Scientific and Technical Periodicals, 1665–1895*, prepared by the American chemist and bibliographer Henry Bolton, which includes more than 800 British titles, rising from 36 in 1800 to 330 in 1900. Bolton's list, however, includes numerous general titles and many titles that reflect a very inclusive definition of "scientific and technical."[3] This inclusivity becomes quite unmanageable in John North's *Waterloo Directory of Newspapers and Periodicals, 1800–1900*, series two of which (out of a planned five) lists more than five thousand titles with the subject key words "science" or "medicine," but includes under these key words such general titles as *Acworth's Ealing Illustrated Magazine and General Advertiser* and the *Agnostic Annual*.[4]

The growth and diversification of scientific, medical, and technical periodicals in nineteenth-century Britain is clearly in part a reflection of larger patterns in periodical production during this age of the industrialization of print. Figures from the *Waterloo Directory* suggest something in the order of a tenfold increase in the number of periodical titles across the century, while average circulation also increased very markedly. It is not surprising that from early on in the nineteenth century, commentators felt themselves to be living in an age in which periodicals exerted a new dominance. That dominance, relative to other types of publication, only increased as the century progressed.[5] Statistical analysis of the entries in LeFanu's list of medical periodicals and Bolton's list of scientific periodicals suggests that the pattern here was broadly comparable (fig. 1.1). Indeed, scientific, medical, and technical periodicals were affected by many of the factors that caused the growth of periodical literature more generally, including the expansion and diversification of reading audiences and the wholesale industrialization of print manufacture and distribution. In what follows, we connect the history of scientific periodicals to that larger history, exploring features that are common as well as those that are distinct.

This chapter begins by reviewing the surprisingly large range of scientific, medical, and technical periodicals that existed in Britain at the turn of the nineteenth century, including a growing array of both society publications and commercial journals, before examining the rapid expansion in types of periodicals after 1815. In these years, the increased commercialization and mechanization of the book trade, together with the growth of reading audiences, underpinned a variety of initiatives to produce new types of scientific periodicals, including cheap weeklies and natural history monthlies. Moreover, learned societies began to emulate the commercial press with new forms of society publication. The second section examines the middle years of the century, when many of the efforts to produce commercial periodicals on particular scientific subjects met with financial disaster but society publications proliferated, and when an increasing range of commercial periodicals were directed at

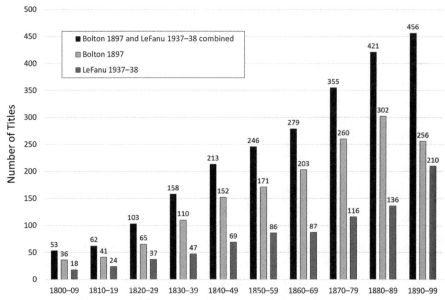

FIG. 1.1. Scientific, medical and technical periodical titles listed in Henry Carrington Bolton's *A Catalogue of Scientific and Technical Periodicals, 1665–1895, Together with Chronological Tables and a Library Check-List*, 2nd ed. (Washington: Smithsonian Institution, 1897); and W. R. LeFanu's "British Periodicals of Medicine: A Chronological List. Part 1: 1684–1899," *Bulletin of the Institute of the History of Medicine* 5 (1937): 735–61, 827–55. The chart displays the number of journal titles in publication in each decade of the nineteenth century. The entries in Bolton's and LeFanu's lists were cross-referenced with entries in COPAC (www.copac.ac .uk), Zeitschriftendatenbank (www.zeitschriftendatenbank.de), the Hathi Trust Digital Library (www.hathitrust.org), and the US National Library of Medicine's PubMed catalogue (www .ncbi.nlm.nih.gov/pubmed/) to establish the exact publication dates of each title. Where this was not possible, or where information was inconsistent, journal titles were excluded from the data sample. The "combined" category records titles that are listed in Bolton's list, LeFanu's list, or in both, and is hence smaller than the sum of the two component parts. We are grateful to Konstantin Kiprijanov for compiling the data and producing this chart; the original data are available for download at conscicom.web.ox.ac.uk.

occupational groups, including not only medical practitioners but also miners, engineers, architects, and pharmaceutical chemists. The final section explores the changed circumstances of the periodical press in the late Victorian period, following the removal of taxes in the 1850s and 1860s and including further technological developments in book production and distribution. It charts the establishment of commercially sustainable journals in a growing range of scientific fields, and the growth of scientifically important "popular science" journals—a context out of which *Nature* emerged—while also showing that the period witnessed a continuing growth of occupationally oriented journals, not least in relation to medicine. Finally, it explores the emergence of university-oriented professional journals in the last decades of the century.

EARLY NINETEENTH-CENTURY PERIODICALS

While the number and range of scientific, medical, and technical periodicals published in Britain at the dawn of the nineteenth century was distinctly restricted by the standards of that century, close inspection reveals it to have been greater than historians might expect. The form of scientific periodical of the longest standing was the volume of transactions, published by a learned society. Of these, the oldest was of course the Royal Society of London's *Philosophical Transactions*, which, after being a private endeavor for almost a century, was taken firmly under the society's control in 1752, becoming formally tied to the society's other activities.[6] Over the final years of the century the practice of issuing transactions was emulated by new learned societies elsewhere in the kingdom—namely, the Manchester Literary and Philosophical Society in 1785, the Royal Irish Academy in 1787, the Royal Society of Edinburgh in 1788, the Society of Antiquaries of Scotland in 1792, and the Dublin Society in 1799—as well as by the Asiatic Society of Bengal in 1788. They were also emulated in the capital by new, more specialist bodies—notably, the Royal College of Physicians in 1768, the Society of Antiquaries in 1770, the Society of Arts in 1783, the Medical Society in 1787, and the Linnean Society in 1791. These publications, typically in large (quarto) format, were rooted in the learned societies they served, and were not typically produced with a view to financial return, though they might find themselves subject to financial constraints.[7] And as Alex Csiszar discusses in chapter 3 of this volume, their lengthy and "polished memoirs" were far removed from the "articles" of commercial journalism.

Alongside the learned transactions, a small but increasing number of commercial medical, technical, and scientific periodicals began to appear in Britain over the closing decades of the eighteenth century, with more than a dozen in production by 1800. These ranged from well-established medical, agricultural, and astronomical periodicals to more recently launched titles in natural history, natural philosophy, and the practical arts. As Thomas Broman has shown, these commercial scientific periodicals need to be seen in relation to a larger transformation. The rapid growth of the periodical press in the eighteenth century contributed to the development of a novel sense among the educated of their participation in "a new collectivity, the 'public,'" that was altogether more open and critical than were the learned academies, which were typically rooted in the patronage structures of a monarchical and aristocratic state, though still fairly small and restricted, and quite distinct from the mass reading public that arose in the following century.[8] At first, scientific, medical, and technical subjects were topics of extensive discussion in the new general magazines and reviews, such as the *Gentleman's Magazine* (1731–1907) and the *Monthly Review* (1749–1844).[9] It was not long, however, before editors and booksellers began to exploit the opportunities within this public sphere to target particular sets of readers with periodicals that were more selective in their content.

As contemporaries observed, Britain lagged somewhat behind leading Continental countries—notably Germany, where universities supplied both editors and readers—in developing such periodicals.[10] Yet certain better-defined groups of consumers seemed worth the speculation in a print market that was increasing in size and entrepreneurial ambition. Arguably the most promising and certainly one of the earliest markets targeted was the one for medical periodicals, where recognizable occupational groups existed whose interests were quite distinct. An increasing range of titles began to appear from the 1760s onwards, several of which secured lasting success, including Andrew Duncan's *Medical and Philosophical Commentaries* (1773–95), the *London Medical Journal* (1781–90), and the *Medical and Chirurgical Review* (1794–1808).[11] Another early market was the one relating to agriculture, where the many improving landowners helped to sustain periodicals such as Arthur Young's *Annals of Agriculture* (1784–1815) and the *Commercial and Agricultural Magazine* (1799–1816).[12] A further occupational group, mariners, provided a market for a very particular type of periodical, the Board of Longitude's *Nautical Almanac* (1766–), and there continued to be a steady market for the astronomical and meteorological data in a range of widely sold almanacs. Beyond the functional, the annual *Gentleman's Diary* (1741–1840) and *Ladies' Diary* (1704–1840) were striking in how they sustained readerships over many decades with their annual offering of entertaining mathematical problems.[13] A handful of other more strictly mathematical periodicals were more ephemeral, but around the turn of the century the *Mathematical, Geometrical and Philosophical Delights* (1793–98) was one of a number of more broadly based periodicals, containing mathematical and other problems, that were produced for schoolchildren and others seeking to educate themselves.[14]

The last two decades before the turn of the century saw significant innovation in the production of periodicals in natural history and natural philosophy. In 1787, the first and the longest-lived of a group of serialized natural history publications appeared in the form of the monthly *Botanical Magazine* of the apothecary turned naturalist William Curtis. With its colored plates and botanical descriptions, Curtis's magazine found a steady market among both naturalists and connoisseurs. Numerous periodicals emulated it over succeeding decades, including zoological as well as botanical titles, as in the case of the *Naturalist's Miscellany* (1789–1813).[15] In the 1790s, several speculators finally concluded that enough of a market might now exist to sustain a "philosophical" or "scientific journal" like those that had begun to appear on the Continent in recent decades. In the space of just eight years, no fewer than six journals were commenced, most by projectors connected as much with the practical arts as with learned natural philosophy. The periodicals they began—including John Wyatt's *Repertory of Arts and Manufactures* (1794–1862), William Nicholson's *Journal of Natural Philosophy* (1797–1813), and Alexander Tilloch's *Philosophical Magazine* (1798–)—were established with the claim that learned philosophers and practical men alike needed cheaper and easier access to the details of the progress of science than could be obtained from Europe's learned transactions. They consequently offered anthologies of discovery that included some original matter, but laid particular store by the reprinting, abstracting, and translation of materials published elsewhere. This approach had the desirable effect of maximizing the market for pioneering products that did not, in Britain, map on to a sizeable occupational group. The editors' imagined audience drew together university professors, manufacturers, and even "young persons" with the offer of an anthology of scientific discovery that met their divergent interests. Moreover, their vision yielded journals that were not only financially sustainable but valued by a range of those with scientific interests.[16]

The success of these new commercial philosophical journals, with their broad appeal to "public" authority, soon created concerns for the learned societies. Offering to public scrutiny an alternative and sometimes much earlier means of access to the activities of such societies, the journals prompted the ire of Royal Society president Sir Joseph Banks, who sought to wrest back control.[17] Before long, however, some of the recently founded more specialist societies began to cooperate with the commercial philosophical journals to offer their own authorized accounts of their proceedings, as is explored in detail in chapter 3 of this book. The new societies founded to cater to emerging scientific disciplines, such as the Geological (f. 1807) and the Astronomical (f. 1820), continued to emulate the Royal Society in accumulating volumes of transactions or memoirs. In the 1820s, however, both of these societies cooperated with their printer, Richard Taylor, to include accounts of their proceedings in his *Philosophical Magazine*. By 1827, members of each were receiving specially offprinted "proceedings" that soon became separate journals in their own right. Moreover, this new mode of appeal to the "public" developed apace. Old societies, notably the Royal Society in 1831, adopted the form alongside existing transactions; new societies, such as the Zoological Society (f. 1826), could choose whether to publish transactions, containing longer memoirs, as well as proceedings, offering much more succinct accounts of scientific findings.[18]

Meanwhile, others had perceived opportunities in the new form of scientific periodical. The early "scientific journals" had been produced largely at the initiative of entrepreneurial editors, for whom they had generated a healthy income. However, even as early as 1805, fashionable and established publishers such as John Murray, Archibald Constable, Longman, and Richard Phillips were interested in purchasing Tilloch's journal as a successful and highly regarded publication. In the dozen years between 1813 and 1825, these commercial publishers joined up with more "learned" and "philosophical" editors to publish titles with larger ambitions, namely chemist Thomas Thomson's *Annals of Philosophy* (1813–26), chemist William Thomas Brande's *Journal of Science and the Arts* (1816–30), natural philosopher David Brewster and naturalist Robert Jameson's *Edinburgh Philosophical Journal* (1819–64), and Brewster's breakaway *Edinburgh Journal of Science* (1824–32). At a moment when the encyclopedic monthly magazine of the eighteenth century was morphing into the self-consciously literary magazine, of which *Blackwood's Edinburgh Magazine* (1817–1980) was the type, publishers and editors alike expected to be able to reach a similarly large readership with well-financed and reputable magazines that were exclusively "scientific," though it would entail adopting a more "popular" mode of address. The publishers threw large sums at their speculations, only to discover that their sought-for market was nothing like so large as they had believed. By 1832 the total number of the general "scientific journals" surviving was just two: the *Philosophical Magazine* and Jameson's *Edinburgh New Philosophical Journal*.[19]

One of the reasons for the failure of these periodicals was the further diversification of periodical literature that took place at this period. At the wealthier end of the market, new weekly journals such as the *Literary Gazette* (1817–63) and the *Athenaeum* (1828–1921) sought to combine literary with scientific news. Moreover, working-class entrepreneurs began to exploit the market for cheap weekly periodicals developed by the postwar radical press before a terrified government introduced the "Six Acts" of 1819 to suppress it through the extension of libel laws and stamp duties. The new form of the cheap apolitical weekly—popularized by the *Mirror of Literature, Amusement and Instruction*

(1822–47)—was soon taken up by the Edinburgh-educated radical journalist Thomas Hodgskin and the Scots patent agent and writer Joseph Clinton Robertson to produce their three-penny weekly *Mechanics' Magazine* (1823–72), intended for "that numerous and important portion of the community, the Mechanics or Artisans," who, it was intended, would be able to say, "This is ours and for us."[20] Not only did the new magazine contribute significantly to the development of the mechanics' institute movement, but it soon spawned a number of similar titles, including the *Glasgow Mechanics' Magazine* (1824–26) and the *London Mechanics' Register* (1824–28), as well as Hodgskin's own *Chemist* (1824–25). The manner in which the cheap weeklies used the new technologies of the increasingly mechanized book trades, such as machine-made paper and stereotype, to cheapen the price of print and expand the market was soon emulated by others, including the Society for the Diffusion of Useful Knowledge, so that contemporaries felt that theirs had become the era of the "march of mind."[21] By the early 1830s, the cheap weekly journal had reached new heights of success in the form of periodicals such as the *Penny Magazine* (1832–45) and *Chambers's Edinburgh Magazine* (1832–1956), which mixed scientific content with a broad array of other matter.

The format of the cheap weekly was also adopted by the London surgeon Thomas Wakley to produce a radically new kind of medical periodical. With more than twice the number of pages, and costing sixpence, Wakley's *Lancet* (1823–) nevertheless used the production strategies of the other new cheap weeklies produced in the same neighborhood of the Strand (fig. 1.2). It also emulated the journalism of postwar political weeklies, notably *Cobbett's Weekly Political Register* (1802–36), in its pugnacious engagement with the medical politics of the day.[22] A number of new medical monthlies and quarterlies had been founded over the preceding two decades, many produced by the emerging specialists in the increasingly vigorous world of medical publishing and edited by ambitious medical practitioners.[23] However, the *Lancet* marked a radical departure from the established formats, offering above all a vibrant sense of rapid progress in both medical knowledge and the medical community. Its ambition to reach a much wider medical readership than previous titles, including medical students and the great bulk of general practitioners, was soon rewarded. In 1824, Wakley claimed more than ten thousand readers, though circulation reportedly settled at around four thousand.[24] Unsurprisingly, Wakley's combative approach encouraged the conservative medical placeholders to find their own vehicle, and the *London Medical Gazette* (1827–51) became one of a series of competitors with the *Lancet* in the weekly market over succeeding decades.

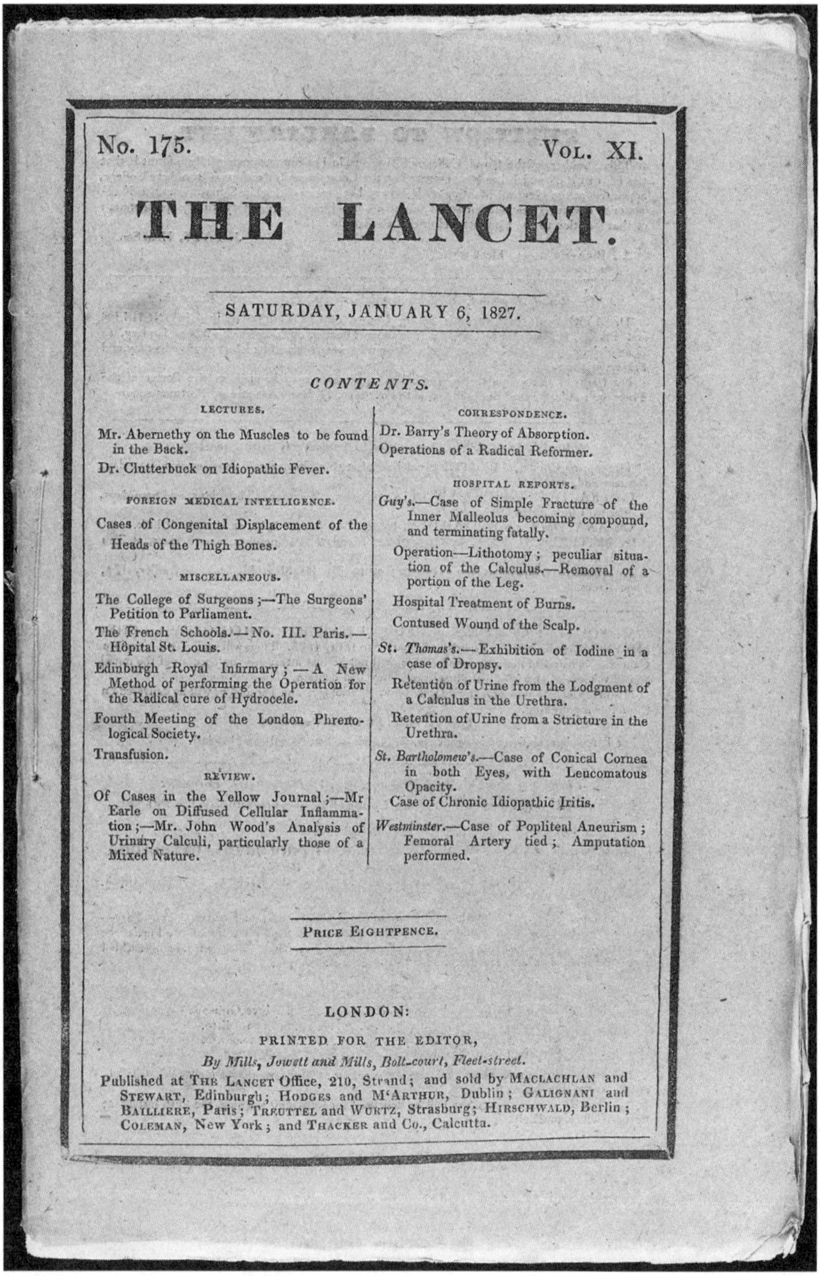

No. 175. Vol. XI.

THE LANCET.

SATURDAY, JANUARY 6, 1827.

CONTENTS.

LECTURES.

Mr. Abernethy on the Muscles to be found in the Back.

Dr. Clutterbuck on Idiopathic Fever.

FOREIGN MEDICAL INTELLIGENCE.

Cases of Congenital Displacement of the Heads of the Thigh Bones.

MISCELLANEOUS.

The College of Surgeons;—The Surgeons' Petition to Parliament.

The French Schools.—No. III. Paris.—Hôpital St. Louis.

Edinburgh Royal Infirmary; —A New Method of performing the Operation for the Radical cure of Hydrocele.

Fourth Meeting of the London Phrenological Society.

Transfusion.

REVIEW.

Of Cases in the Yellow Journal;—Mr Earle on Diffused Cellular Inflammation;—Mr. John Wood's Analysis of Urinary Calculi, particularly those of a Mixed Nature.

CORRESPONDENCE.

Dr. Barry's Theory of Absorption.

Operations of a Radical Reformer.

HOSPITAL REPORTS.

Guy's.—Case of Simple Fracture of the Inner Malleolus becoming compound, and terminating fatally.

Operation—Lithotomy; peculiar situation of the Calculus.—Removal of a portion of the Leg.

Hospital Treatment of Burns.

Contused Wound of the Scalp.

St. Thomas's.—Exhibition of Iodine in a case of Dropsy.

Retention of Urine from the Lodgment of a Calculus in the Urethra.

Retention of Urine from a Stricture in the Urethra.

St. Bartholomew's.—Case of Conical Cornea in both Eyes, with Leucomatous Opacity.

Case of Chronic Idiopathic Iritis.

Westminster.—Case of Popliteal Aneurism; Femoral Artery tied; Amputation performed.

Price Eightpence.

LONDON:

PRINTED FOR THE EDITOR,

By Mills, Jowett and Mills, Bolt-court, Fleet-street.

Published at The Lancet Office, 210, Strand; and sold by Maclachlan and Stewart, Edinburgh; Hodges and M'Arthur, Dublin; Galignani and Bailliere, Paris; Treuttel and Würtz, Strasburg; Hirschwald, Berlin; Coleman, New York; and Thacker and Co., Calcutta.

FIG. 1.2. *Lancet*, 6 January 1827, cover. Wellcome Collection. Creative Commons attribution license.

A further important innovation at the end of the 1820s was the appearance of the first scientific periodical of any note devoted to a particular scientific field. From the 1780s onwards, a growing number of periodicals had offered naturalists, collectors, and gardeners regular depictions and descriptions of plant and animal species, including such new titles as the *Botanist's Repository* (1797–1815), the *Botanical Register* (1815–47), the *Botanical Cabinet* (1817–33), the *British Flower Garden* (1823–38), and the *Botanic Garden* (1824–51). In many ways, however, these were more similar to other serialized part-works in natural history than to periodicals, in that they provided next to no space for a temporally extended conversation. John Loudon's *Magazine of Natural History* (1828–40), with which the introduction to this volume began, offered something altogether more ambitious. As chapters 2 and 5 discuss, this magazine and Loudon's earlier *Gardeners' Magazine* (1826–44), were notable in finding a new readership for such subject-specific fare, leading to a decade of vigorous experimentation that was, nevertheless, rather limited in its success.

MID-NINETEENTH-CENTURY PERIODICALS

One of the great novelties of the 1830s, then, was the rush of editors and publishers to exploit the market that Loudon had found for periodicals on natural history and horticulture. Over the course of the decade, around twenty new periodicals were commenced focusing on these sometimes overlapping subjects, which exhibited a significant variety of approaches in terms of subject matter and intended audience, ranging from the learned bimonthly *Magazine of Zoology and Botany* (1836–38), priced at three shillings and six pence, to the populist monthly *Edinburgh Journal of Natural History and of the Physical Sciences* (1835–40), issued at one shilling in the large format of *Chambers's Edinburgh Magazine* and including Cuvier's *Animal Kingdom* in serialized parts.[25] The monthly and then quarterly *Analyst* (1834–40), which combined significant natural history content with other topics, was the groundbreaking production of the editor of the *Worcester Herald*, William Holl, aimed at a primarily provincial audience among the "intellectual residents" of the West Midlands and interacting significantly with local societies.[26] The sobering reality, however, was that, despite the growing market for print and the increasing use of such technological novelties as machine-made paper and wood engraving, managing the finances for such specialized periodicals was challenging, especially when costs were significantly increased by the notorious "taxes on knowledge" (paper, advertisement, and stamp duties) and by postal charges, which affected provincial ventures disproportionately.[27] Only five of the new titles survived beyond 1844. Of these, two were monthlies directed at gardeners—the hugely successful sixpenny *Floricultural Cabinet* (1833–1916) and the more upmarket *Paxton's Magazine of Botany* (1834–49)—and one was an innovative weekly horticultural newspaper, the *Gardeners' Gazette* (1837–80), probably made possible by the reduction of stamp duty in 1836 from four pence to one penny. Among the new journals, William Jackson Hooker's *Journal of Botany* (1829–57; begun as the *Botanical Miscellany* and with further name changes) and the *Annals and Magazine of Natural History* (1838–1966; begun as the *Annals of Natural History*) alone endured as publications of scientific repute.[28]

In addition to these commercial ventures, however, the 1830s also witnessed the beginnings of what grew to become a nationwide abundance of natural history societies, many of which issued periodicals. Thus, alongside the proceedings (1830–1965) and transactions (1835–1984) of the Zoological Society, the transactions (1834–1932) of the Entomological Society, and the annual report and proceedings (1837–44) of the Botanical Society of Edinburgh, appeared the transactions (1831–38) of the Natural History Society of Northumberland and the proceedings (1834–) of the Berwickshire Naturalists' Club. More generally, the number of specialist scientific societies in London and elsewhere that were issuing periodicals continued to increase, with new metropolitan titles including publications from the Geographical Society (1832–80), the Statistical Society (1834–37, 1838–), the Institution of Civil Engineers (1836–42, 1837–), the Pharmaceutical Society (1841–43), the London Electrical Society (1841), and the Chemical Society (1843–48). Likewise, the number and range of observatory publications continued to grow, with publications commencing at Cambridge in 1829 and Edinburgh in 1834, and with the Greenwich Observatory issuing *Results of Magnetical and Meteorological Observations* from 1840. A particular novelty was the publication from 1833 of the *Report* of the annual meeting of the British Association for the Advancement of Science, offering synoptic reports on the state of several sciences, as well as highly condensed reports on a selection of papers.

In the commercial arena, however, the bitter experience of previous decades and the ever-increasing number of society publications meant that the *Philosophical Magazine* and the *Edinburgh New Philosophical Journal* suffered little further competition as general science journals. Prospective editors and publishers seem to have been chastened by a sense of the smallness of the potential readership for periodicals serving the increasingly arcane demands of the special sciences, in relation to the high costs of production. In 1837, the printer and coeditor of the *Philosophical Magazine*, Richard Taylor, asserted that the journal had never covered its expenses, and that the failure of scientific journals was solely due to the costs incurred.[29] As printer of a large proportion of the publications of learned societies, and publisher of the *Philosophical Magazine* and the *Annals and Magazine of Natural History*, Taylor was the dominant figure in the production of science periodicals. Moreover, he was one of the few innovators at this period, and his business, continued by his illegitimate son, William Francis, maintained a dominant role throughout the century. Taylor's *Scientific Memoirs* (1837–52) was an earnest endeavor to bring British men of science up to date with Continental work by translating selected articles; with a sale of fewer than two hundred copies, however, it was a loss-making "altruistic enterprise."[30] Another notable novelty was the *Cambridge Mathematical Journal* (1837–54), the editors of which sought out a small but sustainable readership rooted in the university's mathematical community, again seeking to make Continental work available.[31] Striking in its singularity was William Sturgeon's *Annals of Electricity* (1836–43). A prominent electrical lecturer, demonstrator, and inventor, Sturgeon struggled against his status as an outsider to London's learned circles, and commenced his journal having had his first paper on electromagnetic machines turned down for publication in the *Philosophical Transactions*. Like the London Electrical Society he was instrumental in founding, the new journal was in part his attempt to "forge a constituency for himself." With subscriptions remaining low, however, it lasted only a few years.[32]

A very notable development in the 1830s was the growth of periodicals related to particular occupational and professional interests that had some bearing on scientific subjects. The 1820s had seen the establishment of additional technical periodicals (e.g., the *London Journal of Arts and Sciences*, 1820–67) and agricultural titles (e.g., the *Quarterly Journal of Agriculture*, 1828–68). Now, however, such work-oriented periodicals diversified significantly. Notable examples of this trend include the monthly *Veterinarian* (1828–1902), *United Service Journal* (1832–1920), *Architectural Magazine* (1834–39), *Railway Magazine* (1835–1903), and *Civil Engineer and Architect's Journal* (1837–67), the quarterly *Mining Review* (1830–40), and the annual *Papers on Subjects Connected with the Duties of the Corps of Royal Engineers* (1839–1918). Some of these occupational journals were even issued as weeklies, notably the *Mining Journal* (1835–) and the *Railway Times* (1838–1918), a development aided from 1836 by the reduction of stamp duties to one penny and the complete removal of duty for local delivery from what were now termed "class" journals. Aimed at restricted groups of readers with very specific interests, these journals became increasingly familiar over succeeding years, and among them the number and range of work-related titles continued to grow apace.[33] The booming railways, building, and engineering offered especially promising markets (e.g., the weekly *Railway Record* [1844–1901], *Economist* [1843–], and *Builder* [1843–1966], and the *Quarterly Papers on Engineering* [1843–49]); but, decade by decade, new markets developed.[34] From 1849, for instance, the gas supply industry supported the weekly *Journal of Gas Lighting* (1849–1972), which later contributed significantly to the debate concerning the introduction of domestic electricity.[35] Such weeklies especially benefited in the following decade as advertising duty (1853) and stamp duty (1855) were finally repealed in the face of continued campaigning.

Most of these practically oriented titles remain little studied, especially from the perspective of the history of science. However, one notable growth area in the 1840s relating to pharmacy and chemistry has attracted more attention, not least because of its obvious bearing on learned discussions. The first of a series of successful publications was Charles and John Watt's sixpenny monthly *Chemist* (1840–58), which promised readers an account of "the various discoveries and improvements in chemistry, chemical manufactures and pharmacy," while "protecting the rights of the chemist, the chemical manufacturer and the druggist . . . from the schemes of ignorance and imposture."[36] The journal's contents engaged to a significant extent with contemporary theoretical discussions, and it was soon joined by a more ambitiously theoretical title, Richard Taylor's fortnightly *Chemical Gazette* (1842–59). While still directed to the manufacturer and the "pharmaceutist," this was a spin-off from the *Philosophical Magazine*, placing great emphasis on the abstracting of Continental work, and was edited by the German-educated William Francis (Taylor's son) and Henry Croft. Sales never rose above a few hundred, but the journal survived until its incorporation into William Crookes's *Chemical News* (1859–1932), as discussed below.[37]

The manner in which the larger occupational market (and Taylor and Francis's strong personal commitments) sustained these two chemical periodicals in the 1840s stands in stark contrast to other special sciences. As Gowan Dawson shows in chapter 4 of this book, for instance, attempts in the 1840s to commence commercial periodicals in geology demonstrated that the market was not viable, and the same applied in the case of microscopy.[38] It was in natural history alone that new titles were founded with some success. The printer and naturalist Edward Newman, who had edited the *Entomological Magazine* (1832–38), associated with the Entomological Club, began his own *Entomologist* (1840–42) believing there was a need for more specialized natural history periodicals. Shortly, however, he closed that title to begin the one-shilling monthly *Zoologist* (1843–1916), a "popular miscellany of natural history" similar to Loudon's earlier magazine and published by the natural history specialist John Van Voorst. It was soon a going concern. But while Newman's more specialized *Phytologist* (1841–63) also endured, sales were very low, and it was probably cross-subsidized.[39] Likewise, as chapter 6 of this volume shows, the *Entomologist's Weekly Intelligencer* (1856–61), which Newman later published, was produced by its editor without financial gain in view. Yet, while the natural history market continued to be financially strained, several horticultural periodicals that found success were of considerable significance in natural history. Especially notable was the sixpenny *Gardener's Chronicle* (1841–), founded as a "stamped newspaper of rural economy and general news" by botanist and horticulturalist John Lindley (who edited the horticultural part), horticulturalist Joseph Paxton, and journalist Charles Wentworth Dilke; while at the cheaper end of the market, the gardening writer George W. Johnson's *Cottage Gardener* (1848–1915) appeared in a smaller weekly format at two pence, or three pence stamped for postal delivery.[40]

The market for medical periodicals also grew and diversified across the middle years of the century. One notable feature was the commencement of new titles produced outside the leading medical centers of London and Edinburgh, including the *Glasgow Medical Journal* (1828–1955) and the *Dublin Journal of Medical and Chemical Science* (1832–1922). Especially notable were the endeavors of the Provincial Medical and Surgical Association to support the interests of the growing numbers of English practitioners outside London, which included the issuing of transactions (1833–53) but also prompted a weekly commercial speculation in the shape of the *Provincial Medical and Surgical Journal* (1840–57). Always closely linked with the association, the new journal was soon taken over by it, becoming the *British Medical Journal* (1857–) after the association became national and changed its name to the British Medical Association.[41] Ireland also secured a weekly in the shape of the *Dublin Medical Press* (1839–65), which became the more broadly based *Medical Press and Circular* (1866–1961) after merging with the London *Medical Circular* (1852–66).[42] The *Lancet*'s dominance was also challenged by a further long-lasting weekly founded in London, namely the surgeon and journalist Frederick Knight Hunt's *Medical Times* (1839–85). As professional standards became more stringent, another important innovation was the commencement of periodicals pitched at allowing practitioners to keep abreast of developments, including the *Retrospect of Practical Medicine and Surgery* (1840–1901) and the *Half-Yearly Abstract of the Medical Sciences* (1845–73).

The 1830s also witnessed the commencement of the first long-lived hospital journals, in the shape of *Guy's Hospital Reports* (1836–1974). By the following decade, the first successful periodicals focused on special medical subjects were beginning to appear. The transactions (1846–1907) of the Pathological Society of London were soon followed over the succeeding decade by those of the Epidemiological Society (1855–1907), the Odontological Society (1856–1907), and the Obstetrical Society (1859–1908), and by the *Ophthalmic Hospital Reports and Journal of the Royal London Ophthalmic Hospital* (1857–1917). Strikingly, however, it was in the distinctive field of psychological medicine that a successful commercial journal was first established in 1848. The asylum owner and physician Forbes Winslow's *Journal of Psychological Medicine and Mental Pathology* (1848–83) soon had competition from the *Asylum Journal of Mental Science* (1853–), produced by the Association of Mental Officers of Asylums and Hospitals for the Insane, but the market was sufficiently robust for the two to continue together over several decades. Similarly, the notably early *British Journal of Dental Science* (1856–1935) continued alongside the related society publications.

Such specialist titles became ever more frequent in the later part of the century as the medical profession itself became specialized, but, as Sally Shuttleworth shows in chapter 10 of this volume, another early development was the commencement of commercial journals concerning public health, typically aimed at a readership including (but extending beyond) medical practitioners, such as the *Journal of Public Health and Sanitary Review* (1855–58). A smattering of medical periodicals aimed at informing members of the public about their own health and treatment also appeared, including the penny weekly *Doctor* (1832–37), the eight-penny monthly *Magazine of Health* (1836), and the distinctly unorthodox *Journal of Health* (1848–67). Unorthodox or dissident medical periodicals flowered more generally in the middle decades of the century, as the long-standing *Phrenological Journal* (1823–47) was joined by a wide variety of other titles, including the patent medicine vendor James Morrison's *Hygeist* (1842–67), the Mesmerist John Elliotson's *Zoist* (1843–56), the *British Journal of Homeopathy* (1843–84), the *Monthly Homeopathic Review* (1856–1907), the *Water-Cure Journal* (1847–49), the *Vegetarian Messenger* (1849–1958), and the *Anti-Tobacco Journal* (1858–1900).[43]

LATE-NINETEENTH-CENTURY PERIODICALS

The 1860s witnessed rapid growth in the number of periodicals of all kinds, and scientific, medical, and technical journals were an important component of this general expansion of the mid-Victorian literary marketplace. Not even the most specialist titles were immune to the technological innovations in printing and paper production, increasingly efficient railway and postal distribution networks, and the repeal of paper duties in 1861, which, along with larger, better educated, and more affluent readerships, fueled this veritable boom in periodical publishing. By the end of the century, further technological innovations such as halftone photographic illustrations and mechanical typesetting, and the increasing dominance of publishers with close ties to academia—whether family firms like Macmillans or the university presses at Oxford and Cambridge—facilitated the development of more efficient, technically accurate, and specialist scientific journals. In the final decades of the nineteenth century, as James Secord has noted, a "fundamental transformation took place which created, in broad terms, the publishing regime in which British science would operate for the following century."[44]

One of the most notable developments of the period was that it was now finally possible to sustain a commercial journal on a single, discrete field of science. In the late 1850s the *Chemical News* (1859–1932), edited by William Crookes in partnership with the publishers Charles Mitchell and then Griffin and Bohn, joined the monthly *Geologist* (1858–1864) as a "special-class scientific periodical," as they were termed by the latter's editor and proprietor, Samuel Joseph Mackie.[45] While Mackie's designation echoed the legal category for journals exempt from stamp duty, introduced by the 1836 newspaper act, stamp duty had been abolished in 1855 and the concept of "class" journalism was now applied more loosely to the parallel processes of scientific specialization and market segmentation. The new breed of scientific "class" periodicals each dealt with just a single scientific discipline, and as commercial enterprises were wholly independent of the official societies in their respective fields, of which they were often critical. The other key difference was that these new "class" journals could be strikingly successful when the right formula was hit upon. As was noted in the previous section, chemistry had been one of the few areas that was already able to sustain its own specialist periodicals in the 1840s, but while Richard Taylor's *Chemical Gazette* never sold more than a few hundred copies (three hundred in 1848, for instance), Crookes's *Chemical News* was, after overcoming some initial difficulties, selling ten thousand copies each week by the end of the century, generating a considerable income for Crookes, who remained its editor until 1919.[46] If Mackie, as is related in chapter 4, was far from matching Crookes's levels of remuneration, his *Geologist* was bought out by Longman in the mid-1860s and relaunched as the hugely successful *Geological Magazine* (1864–), which, its editor Henry Woodward noted, was another of the new type of profitable "class periodical" that began to become established from the 1860s onwards.[47] Other examples include the *Astronomical Register* (1863–86), discussed in chapter 8 of this book, and the *Entomologist's Monthly Magazine* (1864–), which superseded the earlier short-lived, not-for-profit entomological periodicals explored in chapter 6.[48] It was both the reduction of production and distribution costs and, most significantly, the rapid growth of a literate, scientifically interested, and affluent audience that enabled publishers to focus on differentiated readerships that were now, for the first time, sufficiently large to sustain specialist periodicals in a wide range of scientific fields.

This larger and more literate reading audience was also sufficient to enable publishers to experiment with new forms of scientific journalism, and by the 1860s the market for specialist periodicals that made science accessible to both popular and middlebrow reading audiences had become increasingly crowded. While the numbers of popular science periodicals had, after a brief proliferation in the 1820s, remained unchanged until the 1850s, they more than doubled in the following decade.[49] Titles such as *Recreative Science* (1859–62), the *Intellectual Observer* (1862–68), the *Popular Science Review* (1861–95), the *Quarterly Journal of Science* (1864–78), and *Hardwicke's Science-Gossip* (1865–93) all sought to engage new audiences for science with different formulas that combined education with entertainment at highly competitive prices. *Hardwicke's Science-Gossip*, for example, cost just four pence for twenty-four octavo pages and claimed to be the cheapest scientific journal yet published. Its editor, Mordecai Cubitt Cooke, had initially proposed that the new monthly should be called the *Veil of Isis*, but his more market-savvy publisher, Robert Hardwicke, instead proposed *Science-Gossip* (fig. 1.3). In line with this self-consciously demotic title, Cooke outlined his editorial objective as being to "gossip with our readers, as a man chats to his friend . . . talking of scientific subjects in the language of the fireside, and not as *savans*."[50] With this emphasis on quotidian chat, *Science-Gossip* addressed itself principally to the growing ranks of amateur and plebeian naturalists, whose interests were also addressed by regional journals like the *Naturalist* (1864–), published by the Yorkshire Naturalists' Union. Just as such local periodicals helped, as Samuel Alberti has argued, to define a new role for amateurs in which they remained integral to the production of natural historical knowledge, so the popular appeal of *Science-Gossip* and the other popular science periodicals that proliferated from the 1860s did not mean that they were scientifically insignificant.[51]

After all, Cooke received assistance from professional experts in answering readers' queries, and *Science-Gossip* also occasionally published important articles by leading authorities, such as the dermatologist William Tilbury

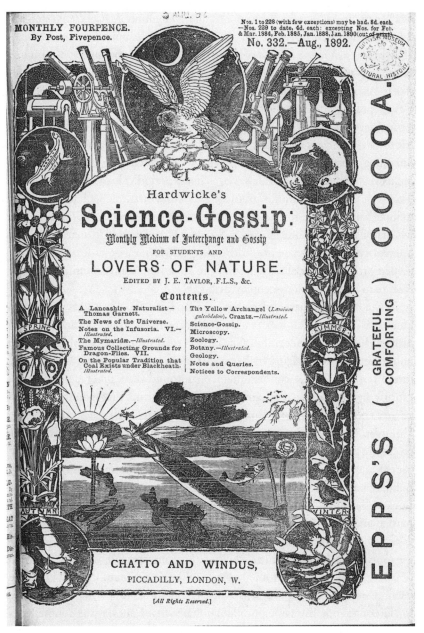

MONTHLY FOURPENCE.
By Post, Fivepence.

Nos. 1 to 228 (with few exceptions) may be had, 8d. each.
—Nos. 229 to date, 4d. each: excepting Nos. for Feb.
& Mar. 1884, Feb. 1885, Jan. 1888, Jan. 1890 (out of print).

No. 332.—Aug., 1892.

Hardwicke's

Science-Gossip:

Monthly Medium of Interchange and Gossip

FOR STUDENTS AND

LOVERS OF NATURE.

EDITED BY J. E. TAYLOR, F.L.S., &c.

Contents.

A Lancashire Naturalist —
Thomas Garnett.
The News of the Universe.
Notes on the Infusoria. VI.—
Illustrated.
The Mymaridæ.—*Illustrated.*
Famous Collecting Grounds for
Dragon-Flies. VII.
On the Popular Tradition that
Coal Exists under Blackheath.
Illustrated.

The Yellow Archangel (*Lamium
galeobdolon*), Crantz.—*Illustrated.*
Science-Gossip.
Microscopy.
Zoology.
Botany.—*Illustrated.*
Geology.
Notes and Queries.
Notices to Correspondents.

COCOA.

EPPS'S (GRATEFUL
COMFORTING)

CHATTO AND WINDUS,
PICCADILLY, LONDON, W.

[All Rights Reserved.]

FIG. 1.3. *Hardwicke's Science-Gossip*, August 1892, cover. Image from the Biodiversity Heritage Library, www.biodiversitylibrary.org. Contributed by Natural History Museum Library, London.

Fox, or by up-and-coming naturalists, like the young E. Ray Lankester, and this brought it new readers from the elite scientific community. Even Charles Darwin was not above contributing to *Science-Gossip*, writing in 1867 on how hedgehogs seemed to use their prickles to carry fruit. Darwin's contribution was in response to previous correspondence and observations from *Science-Gossip*'s readers, and it shows that the participatory nature of such popular periodicals continued to be valued by the most eminent men of science. This affords an important corrective to Susan Sheets-Pyenson's contention that, by the 1860s, a passive top-down popular science had largely replaced the more inclusive "low science" facilitated by science periodicals in earlier decades.[52] Although Cooke, mindful of putting "readers who have not had a scientific training" at their ease, insisted that "it is not our project or ambition to become what is called a 'scientific journal,'" *Science-Gossip*'s mixed format, combining popularity with expertise and addressing a diverse range of readers, actually helped provide a model for what was the most significant scientific journal to emerge in late nineteenth-century Britain.[53]

In the crowded marketplace for popular scientific periodicals in the late nineteenth century, publishers and editors regularly copied formats that had proved themselves successful. Cooke observed in private autobiographical notes that *Science-Gossip* "had many imitators and followers, besides 'Nature.'" It might seem surprising that Cooke would claim that his populist and gossipy periodical had any connection to the journal that would, of course, go on to become the international benchmark for modern science publishing. However, *Nature*, which needed to break even in the commercial marketplace just as much as *Science-Gossip* did (even if it failed to do so for the first two decades), was intended to appeal to both scientific practitioners and the general public when it was launched in 1869, published by Macmillan with Norman Lockyer as editor. Lockyer certainly sought the same cost-effective mix of expertise and popular appeal as Cooke's *Science-Gossip*, and one of *Nature*'s twentieth-century editors, John Maddox, acknowledged that "to begin with, the journal was a gossip sheet."[54] The increasing specialization of science in this period meant that in time *Nature* could only accommodate the needs of its more expert readers, with *Knowledge* (1881–1918), edited by Richard Proctor, later targeting the same popular audience that Lockyer was compelled to abandon from the mid-1870s onward. But *Nature* was not the only one of the late nineteenth century's most authoritative specialist journals to have originally been intended for a more general audience, with the *Geological Magazine* similarly beginning as a new incarnation of the avowedly populist and heterodox *Geologist* and, at least initially, attempting to retain the same readership.

Nature, as Melinda Baldwin has argued, also imitated aspects of Crookes's successful *Chemical News*, and the two weeklies shared an emphasis on speed and brevity in their reporting of news of the latest research that increasingly became a hallmark of science journalism into the twentieth century.[55]

The consolidation of *Nature* as a journal addressed exclusively to expert scientific practitioners that was nonetheless commercially viable helped create a more stratified marketplace for science periodicals in the final decades of the nineteenth century. The increasing specialization of the trade press, with the emergence from the 1860s onwards of highly technical journals like the *Chemist and Druggist* (1859–; fig. 1.4) and the *Electrician* (1878–1952), addressed exclusively to professional practitioners in their respective trades, was one important factor in the process of stratification. New medical periodicals similarly reflected the development of more specialist areas of practice, such as, for instance, the *British Gynaecological Journal* (1885–1907) and the *British Journal of Dermatology* (1888–). Closer to the trade press were a range of new journals addressing audiences engaged in specific medical occupations, with the growing importance of nursing reflected in a number of titles including the *Nursing Record* (1888–1956), the *Nursing Mirror* (1888–1977), and *Nursing Notes* (1891–1945). As Sally Frampton shows in chapter 9 of this book, these new journals, representing the interests of previously marginalized occupational groups, were often resented by established medical periodicals, but they nonetheless contributed to the growing specialization and stratification of the scientific, medical, and technical press in the late nineteenth century.

The process was further augmented by the advent in the 1870s and 1880s of still more specialist journals, financed and published by university presses, that were restricted to small coteries of professional academics. The *Journal of Physiology* (1878–), edited by Michael Foster, was a mouthpiece for the innovative approach to physiological science that Foster was pioneering at Cambridge University, and quickly became viewed as the "'house organ' for the Cambridge School" of experimental physiology, with more than a quarter of its articles deriving from practitioners in the Cambridge laboratory.[56] The *Journal of Physiology* was, like *Nature*, originally published by Macmillans, though after less than two years Foster assumed the role of proprietor, with financial assistance from colleagues at Trinity College, and the increasingly in-house *Journal* was now sold through the Cambridge Scientific Instrument Company. By the mid-1890s the *Journal* was being published by C. J. Clay, official printer to Cambridge University Press, and that press would itself take over publication in the early twentieth century. In 1887, Oxford University Press launched a science periodical of its own, the *Annals of Botany* (1887–),

THE CHEMIST AND DRUGGIST

OUR FRESH START.

WE appear before our readers in a new dress, but not in a new character. We have grown bigger, but all our old subscribers will recognize the features of the CHEMIST AND DRUGGIST of former years. We are what we have always been—the journal of the trade from which we take our name, devoted to the interests of its members.

We have in former numbers explained the reasons which induced us to alter our dress; but we may again state that the chief reason was the uncomfortable tightness of the old one. We found that the book-form prevented our free growth, so we discarded it and adopted the habit of the leading class journals. We have now not merely a larger surface of paper to work upon, but a form in which printed matter can be greatly compressed without appearing heavy.

We do not think it necessary to describe the scope and objects of the CHEMIST AND DRUGGIST, as they are plainly revealed by the contents of the present number.

NOTE ON INSECT WAX.

BY BARNARD S. PROCTOR.

IN the August number of last year's CHEMIST AND DRUGGIST, I drew attention to several materials which have been used as substitutes for wax. I have now to correct an error into which I had fallen with regard to the sample supplied to me as "Insect Wax of China." I have since seen reason to think that the sample thus described was Japan Tree Wax, as it corresponded very closely with that article in all its properties, and "China Wax" is one of the several names under which Japanese wax is known in commerce. I am informed that the true insect wax is not to be found in the market; but through the kindness of Mr. D. Hanbury, I am enabled to note the properties of an authentic specimen.

It greatly resembles spermaceti in appearance, but is not quite so white, and the fracture, which is crystalline and sparkling, is finer grained; it is also much harder and more brittle than spermaceti, and its melting point may be stated

VOL. V. 1864. No. 53.

as 178° Fahrenheit, it having melted at 180°, and congealed at 176°, thus corresponding pretty closely with the melting point which I quoted from Miller. As the substance is not now to be obtained commercially, I have not ascertained for what purposes it is fitted; a further description will be found in an article by Mr. D. Hanbury, in the *Pharmaceutical Journal*, Vol. 12, page 476.

11, Grey-street, Newcastle-on-Tyne.

ON A NEW METHOD OF ANALYSING OIL CAKE.

BY W. B. TEGETMEIER.

SOME time since I was requested to examine the products and report upon the process of a new method of extracting oil from crushed seeds by means of bisulphide of carbon in place of pressure. I found that the oils obtained by this plan were of very superior quality, bright, free from albuminous and mucilaginous matters, and destitute of the slightest trace of the bisulphide.

The residuary mass of ground seed was also much more free from oil than that left after pressure.

Being desirous of verifying my own opinion, I forwarded a sample of the crushed seeds after extraction to Professor A. Church for his analysis. His report of its composition was as follows :—

Water	.. ..	10·28
Oil	.. ..	4·70
Starch, mucilage	.. ..	35·63
Albuminous compounds, containing 5·155 of nitrogen		32·29
Indigestible fibre, cellulose	.. .. ..	9·45
Ash..	.. ..	7·70
		100·00

Thus proving that much more oil was extracted by the chemical than by the mechanical process in general use, ordinary rape cake containing sometimes as much as 12 per cent. of oil, and linseed cake a much larger quantity.

So efficacious is the bisulphide in extracting oil from organic matters, that it is even superior to ether in this respect, and Professor Church informs me that he has adopted it as a cheaper and preferable means of analysis.

Thus the same sample of linseed cake under the old ether process gave 16·57 per cent of oil. But treated in a precisely similar manner with bisulphide of carbon it afforded 16·79 per cent of oil.

The cost of ether in the analyses of oil cakes is so great that the suggestion of this efficacious substitute by Professor Church is one of very great practical importance, and I have therefore much pleasure in bringing it under the notice of the readers of the CHEMIST AND DRUGGIST, as there can be no doubt but that it is equally applicable to determining the amount of oil in any other seeds or in organic matters generally.

FIG. 1.4. *Chemist and Druggist*, 15 January 1864, 1. Reproduced with the permission of Leeds University Library.

with the delegates willing to take on such a publication for the very first time "without any view of securing a profit to the Press," although they insisted on a "guarantee fund" of two hundred pounds, which by 1889 they had drawn on to cover their losses.[57] A lucrative new market was emerging among subscribers in libraries, universities, and technical institutions across the world, however, and the *Annals of Botany* was sufficiently solvent by the end of the following decade for the fund to be returned. Like the *Journal of Physiology*, the *Annals* had established itself as the most authoritative specialist periodical in its field by the close of the nineteenth century. With their innovative processes of peer review, such academically oriented periodicals were, along with more overtly commercial journals like *Nature*, progressively supplanting learned societies as the central institutions where new forms of specialist expertise were adjudicated and guaranteed.[58]

The printed scientific paper increasingly assumed the intellectual authority that was previously the preserve of oral contributions at meetings of learned societies, which might be published only several months or years later, and often in an altered form. The societies had initially responded, as Csiszar shows in chapter 3 of this book, by issuing their own proceedings in formats similar to those of commercial periodicals; however, with the advent of *Nature* and academic journals like the *Journal of Physiology*, these official proceedings could no longer match the speed or specialization of the commercial scientific press. In light of these changes, as well as the concomitant expansion in published papers across the sciences, many societies sought a new role as arbiters and organizers of scientific information. The most conspicuous example of this recalibration of the role of learned societies in relation to the press was the publication of the Royal Society's *Catalogue of Scientific Papers* (1867–1902), which, by including only articles from a select list of validated journals, helped to consolidate the demarcation between academic and popular science periodicals in the final decades of the century. More significantly, the need for such a *Catalogue* also signaled perhaps the most pressing concern in science publishing over the same period.

The massive growth in the number of scientific periodicals over the nineteenth century had inevitably resulted in an exponential increase in the number of papers published. The solution, somewhat counterintuitively, was to create even more science journals, though of a new form pioneered from the 1850s in Germany. These were abstract journals containing systematic digests of the contents of other periodicals in a particular field, with, for instance, the *Zoological Record* (1870–) guiding practitioners with an interest in systematics to the relevant literature in their respective areas. Although originally a commercial enterprise, the *Zoological Record* was, from 1886, published by the Zoological Society; moreover, as with the Royal Society's *Catalogue of Scientific Papers*, the task of collating such vast swaths of information could not be left to the marketplace and instead needed to be subsidized by learned societies. It has been estimated that, by the close of the nineteenth century, there was one abstract journal published for every three hundred conventional science periodicals, with *Science Abstracts* (1898–; renamed *Physics Abstracts* in 1902), published by the Institution of Electrical Engineers, perhaps the most successful and longstanding example, as is discussed in chapter 7 of this book.[59] The volume of serialized information produced each year even in a relatively small subfield like zoological systematics or electromagnetism was so prodigious, however, that abstract journals were often notoriously incomplete and unreliable. Their production was also prohibitively costly: financing *Abstracts of Physical Papers* (1895–98), a short-lived forerunner of *Science Abstracts*, cost the Physical Society far more than their entire annual income from members' subscriptions, and threatened the society's very existence.[60]

Other, more existential hazards were also involved, and in 1896 the British Library's bibliographer, Frank Campbell, warned that the "development of Periodical Literature has been such as to constitute a very considerable danger to the progress of knowledge."[61] The threat of being deluged with a flood of printed paper was at its most acute in the sciences. Indeed, the size, complexity, and heterogeneity of the information to be processed in order to keep up to date with developments in just one particular area was such that the proliferation of science periodicals, which increased exponentially in the period from 1860 to 1900, threatened to overwhelm and even destroy the whole enterprise of scientific research.

The emergence in late-nineteenth-century Britain of scientific periodicals that were connected with the networks and practices of professionalized university-based science was clearly a development of great significance. The purpose of this chapter, however, has been to show that such journals were far from representative of the bulk of scientific, medical, and technical periodicals in the period, and that many other forms of publication were developed over the course of the century that addressed and fostered a wide range of communities of practice. By offering a survey of these periodical forms, this chapter thus underpins the object of the volume: to expose and explore the importance of scientific periodicals in understanding the complex and shifting social and epistemic topography of science in nineteenth-century Britain, both in the preprofessional period and as science became increasingly professionalized.

The survey is inevitably broad-brush, but subsequent chapters explore further the often highly creative initiatives of editors, publishers, societies, and readers in exploiting developments in print culture to develop and maintain scientific communities.

NOTES

1. [Robert Woodhouse], "Review of *Journal of Natural Philosophy*," *Monthly Review*, 2nd ser. 29 (1799), 301–3, on pp. 303, 302.

2. W. R. LeFanu, "British Periodicals of Medicine: A Chronological List. Part 1: 1684–1899," *Bulletin of the Institute of the History of Medicine* 5 (1937): 735–61, 827–55; and idem., *British Periodicals of Medicine: A Chronological List, 1640–1899*, revised edition, ed. Jean Loudon (Oxford, UK: Wellcome Unit for the History of Medicine, 1984).

3. R. M. Gascoigne, *A Historical Catalogue of Scientific Periodicals, 1665–1900: With a Survey of Their Development* (New York: Garland, 1985); Samuel H. Scudder, *Catalogue of Scientific Serials of All Countries Including the Transactions of Learned Societies in the Natural, Physical and Mathematical Sciences, 1633–1876*, reprint (New York: Kraus Reprint, 1965 [1879]); and Henry Carrington Bolton, *A Catalogue of Scientific and Technical Periodicals, 1665–1895, Together with Chronological Tables and a Library Check-List*, 2nd edition (Washington: Smithsonian Institution, 1897).

4. John S. North, *The Waterloo Directory of English Newspapers and Periodicals, 1800–1900*, series 1 and 2, 20 vols., (1994–).

5. Gowan Dawson, Richard Noakes, and Jonathan R. Topham, introduction to *Science in the Nineteenth-Century Periodical: Reading the Magazine of Nature* (Cambridge: Cambridge University Press, 2004), 1–34, esp. pp. 7–10.

6. Aileen Fyfe, Julie McDougall-Waters, and Noah Moxham, "350 Years of Scientific Periodicals," *Notes and Records of the Royal Society* 69 (2015): 227–39.

7. Aileen Fyfe, "Journals, Learned Societies and Money: *Philosophical Transactions* ca. 1750–1900," *Notes and Records of the Royal Society* 69 (2015): 277–99; and Jonathan R. Topham, "Scientific and Medical Books, 1780–1830," in *The Cambridge History of the Book in Britain*, vol. 5, *1695–1830*, ed. Michael Turner and Michael Suarez (Cambridge: Cambridge University Press, 2009), 827–33.

8. Thomas Broman, "Periodical Literature," in *Books and the Sciences in History*, ed. Marina Frasca-Spada and Nick Jardine (Cambridge: Cambridge University Press, 2000), 225–38, on p. 230. See also Alex Csiszar, *The Scientific Journal: Authorship and the Politics of Knowledge in the Nineteenth Century* (Chicago: University of Chicago Press, 2018).

9. See Porter, Roy, "Lay Medical Knowledge in the Eighteenth Century: The Evidence of the *Gentleman's Magazine*," *Medical History* 29 (1985): 138–68; idem., "Laymen, Doctors and Medical Knowledge in the Eighteenth Century: The Evidence of the *Gentleman's Magazine*," in *Patients and Practitioners: Lay Perceptions of Medicine in Pre-Industrial Society*, ed. Roy Porter (Cambridge: Cambridge University Press, 1985), 283–314; and Jonathan R. Topham, "Anthologizing the Book of Nature: The Circulation of Knowledge and the Origins of the Scientific Journal in Late Georgian Britain," in *The Circulation of Knowledge between Britain,*

India and China: The Early-Modern World to the Twentieth Century, ed. Bernard Lightman, Gordon McOuat, and Larry Stewart (Leiden and Boston: Brill, 2013), 119–52.

10. See, in this context, the striking statistics compiled by Fielding H. Garrison in his "The Medical and Scientific Periodicals of the Seventeenth and Eighteenth Centuries," *Bulletin of the Institute of the History of Medicine* 2 (1934): 285–343, on p. 300.

11. On these early medical journals, see David A. Kronick, "Medical 'Publishing Societies' in Eighteenth-Century Britain," *Bulletin of the Medical Library Association* 82 (1994): 277–82; Roy Porter, "The Rise of Medical Journalism in Britain to 1800," in *Medical Journals and Medical Knowledge: Historical Essays*, ed. W. F. Bynum, S. Lock, and R. Porter (London and New York: Routledge, 1992), 6–28; and Iain Chalmers, Ulrich Tröhler, and John Chalmers, "*Medical and Philosophical Commentaries* and Its Successors," in *Andrew Duncan, Senior: Physician of the Enlightenment*, ed. John Chalmers (Edinburgh: National Museums Scotland, 2010), 36–55.

12. See G. E. Fussell, "Early Farming Journals," *Economic History Review* 3 (1932): 417–22; Nicholas Goddard, "The Development and Influence of Agricultural Periodicals and Newspapers, 1780–1880," *Agricultural History Review* 31 (1983): 116–31; Bernard A. Cook, "Agriculture," in *Victorian Periodicals and Victorian Society*, ed. J. Don Vann and Rosemary T. VanArsdel (Toronto: University of Toronto Press, 1995), 235–48; and F. A. Buttress, *Agricultural Periodicals of the British Isles, 1681–1900, and Their Location* (Cambridge: University of Cambridge School of Agriculture, 1950).

13. Mary Croarken, "Tabulating the Heavens: Computing the *Nautical Almanac* in 18th-Century England," *IEEE Annals of the History of Computing* 25 (2003): 48–61; idem., "Human Computers in Eighteenth- and Nineteenth-Century Britain," in *The Oxford Handbook of the History of Mathematics*, ed. Eleanor Robson and Jacqueline Stedahll (Oxford, UK: Oxford University Press, 2009), 375–403; Shelley Costa, "The *Ladies' Diary*: Gender, Mathematics and Civil Society in Early Eighteenth-Century England," *Osiris* 17 (2002): 49–73; Joe Albree and Scott H. Brown, "'A Valuable Monument of Mathematical Genius': The *Ladies' Diary* (1704–1840)," *Historia Mathematica* 36 (2009): 10–47; and Sloan Evans Despeaux, "Mathematical Questions: A Convergence of Mathematical Practices in British Journals of the Eighteenth and Nineteenth Centuries," *Revue d'histoire des mathématiques* 20 (2014): 5–71.

14. On mathematical periodicals, see Sloan Evans Despeaux, "International Mathematical Contributions to British Scientific Journals, 1800–1900," in *Mathematics Unbound: The Evolution of an International Mathematical Research Community*, ed. Karen Hunger Parshall and Adrian C. Rice ([Providence, RI]: American Mathematical Society and London Mathematical Society, 2002), 61–88; and idem., "A Voice for Mathematics: Victorian Mathematical Journals and Societies," in *Mathematics in Victorian Britain*, ed. Raymond Flood, Adrian Rice, and Robin Wilson (Oxford: Oxford University Press, 2011), 155–74.

15. W. Botting Hemsley, *A New and Complete Index to the Botanical Magazine from Its Commencement in 1787 to the End of 1904, Including the First, Second, and Third Series; To Which Is Prefixed a History of the Magazine* (London: Lovell Reeve, 1906); and Ray Desmond, *A Celebration of Flowers: Two Hundred Years of Curtis's Botanical Magazine* (Kew, UK: Royal Botanic Gardens, in association with Collingridge, 1987).

16. Topham, "Anthologizing"; Iain Watts, "'We Want No Authors': William Nicholson and the Contested Role of the Scientific Journal in Britain, 1797–1813," *British Journal for the History of Science* 47 (2014): 397–419; W. H. Brock and A. J. Meadows, *The Lamp of Learning:*

Taylor & Francis and the Development of Science Publishing, 2nd edition (London: Taylor and Francis, 1998 [1984]); and S. Lilley, "'Nicholson's Journal,' 1797–1813," *Annals of Science* 6 (1948–50): 78–101.

17. Watts, "We Want No Authors."

18. James A. Secord, "Science, Technology, and Mathematics," in *The History of the Book in Britain*, vol. 6, *1830–1914*, ed. David McKitterick (Cambridge: Cambridge University Press, 2009), 443–74, esp. 451–56; and Csiszar, *The Scientific Journal*.

19. Jonathan R. Topham, "The Scientific, the Literary and the Popular: Commerce and the Reimagining of the Scientific Journal in Britain, 1813–1825," *Notes and Records of the Royal Society* 70 (2016): 305–24; and Brock and Meadows, *Lamp of Learning*.

20. James Mussell, "'This Is Ours and For Us:' The *Mechanic's Magazine* and Low Scientific Culture in Regency London," in *Repositioning Victorian Sciences*, ed. David Clifford et al. (London: Anthem Press 2006), 107–18; and Jonathan R. Topham, "John Limbird, Thomas Byerley, and the Production of Cheap Periodicals in the 1820s," *Book History* 8 (2005): 75–106.

21. Jonathan R. Topham, "Publishing 'Popular Science' in Early Nineteenth-Century Britain," in *Science in the Marketplace: Nineteenth-Century Sites and Experiences*, ed. Aileen Fyfe and Bernard Lightman (Chicago and London: University of Chicago Press, 2007), 135–68. For further titles, see Susan Sheets-Pyenson, "Popular Science Periodicals in Paris and London: The Emergence of a Low Scientific Culture, 1820–1875," *Annals of Science* 42 (1985): 549–72.

22. Michael Brown, "'Bats, Rats and Barristers': The 'Lancet', Libel and the Radical Stylistics of Early Nineteenth-Century English Medicine," *Social History* 39 (2014): 189–209; Topham, "John Limbird," and Mary Bostetter, "The Journalism of Thomas Wakley," in *Innovators and Preachers: The Role of the Editor in Victorian England*, ed. Joel Wiener (Westport, CT: Greenwood, 1985), 275–92.

23. W. F. Bynum and Janice C. Wilson, "Periodical Knowledge: Medical Journals and Their Editors in Nineteenth-Century Britain," in *Medical Journals and Medical Knowledge: Historical Essays*, ed. W. F. Bynum, S. Lock, and R. Porter (London and New York: Routledge, 1992), 6–28.

24. Brown, "Bats, Rats, and Barristers," 183n4.

25. Susan Sheets-Pyenson, "From the North to Red Lion Court: The Creation and Early Years of the Annals of Natural History," *Archives of Natural History* 10 (1981): 221–49; and *Examiner*, 31 January 1836, p. 80.

26. *Analyst* 1 (1834), "Advertisement" and [iii].

27. W. H. Brock, "The Development of Commercial Science Journals in Victorian Britain," in Meadows, *Development of Science*, 95–122; and David E. Allen, "The Struggle for Specialist Journals: Natural History in the British Periodicals Market in the First Half of the Nineteenth Century," *Archives of Natural History* 23 (1996): 107–23.

28. Allen, "The Struggle for Specialist Journals"; Susan Sheets-Pyenson, "A Measure of Success: The Publication of Natural History Journals in Early Victorian Britain," *Publishing History* 9 (1981): 21–36; Sheets-Pyenson, "From the North to Red Lion Court"; and Ray Desmond, "Loudon and Nineteenth-Century Horticultural Journalism," in *John Claudius Loudon and the Early Nineteenth Century in Great Britain,* ed. Elisabeth B. MacDougall (Washington: Dumbarton Oaks Trustees for Harvard University, 1980), 77–97.

29. *First Report from the Select Committee on Postage; Together with the Minutes of Evi-*

dence, and Appendix, House of Commons Parliamentary Papers, Session 1837–38, 33: 1–516, on pp. 319 and 325.

30. Brock and Meadows, *Lamp of Learning*, 103.

31. Gowan Dawson, "Thomson, William (1824–1907)," in *Dictionary of Nineteenth-Century Journalism*, ed. Laurel Brake and Marysa Demoor (London: British Library, 2009), 625; Crosbie Smith and M. Norton Wise, *Energy and Empire: A Biographical Study of Lord Kelvin* (Cambridge: Cambridge University Press, 1989); Sloan Evans Despeaux, "Launching Mathematical Research without a Formal Mandate: The Role of University-Affiliated Journals in Britain, 1837–1870," *Historia Mathematica* 34 (2007): 89–106; and Tony Crilly, "The *Cambridge Mathematical Journal* and Its Descendants: The Linchpin of a Research Community in the Early and Mid-Victorian Age," *Historia Mathematica* 31 (2004): 455–97.

32. Iwan Rhys Morus, *Frankenstein's Children: Electricity, Exhibition, and Experiment in Early-Nineteenth-Century* (London: Princeton University Press, 1998), 46.

33. Andrew King, "'Class' Publications," in *Dictionary of Nineteenth-Century Journalism*, ed. Laurel Brake and Marysa Demoor (London: British Library, 2009), 126. See also Martin Hewitt, *The Dawn of the Cheap Press in Victorian Britain: The End of the "Taxes on Knowledge," 1849–1869* (London: Bloomsbury Academic, 2014).

34. See Ben Marsden and Crosbie Smith, *Engineering Empires: A Cultural History of Technology in Nineteenth-Century Britain* (London: Palgrave Macmillan, 2005), 240–41; Ruth Richardson and Robert Thorne, *The "Builder": Illustrations Index* (London: Builder Group and Hutton, 1994); Ruth Richardson, "'Notorious Abominations:' Architecture and the Public Health in 'The Builder,' 1843–83," in *Medical Journals and Medical Knowledge: Historical Essays*, ed. W. F. Bynum, S. Lock, and R. Porter (London and New York: Routledge, 1992), 90–107; Michael Brooks, "*The Builder* in the 1840s: The Making of a Magazine, the Shaping of a Profession," *Victorian Periodicals Review* 14 (1981): 86–93; "*The Economist," 1843–1943: A Centenary Volume* (London: Oxford University Press, 1943); Ruth Richardson and Robert Thorne, "Architecture," Albert Tucker "Military," John E. C. Palmer and Harold W. Paar, "Transport," and David J. Moss and Chris Hosgood, "The Financial and Trade Press," in *Victorian Periodicals and Victorian Society*, ed. J. Don Vann and Rosemary T. VanArsdel (Toronto: University of Toronto Press, 1994), 45–61, 62–80, 179–98, 199–218; and Matthew Taunton, "Mining Press," Christian Wolmar, "Railway Press," and Ana Parejo Vadillo, "Transport Press," in *Dictionary of Nineteenth-Century Journalism*, ed. Laurel Brake and Marysa Demoor (London: British Library, 2009), 412, 527, 639–40.

35. Graeme Gooday, *Domesticating Electricity: Technology, Uncertainty and Gender, 1880–1914* (London: Pickering and Chatto, 2008).

36. *Chemist* 1 (1840), 1.

37. Brock and Meadows, *Lamp of Learning*, 99–100, 130–31; William H. Brock, "The *Chemical News*, 1859–1932," *Bulletin of the History of Chemistry* 12 (1992): 30–35; idem., "The Making of an Editor: The Case of William Crookes," in *Culture and Science in the Nineteenth-Century Media*, ed. Louise Henson et al. (Aldershot, UK: Ashgate, 2004), 189–98; and idem., *William Crookes (1832–1919) and the Commercialization of Science* (Aldershot, UK: Ashgate, 2008).

38. W. H. Brock, "Patronage and Publishing: Journals of Microscopy 1839–1989," *Journal of Microscopy* 155 (1989): 249–66.

39. Allen, "The Struggle," 115–18.

40. William Thomas Stearn, "The Life, Times and Achievements of John Lindley, 1799–1865," in *John Lindley, 1799–1865: Gardener, Botanist, and Pioneer Orchidologist*, ed. William T. Stearn (Woodbridge, UK: Antique Collectors' Club in association with the Royal Horticultural Society, 1999), 15–72, on pp. 58–59.

41. P. W. J. Bartrip, *Mirror of Medicine: A History of the "British Medical Journal"* (Oxford, UK: British Medical Journal and Clarendon Press, 1990); and idem., "The *British Medical Journal*: A Retrospect," in *Medical Journals and Medical Knowledge: Historical Essays*, ed. W. F. Bynum, S. Lock, and R. Porter (London and New York: Routledge, 1992), 126–45.

42. Robert J. Rowlette, *The Medical Press and Circular, 1839–1939: A Hundred Years in the Life of a Medical Journal* (London: [The Medical Press and Circular], 1939).

43. Andrew King, "Medical Journals: Alternative, Complementary, Fringe," in *Dictionary of Nineteenth-Century Journalism*, ed. Laurel Brake and Marysa Demoor (London: British Library, 2009), 406–7; and Jennifer Ruth, "'Gross Humbug' or 'The Language of Truth'? The Case of the *Zoist*," *Victorian Periodicals Review* 32 (1999): 299–323. See also Olwen C. Niessen, "Temperance," in *Victorian Periodicals and Victorian Society*, ed. J. Don Vann and Rosemary T. VanArsdel (Aldershot, UK: Scolar Press, 1994), 251–77.

44. Secord, "Science, Technology, and Mathematics," 456–57.

45. [Samuel Joseph Mackie], "Notice to Subscribers, Contributors, & Advertisers," *Geological and Natural History Repository* 1 (1865–67): [iii]; and "The Geologist," *Geologist* 1 (1858): 1–5, on p. 1.

46. See William H. Brock, "Chemical News," in *Dictionary of Nineteenth-Century Journalism*, ed. Laurel Brake and Marysa Demoor (London: British Library, 2009), 110–11.

47. Henry Woodward, "The 'Coming of Age' of the *Geological Magazine*," *Geological Magazine* n.s. 3 (1886): 45–48, on p. 48.

48. See also Bernard Lightman, "The Mid-Victorian Period and the *Astronomical Register* (1863–1886): 'A Medium of Communication for Amateurs and Others,'" *Public Understanding of Science* 27 (2018): 629–36.

49. Ruth Barton, "Just before *Nature*: The Purposes of Science and the Purposes of Popularization in Some English Popular Science Journals of the 1860s," *Annals of Science* 55 (1998): 1–33, on p. 2.

50. [Mordecai Cubitt Cooke], "Science Gossip," *Hardwicke's Science-Gossip* 2 (1866): 1.

51. Samuel J. M. M. Alberti, "Amateurs and Professionals in One County: Biology and Natural History in Late Victorian Yorkshire," *Journal of the History of Biology* 34 (2001): 115–47.

52. Susan Sheets-Pyenson, "Popular Science Periodicals in Paris and London: The Emergence of a Low Scientific Culture, 1820–1875," *Annals of Science* 42 (1985): 549–72, on p. 555.

53. [Mordecai Cubitt Cooke], "Our Compliments to Our Readers," *Hardwicke's Science-Gossip* 4 (1868): 1–2.

54. Quoted in Mary P. English, *Mordecai Cubitt Cooke: Victorian Naturalist, Mycologist, Teacher and Eccentric* (Bristol, UK: Biopress, 1987), 112.

55. Melinda Baldwin, *Making "Nature": The History of a Scientific Journal* (Chicago: University of Chicago Press, 2015), 27–28.

56. Gerald L. Geison, *Michael Foster and the Cambridge School of Physiology* (Princeton, NJ: Princeton University Press, 1978), 188.

57. Jonathan R. Topham, "Science, Mathematics, and Medicine," in *The History of Oxford University Press: Volume II: 1780 to 1896*, ed. Simon Eliot (Oxford, UK: Oxford University Press, 2013), 513–57, on p. 553.

58. On the emergence of peer review, see Alex Csiszar, "Troubled from the Beginning," *Nature* 532 (2016): 306–8.

59. William H. Brock, "Science," in *Victorian Periodicals and Victorian Society*, ed. J. Don Vann and Rosemary T. VanArsdel (Aldershot, UK: Scolar Press, 1994), 81–96, on p. 87.

60. See A. J. Meadows, "Access to the Results of Scientific Research: Developments in Victorian Britain," in *The Development of Science Publishing in Europe*, ed. A. J. Meadows (Amsterdam: Elsevier, 1980), 43–62, on p. 52.

61. Quoted in Alex Csiszar, "Seriality and the Search for Order: Scientific Print and Its Problems during the Late Nineteenth Century," *History of Science* 48 (2010): 399–434. on p. 405.

CHAPTER 2

Redrawing the Image of Science: Technologies of Illustration and the Audiences for Scientific Periodicals in Britain, 1790–1840

Jonathan R. Topham

In works devoted to [scientific] subjects, representations of physical objects are indispensable; and this cannot be better effected than by wood-cuts, which are now executed with much beauty, and, besides, combine conveniency with cheapness. Science, therefore, as well as literature, lies under deep obligations to the individuals who have carried this art to so high a degree of perfection.

Chambers's Edinburgh Journal (1836)[1]

I have witnessed in my own recollection a failure of all the scientific journals almost that have been set on foot . . . they have all of them failed from an inability to cover their expense, and it is almost an impracticable thing to keep a scientific journal alive in this country.

RICHARD TAYLOR (1838)[2]

Over recent years, historians have highlighted the significant role that the transformation in the manufacturing processes, products, and markets for print played in shaping the identity and practice of the sciences in early-nineteenth-century Britain. The mechanization of paper manufacture, printing, and binding in the half century between 1790 and 1840 underpinned the emergence of much cheaper scientific publications that were much more widely accessible than previously, enabling the production of educational books, scientific journals, and works of popular science that fundamentally altered the place of science in society and the character of scientific knowledge.[3] However, a crucial aspect of this transformation has been largely neglected by science historians—namely, the fundamental alteration that took place in the technologies of illustration. In the 1790s, publications of quality were almost exclusively illustrated with copper-plate engravings or etchings, while woodcuts were largely reserved for cheaper works, such as school books, and were typically of poor quality. In the 1830s, by contrast, illustration was dominated by wood engravings, some of exquisite quality, alongside a range of competing technologies, including not only copper-plate engravings but also lithographs and engravings on steel. Moreover, wood engravings could be incorporated within the new technologies of mass production, including the steam rotary press and stereotyping.[4]

As historians of journalism have shown, these developments—alongside the more general changes in printing practice—were pivotal in transforming the audiences for printed matter, above all because the cost reductions they offered enabled editors and publishers to use visual matter to appeal to new groups of readers. This was notably the case in the production of new mass-circulation periodicals intended to be attractive to working-class readers, such as the *Penny Magazine* (1832–45) of the Society for the Diffusion of Useful Knowledge, although it is the *Illustrated London News* (1842–2003) that is often seen as marking the apotheosis of the illustrated journal.[5] As I will explore in this chapter, however, such popular publications were by no means the only periodicals affected by the transformation in illustrative technology. The changes also had important consequences for the burgeoning periodical literature of the sciences and medicine, and indeed for scientific and medical books more generally. From the perspective of this volume, the most important of those consequences relate to the effect that the technological changes had on the economics of scientific publishing. As the epigraphs above make clear, the choice of imaging technology had the potential to radically reduce the cost of scientific illustration in a market where scientific periodicals struggled to cover their costs. As editors, publishers, and societies sought to reach new, larger, or more specialized audiences with scientific periodicals, the changing technologies and economics of illustration formed a key element in their decision making. Thus, the kinds of communities of scientific practice that could be fostered by the burgeoning periodical press depended in no small measure on these underlying changes in periodical manufacture.

It was not, however, merely the economics of the changing technologies that were of importance for the development of scientific communities. The technological transformations also had significant consequences for the content of scientific communication, offering new graphic possibilities and challenging the expectations of authors, illustrators, and readers. In recent years, historians have shown increasing interest in the visual culture of science in early-nineteenth-century Britain. However, while several major studies have taken cognizance of the effects of the changing technologies of illustration, none has placed such technologies in the foreground of the analysis.[6] We still know relatively little about when particular technologies began to become practically available and how quickly they were adopted. There is also much to learn about how those involved—authors and artists, publishers and editors, and readers and observers—viewed the graphic qualities of the different technologies, and what the concerns and difficulties were that affected their choice of technology. Important questions remain concerning the effect of their decisions on the processes by which images were produced and the identity of those involved in producing them. Finally, of course, the qualities and conventions of the images produced, and the reactions of those who viewed them, need to be better understood in relation to the technologies employed. Addressing such questions promises significantly to enrich the history of the visual culture of science. In addition, however, it promises to enrich our understanding of how those involved conceived of the audiences for scientific periodicals and other publications, as they considered whether particular kinds of illustration, produced using particular technologies, better served particular readerships.

As the foregoing implies, the technological changes also affected the communities of science in that they affected who was engaged in the scientific work of visual representation. The technicians of print responsible for printed illustrations—significant numbers of women as well as men—have suffered substantial historical neglect. Many cultivators of the sciences were skilled artists, and some were also skilled in the associated printing technologies, but the changes in technology affected the relationships between authors, artists, and print technicians. In the introduction of both wood engraving and lithography, the availability of appropriately skilled designers, engravers, and printers was key. Such individuals often developed close relationships with scientific practitioners, and recovering an understanding of their skills and working practices is highly pertinent to the attempt to understand the development of scientific imagery in the changing technological context of the nineteenth century. The conventions established in the emerging visual languages of the several sciences were conditioned by these practical aspects of image making, as well as by the economic and political implications of choosing particular technologies.

My object in this chapter is to begin to address some of these fundamental questions about the changes that took place in illustrative technologies in the period between 1790 and 1840, with the intention of shedding light on their consequences for the production of scientific periodicals and for the communities that those periodicals fostered and served. My central claim is that the transformation was a pivotal and somewhat overlooked element in the cheapening of scientific periodicals, and in the widening of their appeal in the years before Victoria's accession. At the start of that period, scientific periodicals in Britain were restricted to a handful of learned transactions, illustrated luxuriously with copper-plate figures. By the end, there was a plethora of competing titles, very diverse in price and appearance and illustrated in a variety of ways. These included society proceedings as well as transactions (now illustrated in increasingly diverse ways) and a range of general scientific journals, such as the *Philosophical Magazine* (1798–). They also, however, included cheap technical journals for mechanics, self-proclaimed magazines of popular science, and an increasing range of magazines on natural history and gardening, many of which took advantage of changes in the technologies of illustration—and especially the growth of wood engraving—to offer products that were accessible in form and price to the expanding reading audiences of the industrial age. Such periodicals, which helped to foster wider engagement in the sciences and in new communities of science, such as mechanics institutes and natural history clubs, thus depended on the transformation of illustrative technology alongside the other changes in print manufacture.

The chapter falls into four sections. I begin by considering the strikingly slow take-up of wood engraving for scientific purposes, and show that, while the new technology seemed to many to offer important advantages, it was ill-suited to the high-price model that dominated scientific publishing in the first quarter of the century. The next two sections examine parallel developments in the 1820s. First I outline the emergence of lithography in Britain at the end of the 1810s, showing that, in the high-prestige periodicals of learned societies, the new technology began to be used as a means of saving money while maintaining an air of gentlemanly opulence. Next I show that it was the new cheap journals of the 1820s—notably the *Mirror of Literature* (1822–47) and the *Mechanics' Magazine* (1823–72)—that pushed forward the adoption of wood engraving as part of a concerted program of instruction, but with an eye also to entertainment. Similar motives actuated the innovative horticulturalist and journalist John Claudius Loudon in his application of wood engraving in producing his innovative *Gardener's Magazine* (1826–44) and *Magazine of Natural History* (1828–40). In the final section, I briefly examine how the growing adoption of the new technologies in the scientific journals of the 1830s fueled a debate about how images should operate in relation to the work of science and the character of the communities involved, highlighting some of the questions that remain concerning the grounds on which choices were made regarding the use of illustrative technologies.

While this chapter thus focuses primarily on the importance of illustrative technologies for the history of scientific periodicals, it also contributes to establishing a wider agenda in the history of the visual and print culture of the sciences in the period. In the course of the chapter, I offer something of an overview of the transformation of illustrative technologies in relation to the sciences generally. To a significant extent, of course, the developments in scientific periodicals paralleled those in scientific books. With research on the subject still in its infancy, this chapter provides a framework on which future researchers might build, and it opens up important new questions about how distinctive the use of the new technologies was in periodicals as opposed to books, and in scientific publications as opposed to other publications. More generally, the chapter identifies some of the key ways in which closer attention to the practicalities, economics, and workers involved in printed illustrations in early-nineteenth-century science can reinvigorate the science and visual culture research agenda by focusing attention on the reasons why certain technologies were preferred to others. I return to these points briefly at the end.

THE REVIVAL OF WOOD ENGRAVING AND THE PERSISTENCE OF INTAGLIO PLATES, 1790–1820

At the end of the eighteenth century, the standard technology of illustration used in scientific publications depended upon making incisions in a copper plate into which ink was introduced. These were usually made with a sharp implement (in engraving) or by using acid to cut away parts of the plate exposed through a "ground" (in etching), though less commonly used variants (mezzotint and aquatint) produced tonal effects. These were all highly skilled and labor-intensive processes, typically involving a specialist technician. The resultant plates were sometimes known as "intaglios," from the Italian verb *intagliare* (to cut in), and all were printed in much the same way. It took the intense pressure of a roller press to lift the ink out of the incisions and onto the paper, so that intaglio plates had to be printed separately from the letterpress, typically on good quality paper. Moreover, the intense pressure meant that the copper—attractive for its relative softness and ease of working—gradually deformed, limiting to a few hundred the number of copies that could be printed before the plate needed repair or replacement. Any color was usually applied manually, after printing. Thus, while the illustrations produced were often sophisticated and aesthetically rewarding, they were expensive, and added very markedly to the cost of the publication. Not surprisingly, then, such illustrations were typically used sparingly. Yet, as transactions began to be issued by the new learned societies established outside London (notably in Dublin, Edinburgh, and Manchester) and by those designed to serve sectional interests (such as the Society of Arts and the Linnean Society), those publications followed the *Philosophical Transactions* of the Royal Society (1665-) in including occasional plates. The commercial scientific magazines that began to be issued for the first time in the 1790s, such as the *Journal of Natural Philosophy* (1797-1813) and the *Philosophical Magazine*, similarly included a small number of copper-plate illustrations.

This was the context in which the revival of fine wood engraving began. Wood blocks had been used to produce such important scientific works as Vesalius's *De humani corporis fabrica* (1543), but the technology had shortly afterwards been supplanted for purposes of fine illustration by copper-plate engraving. By the eighteenth century, wood was used chiefly to produce cheap and often crude illustrations, "of little use but to embellish half-penny ballads and school-books for little children."[7] These were chiefly what are sometimes distinguished as "woodcuts," in which the image was produced by cutting into the long grain of a wood block to leave the drawing standing proud in a way that could be printed alongside the letter press.[8] This relief process had many advantages over copper-plate engraving: the block was not only easier to print and cheaper to prepare, but it was also much more durable. However, the quality of the image produced on long-grain blocks was markedly inferior, especially in regard to the fineness of line. Better results could be achieved with wood engraving, where the blocks were cut across the hard end grain of the wood, providing a surface in which finer lines could be produced. While he did not originate it, this was the process that the provincial engraver Thomas Bewick took to a new level of sophistication in the last decade of the century, leading contemporaries to consider that a new epoch had opened in the history of wood engraving.[9]

Bewick was first introduced to wood engraving when, as an apprentice in Newcastle-upon-Tyne, he was set to engrave geometrical drawings for the local mathematician Charles Hutton's *Treatise on Mensuration* (1768). However, it was his *General History of Quadrupeds* (1790) that brought the possibilities of wood engraving to public prominence, with the book running through three editions in as many years. Bewick had developed a great love for natural history as a child, and he harbored an ambition of offering a work for children with illustrations of animals superior to the woeful copper engravings found in the standard trade work he had encountered in his youth, Thomas Boreman's *A Description of Three Hundred Animals* (1730; 11th ed., 1774). His *History of Quadrupeds* was planned with his former master, Ralph Beilby, who agreed to provide descriptive text, and as the quality of the illustrations became clear, the project became more ambitious.[10] Bewick brought a fineness, delicacy, and artistry to wood engraving—as well as a feel for nature—using an expressive white-line technique in which the picture was produced by cutting out lines and larger areas of white from a black background, rather than merely cutting away ("blocking out") a white background in order to reproduce the black lines of a drawing (fig. 2.1a). Contemporaries were quick to appreciate the novelty. Woodblock illustrations had been despised since copper engraving became the pinnacle of technique, the *Critical Review* noted, but Vesalius's wood engravings had "a force, a spirit, and an expression" unequaled by Boerhaave's later edition of *De humani corporis fabrica* using copper engravings. The wood engravings in Bewick's work—"executed on a new principle"—also had an uncommon "delicacy and clearness." They were, according to the *General Magazine*, "beyond all comparison—the chef d'oeuvre of the art of wood engraving."[11]

Reviewers considered that the new technique was especially successful—indeed, superior to copper engraving—in picturing animals (figs. 2.1a and 2.1b). Reviewing Bewick's *History of British Birds* (1797–1804), the *Annual Review* reported that the engraver had

> soon found that the yielding consistence of wood is better fitted to express the ease, freedom, and spirit which ought to characterize portraits of animated beings than the stubborn surface of a metallic substance. . . . There is in [Bewick's engravings] a boldness of design, a correctness of outline, an exactness of attitude, and a discrimination of general character, conveying at the first glance a just and lively idea of each different animal, to which nothing in modern times has ever aspired, and which the most eminent old artists have not surpassed.[12]

Bewick's ability to capture the distinctive character of animals was in part attributed to his special knowledge and love of nature, and his successors were often considered to be inferior in that regard. Yet, informed writers considered that wood engraving offered special opportunities to the artist. An article on "wood-cuts" for the 1801 supplement to the third edition of the *Encyclopædia Britannica* claimed that, while copper plates were superior "in point of delicacy and minuteness," wood engravings were equally superior "in regard to

(a)

(b)

FIG. 2.1. The comparative value of wood engraving in conveying a "just and lively idea" of animals. Illustrations of the squirrel from (a) Thomas Bewick's *General History of Quadrupeds* (Newcastle-upon-Tyne: S. Hodgson, R. Beilby, and T. Bewick; London: G. G. J. Robinson and C. Dilly, 1790), 333 (wood engraving by Thomas Bewick), and (b) Oliver Goldsmith's *History of the Earth, and Animated Nature*, vol. 4 (London: F. Wingrave, 1791), opposite 23 (copperplate engraving by Isaac Taylor), showing the superiority of the former. Reproduced with the permission of Special Collections, Leeds University Library.

strength and richness." The blacks and whites produced were unsurpassed, and the technique lent itself to the chiaroscuro effects so much in demand, with strong contrast between light and shade. William Chatto later claimed, in his highly regarded *Treatise on Wood Engraving* (1839), that this was the "greatest advantage" of wood engraving over copper. He also noted that by using "lowering"—in which parts of the block were scraped to reduce the height—a softness of texture could be achieved that was of value in zoological illustration. Similarly, the *Penny Cyclopaedia* (1833–43) claimed that, while wood engraving could not achieve the "extreme neatness, length and sweep of line, and bold outline of the copper," it could equal even mezzotint in the "depth of shadow and effect," only with "more distinctness of detail."[13]

The graphic advantages of wood engraving aside, commentators were quick to notice its practical advantages, and its potential in scientific illustration. A pseudonymous writer in the *Monthly Magazine* ("NM") discussed the virtues of the new wood engraving—its "rich fullness of shade, a mellow softness in their gradations, and a great strength of touch"—suggesting that it was excellent for artistic works when on a larger scale. Its chief use, however, would be in reducing costs for illustration on a smaller scale, notably in relation to scientific subjects. Anatomical illustration was likely to be the most important, he continued, observing:

> I am perfectly satisfied that anatomical plates can be executed on wood with all the precision possible on copper, and, in some particulars, (especially those where the muscles are represented) with much greater elegance and beauty. A set of such plates, if executed from accurate designs, by having the whole civilized globe for a market, (the explanations being easily printed in different languages) could be afforded at a very low price, so as to bring them within the reach of every student of physic; while the undertaker would be insured in a most abundant profit.

Mathematical diagrams and machinery could also be accurately accomplished using wood engraving; Bewick had demonstrated the technique's value in zoology, and a finer effect might yet be expected in producing illustrations of insects, shells, and minerals. The writer claimed to have been told by a knowledgeable informant that, had the plates of the latest edition of the *Encyclopaedia Britannica* been engraved on wood instead of copper, it would have saved ten guineas per plate in printing costs, or a total of four thousand on the work as a whole.[14] The point was echoed in the encyclopedia's own 1801 supplement, which asserted that wood engraving was being underused as an "*economical art* for illustrating mechanics and other subjects of science."[15]

Some scientific publications certainly began to be produced using wood engravings. The precedent had obviously been set for works of natural history, and as early as 1807 an abridged edition of Goldsmith's *Animated Nature* was being issued as "illustrated by nearly two hundred Engravings on Wood, in the Manner of Bewick."[16] Nevertheless, the great bulk of publications continued to be illustrated using copper plates, and this applied particularly to the learned transactions and scientific journals. One might seek to account for this in technical terms. For instance, contemporaries had anticipated difficulties with applying the new technology arising from the problem of securing sufficiently large end-grain blocks, and even in 1839 William Chatto estimated five inches square as the maximum block size to be achieved without joining blocks.[17] Yet, while some journal illustrations required a larger canvas than this, many did not, so that it could only amount to a partial explanation. Similarly, we have seen that copper-plate engraving allowed for the production of much finer lines than could be achieved with wood, and it could be time-consuming and expensive to print engraved wood blocks in such a way (using overlays) as to achieve a delicate variation in tone. However, for many purposes this was hardly relevant, and the potential cost savings offered a considerable incentive to use wood engraving where nothing was to be lost.[18] A further consideration is the limited number of wood engravers available to carry out the work, but some reports suggest that, despite their small numbers, wood engravers at this period "did not meet with constant and regular employment."[19] Perhaps more significant was that draftsmen needed to know how to draw on wood in a way that would get the best out of the technology, and as late as 1839 Chatto claimed that there was only one active artist adept at drawing for wood engraving.[20]

Underpinning these technical considerations, however, was a more general conservatism that affected the uptake of the new process in general, and not least in the scientific periodicals. Wood blocks had hitherto been associated with the production of cheap books, and later commentators also emphasized that wood engraving was "the art of design which is naturally associated with cheap and rapid printing."[21] As we shall see, cheap publishing was indeed the context in which its use first prospered in the 1820s. By contrast, in the high-price book market of the first quarter of the nineteenth century, scientific books and journals continued to be expensive luxury goods. As new specialist societies began to issue transactions alongside those of the Royal Society in the 1800s and 1810s, they took on the form of prestige publications with the associated high production costs. The next section shows that the new technology of lithography offered a way to reduce costs and increase convenience without compromising the sense of luxury.

LITHOGRAPHY AND THE PRECARIOUS FINANCES OF PRESTIGE PUBLICATIONS, 1820-30

The relatively slow advance of wood engraving in Britain finds something of a parallel in the length of time it took for lithography to gain a foothold. Invented by the German actor Alois Senefelder in 1796, the technique was patented by him in London in 1801. However, with many forbidding technicalities to master and with the Napoleonic wars intervening, it was not until the late 1810s that the technology began to be exploited in any systematic way in Britain.[22] As its name implies, lithography involved printing from stone. Instead of the ink being carried in incisions on a metal plate or on the relief surface of a wood block, the process depended on the differential chemical affinities of the ink, such that printing could take place from the flat surface of the fine lithographic limestone. The lithographic artist used waxy crayons or ink to produce the image on the surface of the stone before etching the stone to prepare it to absorb water in the areas not to be printed. The ink consequently adhered only to the image, and could be transferred to paper in a suitable printing press. As a planographic process in which printing took place from a flat surface rather than a raised one, lithography still involved printing the image separately from the book's text, typically on a separate page as a "plate." However, it had many potential advantages over copper plates. The process of preparing the stone was altogether less laborious than that of preparing a plate; moreover, competent draftsmen could learn to draw on stone altogether more easily than they could learn to engrave or etch, thus offering a novel immediacy. In consequence, lithography was both more rapid and cheaper than intaglio processes, and additional cost savings resulted from the much greater durability of the lithographic stone, which could yield tens of thousands of impressions without deterioration of quality. In addition, there were a number of graphic advantages. While it could not offer the same clarity as copper engraving, lithography permitted a particularly wide range of marks to be made, and could be used to produce a distinctive pencil-like tone.

It was not, however, until the late 1810s that British commentators began to voice these claims for lithography. Notable among them was the German-born fine art publisher, Rudolf Ackermann, who set himself to promote lithography, establishing a lithographic press in 1816, and offering specimens of what the new technology could achieve in his fashionable monthly, the *Repository of Arts*.[23] He also published Senefelder's *Complete Course of Lithography* in 1819 and several other manuals (both original and in translation) soon followed. By the early 1820s, London had several active lithographic printers who could offer a range of variations on the basic technique, including transfer lithography, whereby copper-plate engravings could be printed using lithographic stone. Moreover, lithography's advocates were vocal about its potential value for scientific illustration. As early as 1813, one was suggesting its suitability for the easy production of natural history illustrations at low cost, and this was soon echoed in reports from France, and by Ackermann and his lithographic printer Charles Hullmandel.[24]

Lithography recommended itself to scientific practitioners in a number of ways. The process itself fell within the scientific purview of both geologists and chemists, and William Buckland and Michael Faraday were both involved in offering advice to Hullmandel.[25] Furthermore, it offered the prospect of practitioners being enabled to produce their own drawings for publication without the intervention of a craftsman. While some scientific men, such as the surgeon Charles Bell, had learned to etch copper plates to a high standard, lithography offered the prospect of something much less demanding, with a range of possible benefits. Indeed, the only individual known to have experimented with lithography before the publication of substantial manuals was the sixteen-year-old apprentice geologist John Phillips. Phillips was probably acting at the instigation of his uncle, William Smith, who was excited by what the technique might offer in relation to publishing his drawings and producing much-needed income. Learning from several brief accounts, including translations from French periodicals in the *Annals of Philosophy* (1813–26), Phillips was able to set up a short-lived lithographic press in the years 1817 to 1819, and to advertise his services.[26] While Phillips's commercial involvement was unusual, other geologists very rapidly exploited the technique in producing small numbers of copies of their drawings for semiprivate circulation.[27] Moreover, they were impressed by the graphic qualities of the prints produced. For instance, a print of an ichthyosaur lithographed for Henry De La Beche in 1819 prompted one observer to claim that the effect was "far better for fossils than the fine engravings to Sir Everard Home's papers in the Philosophical Transactions."[28]

The convenience, cheapness, and graphic qualities of lithography all recommended it for scientific use, but a further recommendation was that the technology most easily produced illustrations as separate "plates." As we have seen, the great bulk of scientific books and journals had been illustrated in this way over the preceding century, while publications in which the illustration and text were combined were typically cheap books, especially for children. In these years before the new industrial technologies were applied to reduce costs, printed matter remained generally very expensive. Scientific publications were chiefly for the wealthy, and many of those with illustrations were notable for their appearance of luxury and prestige. In this context, lithography offered the prospect of producing books more cheaply without altering the form of the publication. Lithographic plates could replace copper plates while exuding the same air of luxury, and with none of the "cheap" connotations of wood engravings. Rudolf Ackermann had demonstrated as much in his fashionable *Repository of Arts*, and the point was not lost on those producing scientific books, journals, and above all the learned transactions.

This circumstance is well illustrated by the case of the Geological Society. When the society began to produce *Transactions* in 1811, it had strong reasons to desire a prestige publication. To begin with, the society had only just emerged from a battle to establish itself as independent of the Royal Society, not least in the matter of having the right to publish its members' memoirs independently of the *Philosophical Transactions*. Moreover, the society was conscious of the vulnerable status of the nascent science of geology, and the preface to the first volume highlighted the society's independence from controversial debate relating to theories of the earth. Decisions concerning the production of the *Transactions* consequently rendered the new publication highly reminiscent of the *Philosophical Transactions*. Choosing between specimens of type, paper, and form provided for them by the society member and printer William Phillips, they depended on the financial backing of several members to produce a luxurious quarto publication on good paper using ink of the "best quality." To complete the effect, the illustrations, like those in the *Philosophical Transactions*, were intaglio engravings, many of them colored. Whatever else remained to be proved, the geological *Transactions* were properly scientific in form, at least. Moreover, at a price of one pound, twelve shillings to the book trade and to members—the *Transactions* were not included in the fellows' subscription—their audience was distinctly select.[29]

Having a select audience is one thing, but having a vanishing audience is another, and in 1821, after five volumes of the *Transactions* had been produced, the society's council resolved to "take into its early consideration" the publication's "high price."[30] The print run had been 750 copies, but while all but two hundred copies of the first volume had been distributed by June 1822, almost six hundred of the latest part remained, meaning that more than half of the society's members had not bought a copy (see fig. 4.1). Volumes 2 to 5 had been

published at the risk of one of its members, the printer William Phillips, costing £4,500 to produce, but so far only one had broken even.[31] A subcommittee
appointed to enquire into "the most desirable form of publishing" concluded
that cost savings could be made by the society publishing the *Transactions*
on its own account, by making better use of the costly paper by introducing
smaller type and a "fuller page," and by "the substitution, wherever practicable, of Lithographic plates for copper plate engravings."[32] An estimate suggested that lithographed plates would cost just over a third of the price of
engraved plates, though by the time many had been colored, the plates would
still cost significantly more than the letterpress printing and paper combined.[33]
The first half volume under the new regime included just two engravings, in
comparison to twenty-two lithographs, and where the previous half volume
had sold for two pounds, twelve shillings, the new one could be offered at one
pound, five shillings to members and booksellers, and at one pound, eleven
shillings and six pence to the public. Sales consequently revived, and within
three and a half years, 383 copies had been sold, yielding a profit of £134.[34]

From the financial point of view, then, the technological change achieved its
objective, but it was also successful graphically. Reviewing the new volume of
the *Transactions of the Geological Society* in the *Quarterly Review*, Charles Lyell was bullish about the effectiveness of lithography in geological illustration,
observing: "This art, so strongly recommended by its superior cheapness, may
exert a favourable influence on the future progress of science, and particularly
on natural history, which has always been retarded by the unavoidable expense
of engraving."[35] Lithography was especially effective in representing the textured surfaces of rock and fossil specimens, as Charles Hullmandel demonstrated with one of the samples in his *Art of Drawing on Stone* (1824). For these
purposes, the draftsman would usually work with lithographic chalk, which
produced a distinctive textured quality. Here, the Geological Society was fortunate in being able to draw on the skills of Hullmandel's protégé George Jo-
hann Scharf, a draftsman trained at the home of lithography in Munich, whose
skill rapidly became prized by the geologists, for all that they treated him as a
"mechanic" (figs. 2.2a and 2.2b).[36] Other techniques allowed for more definite
lines, as for use in geological sections, but fine lines were more difficult. Consequently, the society still resorted to copper plates for some purposes—notably
in the production of detailed maps, where fine lines were of the essence. Even
here, however, lithography promised assistance. Hullmandel considered that
one of the "most useful applications of lithography" was likely to be the technique by which copper plate impressions could be transferred to stone for
printing.[37] Such transfer lithography grew in importance in following years,

though it soon had to compete with the use of steel engraving—another new process, which produced intaglio plates that lasted much longer than copper.

It was by no means only the Geological Society to whom the financial and graphic qualities of lithography appealed, but that society was the first to adopt the technology, and others only gradually followed suit. While the Royal Society and the Society of Arts persisted with copper plates, some other societies began to experiment with the new technology—notably those with animal and plant specimens to illustrate, where the artistry of copper-plate engraving proved especially demanding and expensive. When the short-lived monthly magazine *Library of the Fine Arts* took stock of the state of lithography in England in 1831, it reported not only that Scharf's "accuracy and neatness" had been "highly appreciated" in the *Transactions of the Geological Society*, but that lithography had been shown more generally to have advantages "in the delineation of subjects of natural history." The young flower painter Valentine Bartholomew—who had lived and worked with Hullmandel during the preceding decade, marrying his sister in 1827—had produced lithographed drawings of flowers that the *Library of the Fine Arts* considered to have proved the technique's "fitness for botanical illustrations, and for any subjects to be afterwards finished in colours; as the softness and richness of tint and delicacy of outline in the lithographic drawing render it when coloured hardly distinguishable from an original drawing." The technique had also been "found particularly effective" for "anatomical subjects, and delineations of morbid parts."[38]

Lithography's utility in anatomical illustration had been demonstrated, the writer reported, by the illustrations in the *Medico-Chirurgical Transactions* (1809–1907) of the Medical and Chirurgical Society of London, which had made the transition from copper to stone abruptly between 1825 and 1827. In natural history the transition was much more tentative. Like the Geological Society, the Linnean Society found the running of its *Transactions* (1791–) to be a major drain on its resources, above all because of the cost of the copper plates, many of which were colored. However, while it began using lithography in 1827, the majority of plates continued to be intaglio.[39] Likewise, while those who founded the society's Zoological Club—soon to become a separate Zoological Society—used transfer lithography to circulate their inaugural resolutions in 1823, the *Zoological Journal* they commenced the following year had the merest brush with the technique, relying almost exclusively on intaglio. It was only in the 1830s, when the Zoological Society began issuing its own *Transactions* (1835–1984), that significant use was made of lithography.[40] Finally, while the Horticultural Society's finances in the 1820s were straightened,

(a)

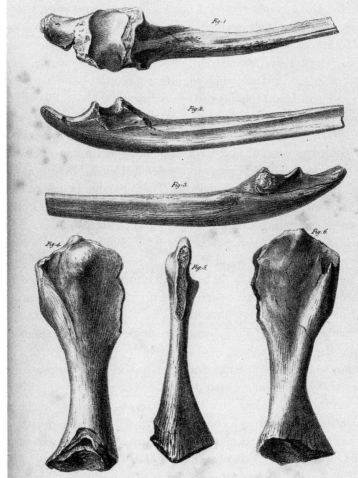

H.T.De la Beche del.   Printed by C.Hullmandel.   G.Scharf.Lithog.

(b)

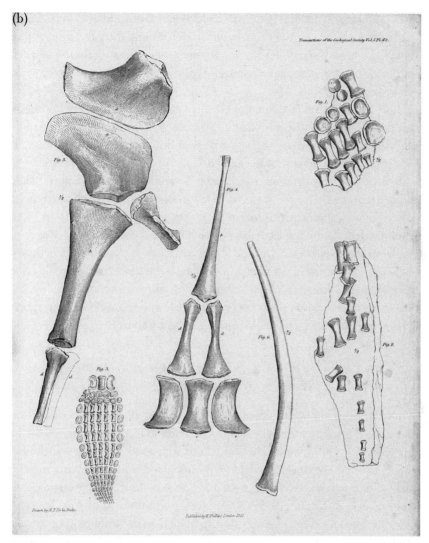

Fig. 1.

Fig. 2.

Fig. 3.

Fig. 4.

Fig. 5.

Fig. 6.

Drawn by H.T. De la Beche.

Published by W. Phillips London 1821.

F I G .   2 . 2 . The comparative value of lithography in providing a powerful sense of the texture of specimens. Two illustrations in *Transactions of the Geological Society*, both based on Gideon Mantell's drawings of the plesiosaurus: (a) new series, vol. 1 (1824), plate 21, produced using lithography by George Scharf, and (b) original series, vol. (1817), plate 42, produced using copper engraving. Images from Biodiversity Heritage Library, www.biodiversitylibrary .org, contributed by (a) California Academy of Sciences Library and (b) Ernst Mayr Library, Museum of Comparative Zoology, Harvard University.

and the luxurious qualities of its *Transactions* (1807–48) added to the financial strain, there was no attempt to cut costs through the application of lithography. Rather, the society experimented much more than others with the use of wood engraving for more mundane illustrations, while also experimenting with some steel intaglio plates.[41]

This conservatism in the natural history transactions deserves further attention, and it is worth noting that lithography was slow to be adopted more generally in the high-prestige, high-cost botanical and zoological part works that were so characteristic of the early decades of the nineteenth century.[42] In particular, the growing number of monthly botanical part works that emulated William Curtis's *Botanical Magazine* (1787–) continued the practice of issuing colored intaglio engravings. Only in 1845, with a new publisher and the lithographic talents of botanical artist Walter Fitch, did the *Botanical Magazine* finally make the change.[43] In such a well-established, extremely skilled, and high-status culture of illustrative printmaking, it is perhaps not surprising that the adoption of lithography was rather slow, despite its attractions in offering plates that were at once relatively conservative in appearance —looking superficially much like intaglio plates—and significantly cheaper to produce.

CHEAP JOURNALS, NEW AUDIENCES, AND THE GROWTH OF WOOD ENGRAVING, 1820-30

While the learned societies continued to be invested in the expensive form of the prestige publication, albeit with costs now sometimes tempered by the introduction of lithography, the early 1820s witnessed a striking transformation as the first cheap scientific journals employed wood engraving to great effect. The emergence in the 1820s of the first commercially successful cheap periodicals—addressing the rapidly increasing numbers of working- and lower-middle-class readers and using some of the new technologies of mass production—was one of the wonders of the age, contributing to a new sense of the "march of mind." Reflecting that the application of wood engraving "for the general purposes of pictorial illustration was comparatively slow" in the three decades after 1790, William Chatto attributed its rapid growth in the 1820s to radical publisher William Hone—whose political satires and popular miscellanies sold very extensively—and to the Society for the Diffusion of Useful Knowledge (SDUK), an organization founded in 1826 with a view to providing high-quality educational works at cheap prices.[44] However, it was the success of the cheap-periodical visionary John Limbird, another former radical publisher, that inspired many, including the SDUK, to adopt the extensive use of wood engraving.

During the 1810s, wood engravings had begun to appear occasionally in the broadly based monthly magazines, such as the *Gentleman's Magazine* (1731–1907), that had survived the previous century, and by the start of the 1820s such magazines were beginning to make them more of a regular feature. Limbird copied this practice in his *Mirror of Literature, Amusement and Instruction*, a sixteen-page weekly miscellany begun in 1822. The magazine was priced at two pence—a price previously only seen in the cheap radical press that had been deliberately suppressed by the imposition of a four and one-half pence newspaper tax in 1819—and Limbird's strategy of giving readers affordable access to snippets of the literature of the day ensured an unprecedented success, with regular sales probably in the tens of thousands. The inclusion of one or two moderately large wood engravings was a particular selling point that others soon emulated. Indeed, in the wake of Limbird's success, a cascade of cheap, mostly short-lived, weekly publications tumbled onto the market, many of them illustrated, such as the *Nic-Nac* (1822–28), the *Portfolio* (1823–29), and the *Olio* (1828–33). It was above all these new cheap periodicals that popularized wood engraving and served to develop a regular workforce.[45]

The *Mirror*'s weekly mélange included a range of scientific and technical illustrations. The first eight numbers, for instance, included illustrations of the treadmill at Brixton Prison, a "mermaid" specimen that was being exhibited in London, the recent eruption of Mount Vesuvius, and the moose on display at William Bullock's Egyptian Hall in Piccadilly. As this suggests, illustrations often related to spectacles of the day, and their sensationalism was obviously designed to draw the eye and to entertain. Yet, while such illustrations were far from providing the kind of technical information needed for scientific purposes, they were broadly informative, and the journal was clear about its dual mission: to instruct as well as to amuse. Introducing the depiction of the moose, for instance, it observed: "Anxious to keep our promise with the public, in rendering our little work a 'MIRROR of Literature, Amusement, and Instruction,' we shall occasionally give engravings of some of the most remarkable subjects of natural history, accompanied by accurate descriptions."[46]

The advantages of wood engraving for scientific purposes were altogether more clearly on display in one of the *Mirror*'s earliest imitators, the *Mechanics' Magazine*. Founded in 1823 by Limbird's associate, the patent agent Joseph Clinton Robertson, this three-penny weekly had the use of wood engravings at the heart of its mission to inform "the British artisan." Offering a "digested selection" from periodicals and books of the day, it promised

> Accounts of all New Discoveries, Inventions, and Improvements, *with illustrative Drawings*, Explanations of Secret Processes, Economical Receipts, Practical Applications of Mineralogy and Chemistry; Plans and Suggestions for the Abridgment of Labour; Reports of the State of the Arts in this and other Countries; Memoirs, and occasionally Portraits, of eminent Mechanics, &c. &c.

The periodical was true to its word: there were "numerous Wood-cuts" in each number, many of which were diagrams providing information about machinery.[47] This informational aspect was a matter of considerable importance. The *Imperial Magazine*, a general monthly aimed at a lower-middle-class dissenting audience, observed that the *Mechanics' Magazine* abounded "with wood engravings, illustrative of the various subjects which required something more than simple description, to render them satisfactorily intelligible to every reader."[48] Not that all the illustrations were purely informational. The first number alone had a portrait of James Watt and a depiction of Icarus, both by "the skilful Sears." Yet, while Matthew Urlwin Sears was a technically accomplished engraver who worked also for Limbird, one contemporary remembered him as "little of an artist, with no taste," and the *Mechanics' Magazine* certainly offered little to please the aesthete.[49]

The large circulation achieved by the *Mirror of Literature* and the *Mechanics' Magazine* prompted both commercially and ideologically motivated publishers to investigate the possibilities of cheap publishing, and the use of wood engraving expanded rapidly in the process. This was above all the case with the publications of the SDUK. Its flagship "Library of Useful Knowledge," a series of fortnightly six-penny treatises launched in March 1827, was closely modeled on the format of the cheap journals. Moreover, while the "preliminary treatise" on "the objects, advantages, and pleasures of science" was unillustrated, those that followed contained a constant supply of largely diagrammatic illustrations. The society was wary of indulging the sensuality of ill-educated workers, and the preliminary discourse—by Henry Brougham— explained that no figures were to be used in the treatise to "assist the imagination," because the object was to appeal to "reason, without help from the senses." As Anne Secord has shown, this severe judgement "reflected widespread concerns about the nature and management of visual pleasure."[50] Yet the informational content of illustrations was highly prized. Indeed, by 1837 the editor of the *Penny Mechanic* (1836–43) could quote as familiar the maxim that "one square inch of wood is worth a page of letter press," pointing out that illustration had enabled him to explain "many complicated pieces of machinery . . . that could not have been described without the assistance of the draughtsman."[51]

Very rapidly, moreover, the SDUK widened its perspective on the role of illustration as it expanded its commitment to the role of pleasure in learning. Reflecting that much of the reading that was done for "mere amusement" might be "made a source of great improvement," the society's first annual report announced its intention to commence a "Library of Entertaining Knowledge" combining "instruction and amusement, comprising as much entertaining matter as can be given along with useful knowledge, and as much knowledge as can be conveyed in an amusing form."[52] These volumes—which came to be published by the cheap-publishing visionary Charles Knight—greatly expanded the society's illustrative ambition and technical prowess in the use of wood engraving. Knight's own early volume *The Menageries: Quadrupeds, Designed and Drawn from Living Subjects* (1829) provides a good example. Here, the depiction of animals within a range of scenes—executed by "rising young men," two of whom were later Royal Academicians—was clearly intended to be quite as pleasurable as instructive.[53] In the years that followed, Knight became the SDUK's sole publisher, and it was his passion for the improving qualities of artistic representation that led to the dominant position of high-quality wood engravings in the *Penny Magazine*, commenced in March 1832.[54] Moreover, the achievement depended on technical experimentation in the preparation of the wood blocks—especially in relation to lowering— that enabled them to be printed using the steam presses of London printer William Clowes in order to deliver the magazine's print runs of up to two hundred thousand copies. Since the 1820s Clowes had been a keen advocate of the new steam technology as a means of cheapening and expanding the market for print, and Knight worked closely with him in achieving a high-quality product.[55]

The degree of success enjoyed by the new cheap journals of the 1820s in the use of wood engraving stands in stark contrast to the continuing conservatism not only of the learned transactions, but also of the established scientific and technical journals, and of such new titles as David Brewster's *Edinburgh Journal of Science* (1824–32) and Thomas Gill's *Technical Repository* (1822–30). However, two commercial journals of the late 1820s offered a striking demonstration of the utility of wood engraving for more learned scientific purposes—the *Gardener's Magazine* (1826–43) and the *Magazine of Natural History* (1829–40). Both were the productions of the Scottish landscape gardener and author John Claudius Loudon, working with the leading London publishers Longman.

Loudon was a farmer's son who had been apprenticed as a nurseryman and who studied at the University of Edinburgh. On moving to London, he had soon established a reputation in horticultural circles, becoming a fellow of the Society of Arts, the Horticultural Society, and the Linnean Society. With support from Sir Joseph Banks, he published the monumental 1,500-page *Encyclopaedia of Gardening* with Longman in 1822. The work was novel in its comprehensive scope, and was soon being described as a "standard book," passing through multiple editions. Moreover, while at two pounds and ten shillings it was very expensive, it was intended for use by "practical gardeners" as well as their patrons, embodying Loudon's Benthamite vision for their professional education.[56] With such a readership in view, Loudon illustrated the work with nearly six hundred wood engravings by one of the capital's leading wood engravers, Robert Branston, which reviewers considered did the artist "the highest credit."[57] The *Gardener's Magazine*, launched in January 1826, offered a periodical continuation of this encyclopedia, providing an account of improvements in gardening that was "accessible to the practical gardener, land-steward, bailiff, and others concerned in country affairs." The work was not cheap, but at three shillings and six pence per quarter it worked out roughly the same as a regular subscription to the *Mechanics' Magazine*, albeit for just over half the number of pages. Once again, it was to be illustrated by wood engravings "where useful," and the first four quarterly issues contained a hundred illustrations between them, which ranged from diagrams of plant dissections and tools, through views of horticultural structures to pictorial landscapes and portraits (see figs. 2.3a and 2.3b).[58]

The woodcuts were a crucial part of the innovative package that Loudon sought to offer, since they helped to keep costs low, and thus permitted a greater range of illustration. Here, his eye was clearly on the dream of a wider audience embodied in the new cheap journals, rather than on investing in lithography to emulate the learned transactions. The editor spelled out his vision more explicitly at the start of the following year, having been piqued by a pointed reference in the preface to the *Transactions of the Horticultural Society* to the practice of some journals of reprinting original matter from that publication. In response, Loudon wanted to emphasize, first, that the original matter in his own magazine was quite equal to that in the *Transactions* ("the same persons, and sorts of persons, write in both works"), second, that his magazine worked out at around one-sixth of the price of the *Transactions*, and, third, that while its articles were "not ornamented by coloured plates, or

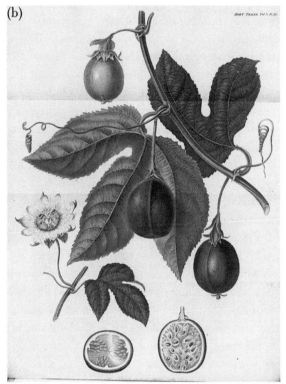

FIG. 2.3. The comparative value of wood engraving for practical purposes. Two illustrations of *Passiflora* species: (a) a wood engraving from *Gardener's Magazine* 1 (1826): 16, to illustrate the "fleshy rays" that the author ("an Amateur") advised should be removed to avoid causing putrefaction in the fruit, and (b) a hand-colored copper engraving from the *Transactions of the Horticultural Society* 3 (1820): plate 3, to illustrate the appearance of the leaves, flowers, and fruit of a particular cultivar for taxonomic purposes. Images from the Biodiversity Heritage Library, www.biodiversitylibrary.org, contributed by (a) Smithsonian Libraries and (b) Peter H. Raven Library, Missouri Botanical Garden.

engravings from copper or steel," they were "illustrated by a greater number of engravings from wood, sufficiently intelligible for all useful purposes, than is the present or any former volume of the Horticultural Transactions." Indeed, Loudon claimed, when the magazine drew on the *Transactions*, it not only offered a usefully abridged account but "frequently illustrated this essence by engravings, which rendered it of more value than the original in its unabridged and unillustrated state"; in the case of a recent number, eleven engravings were "composed expressly for the purpose." Above all, he considered that his added wood engravings provided additional and valuable information, as when he included detailed diagrams to make clearer how the original author suggested grafting rose buds.[59]

Contemporaries agreed that the *Gardener's Magazine* would mark a "new era" in the gardening literature: it was "of incalculable value to working gardeners and farmers," who could not afford to buy "expensive works, such as the Horticultural Transactions, Linnæan Transactions, and other works containing much valuable matter, but not accessible to general readers."[60] Moreover, while the learned societies' transactions proved a drain on their resources, the low cost of engravings and other cost-cutting measures of the *Gardener's Magazine* meant that profits were considerable, with Loudon earning around £500 to £750 annually in the early days.[61] In such circumstances, and with a new *Encyclopaedia of Plants* in preparation, it is not surprising that Loudon now decided to apply the format for a related audience in his three shillings, six pence bimonthly *Magazine of Natural History*, which he alternated with the now bimonthly *Gardener's Magazine*. The new magazine emulated the existing one in its emphasis on providing ready access to the progress of science while seeking to "extend a taste for this description of knowledge among general readers and observers, and especially among gardeners, farmers, and young persons resident in the country." Initial sales of more than two thousand copies suggest that Loudon was reaching a wide audience.[62]

Once again, the use of wood engravings—there were more than two hundred in the first volume—was a key aspect of the magazine's cost-saving formula. On this occasion, however, it was not only the name of the engraver that appeared in the publicity materials. Robert Branston's artistry was admired, but natural subjects were not his forte. However, the members of the London school of wood engravers of which he was considered the head were used to working from others' designs. Loudon arranged to have the botanical drawings carried out by the well-respected naturalist and botanical artist James de Carle Sowerby; the zoological ones by William Harvey, by this time London's premier wood-block draftsman; and the trees by the landscape painter and tree specialist Jacob George Strutt.[63] With such high-profile artists involved, he could hope to convince knowledgeable naturalists that for many purposes his magazine was capable of relaying information graphically that was not inferior to what might be expected in the competing *Zoological Journal* (1824–34), with its copper plates. Indeed, he challenged his readers to compare his wood engraving of Geoffroy's shrike, *Lanius plumatus*, with the copper plate in Edward Griffith's revised edition of Cuvier's *Animal Kingdom* (1827–35) on which it was based (figs. 2.4a and 2.4b). It was, he considered, "nearly as expressive, or, at least, sufficiently so for every useful purpose." He also argued that wood engraving could be used to add additional diagrams offering the osteological information so central to Cuvier's system. Instead of making their work ludicrously expensive with intaglio plates, Griffiths and his publishers should have used wood, Loudon argued.[64] Certainly, Loudon's use of wood engravings in his new magazine helped to keep its finances reasonably manageable, even though sales soon declined. Moreover, as we shall see shortly, it established a model for many competitors. However, the growing prominence of wood engraving in the context of the scientific journal only further fed the debate concerning the relative merits of different kinds of illustrative technology for scientific purposes, and it is to this that we now briefly turn.

ILLUSTRATIVE TECHNOLOGIES, THE POLITICS OF KNOWLEDGE, AND THE PURPOSE OF SCIENTIFIC IMAGERY, 1830–40

By 1830, then, several scientific periodicals had adopted the new illustrative technologies of lithography and wood engraving in preference to intaglio, demonstrating their potential utility for a range of scientific purposes. In both cases, the decision reflected the financial advantages that the new technologies offered. In prestige periodicals, such as the transactions of learned societies, use of lithography could make the difference between financial disaster and triumph. In the new cheaper periodicals of the "useful knowledge" movement of the 1820s, wood engraving offered a means of securing all the attractions of illustration within a tight budget, and consistently with other new technologies of the industrial age, such as stereotyping and the steam press. Yet, while lithography maintained a gentlemanly feel in the luxuriant plates of learned transactions, wood engraving and the cheap periodicals in which it was used spoke directly to the changing politics of knowledge. The technology of illustration became inextricably linked to the audiences for science and the purposes to which illustrations were put. In 1839, William Chatto observed:

(a)

we have numerous genera, of which a single figure is not given of any one of the species; while of other genera, copperplates are engraved of several of the species. This indefinite, unsystematic mode of giving illustrative engravings, publishers will, in time, learn to avoid, from the necessity which they will find of accommodating their productions to the present improved state of the public judgment in books, called forth, in a great measure, by an increased taste for reading, and the diminished means of procuring this gratification. We can see no occasion for engravings on copper to illustrate a work like the present; had the figures been on wood, they would have been nearly as expressive, or, at least, sufficiently so for every useful purpose, as that which we now give of the Geoffroy's Shrike (*fig.* 143.) will show to those who

143

can compare it with the copperplate from which it was copied in Part XV. They could have been printed along with the descriptions, and more readily compared with them; and, had they been limited to the type species of each genus, the work would have been much more complete, and, we should think, not near so high priced. There is also an omission which detracts from the value of the work. As the system of Cuvier is founded on the physiology of animals, there ought to have been plates given of those parts of the osteology of vertebrated animals which form the distinctive characters. They are not given in the original work of Cuvier, because it was intended to form a cheap text-book, and the reader is supposed to have access to the museums of Paris, and to extensive libraries of natural history; but in a work which will be ten times the price of the original, they ought to have been added. We regret to be obliged to make these objections,

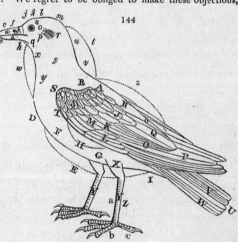

144

and should have been much better pleased to have given the work entire instead of qualified approbation.

As a principal object of the last number of this Magazine was to impress on the mind of the young reader the terminology of birds, we shall here copy from Mr. Griffith's work an engraving which exemplifies that terminology in a very judicious manner. (*fig.* 144)

(b)

GEOFFROY'S SHRIKE.

L. PLUMATUS.

London. Published by G.B.Whittaker. March 1835.

FIG. 2.4. The comparative value of wood engraving for scientific purposes. Two illustrations of Geoffroy's shrike, *Lanius plumatus*, compared by John Claudius Loudon: (a) a wood engraving from the *Magazine of Natural History* 1 (1829): 276; and (b) a copper engraving from Edward Griffith's *The Animal Kingdom, Arranged in Conformity with Its Organization by the Baron Cuvier*, vol. 6 (London: Geo. B. Whittaker, 1827–35), opposite 484. Images from the Biodiversity Heritage Library, www.biodiversitylibrary.org, contributed by (a) Natural History Museum Library, London, and (b) University Library, University of Illinois at Urbana-Champaign.

Wood-engravings are not to be estimated by a comparison with copper-plates; but are to be judged of by the power and significance with which they excite ideas in the mind, with reference to the means employed in their execution, and on a consideration of the thousands whose knowledge is thus extended, and whose pleasure is thus increased, compared with the hundreds who can afford to purchase copper-plate engravings.[65]

In such a climate, the continued employment of more expensive illustrative technologies was a statement about the legitimate users and proper uses of the knowledge texts they adorned.

As Anne Secord has shown, many in this changing world considered the sensory pleasures offered by illustrations to be a key means of recruiting potential practitioners into the enterprise of natural history. The great flowering of illustrated horticultural and natural history periodicals that occurred in the 1830s—Ray Desmond reports that fifteen such periodicals were established in that decade—doubtless contributed to building the large and active community of natural history enthusiasts so characteristic of Victorian Britain.[66] Some, such as the shilling monthly *Horticultural Register* (1831–36) of the Duke of Devonshire's landscape gardener Joseph Paxton, copied Loudon's formula closely, including the use of wood engravings. Many others, however, continued to use other illustrative technologies—including both intaglio (where the more resilient steel was increasingly used to reduce costs) and lithography—despite their higher cost and the precarious finances that led most magazines to fail. Indeed, within a year Paxton's *Horticultural Register* was also including intaglio plates, and when Paxton commenced his two-shilling monthly *Magazine of Botany* in 1833, it included hand-colored lithographs alongside the wood engravings in the text. Even the bargain-basement six-penny monthly *Floricultural Cabinet* (1833–59), of Paxton's erstwhile collaborator Joseph Harrison, shifted from wood engravings to intaglio for its colored botanical plates after five issues.[67] James Rennie's *Field Naturalist* (1833–35) and Richard Owen's *Zoological Magazine* (1833) also combined wood engraving with plates. Most strikingly, the *Magazine of Zoology and Botany*, begun by leading Edinburgh naturalists in 1836, was dominated by its colored plates (three per bimonthly number) rather than its wood engravings. This was the magazine—remodeled by the savvy journal printer Richard Taylor as the *Annals of Natural History* (1838–1966)—that survived the attritional early years of natural history journal publishing, despite the fact that its illustrations cost twice those of the *Magazine of Natural History*.[68] Such choices in illustrative technology clearly reflected perceptions concerning the aesthetic desires of potential purchasers, and, in a related manner, evolving conventions of format. However, choices were also sometimes shaped by judgements concerning the adequacy of wood engraving for the purposes of scientific instruction and communication. In 1835, William MacGillivray—the conservator of the museum of the Royal College of Surgeons in Edinburgh and collaborator with Audubon—began a two-penny fortnightly *Edinburgh Journal of Natural History, and of the Physical Sciences* (1835–40), which contained some wood engravings but also a color plate. MacGillivray explained:

> Within the last few years, various cheap publications on this subject, illustrated by engravings on wood, have led all classes to observe and to enjoy the ever-varied beauties of the creation. But no description, however correct, or no wood-cut, however well executed, can give that complete idea of a natural object, which is effected by an engraving on steel, when coloured with accuracy. The enormous price at which these illustrations of a higher order are usually sold, has alone prevented them from becoming extensively popular. Hitherto, coloured engravings, executed with beauty and correctness, have been accessible to the wealthier classes alone. It is proposed, in this work, to place elegant engravings of the choicest productions of Nature, within the reach of all classes of the community.[69]

To MacGillivray, the desire to expand the social range of those engaged in natural history by using cheaper illustrative technologies risked undermining informational accuracy. Yet, while the use of color certainly conveyed additional information, it is otherwise far from clear where the supposed superiority of intaglio lay (see figs. 2.5a and 2.5b).

Indeed, practitioners were not agreed as to the most informative illustrative technology for the purposes of natural history. The aesthetic qualities associated with the hand-colored intaglio plates that sold botanical part-works—including such periodicals as the *Botanical Magazine*—were considered by some to be fraught with dangers which the new, cheaper technologies might actually obviate. In 1831 the cryptogamic botanist William Wilson wrote to his friend, Glasgow's professor of botany William Jackson Hooker, about the need for "a well conducted work of botanical illustrations, either in Lithograph or Woodcut." He continued:

> If I could draw with facility, I would myself attempt it—accurate dissections, & essential characters of genera & species would I am convinced do more towards the propagation of the science, than all the coloured figures that ever

young and adult garbs of both sexes, and this has induced a still further multiplying of specific names. This bird is a native of the Northern States of America, and migrates southward on the approach of winter. It constructs its nest in the hollow of a tree.

DESCRIPTION OF TWO NEW SHELLS.

The two Shells represented beneath were procured from Orkney, and are now in the Cabinet of William Nicol, Esq., Edinburgh. It is difficult to determine whether they are new species, or only greatly produced varieties of the *Buccinum Anglicanum* and *B. undatum*.

1   2

Fig. 1. agrees in all its characters with the *B. Anglicanum* of Lamarck, except in the spire being much longer and more fusiform, the breadth of the body of the shell being only about a third of its length; while in the *B. Anglicanum*, it measures nearly the half of the shell; it is of a reddish brown colour, fasciated and clouded with darker shades of the same colour. If it be really distinct from the *B. Anglicanum*, it might with propriety be distinguished by the name of *B. elongatum*. But until we have seen and examined the animal, the name must remain in abeyance.

Fig. 2. has all the characters and appearance of *B. undatum*, except in its greatly elongated shape, and if the animals really differ, it might be distinguished by the specific name of *nautinitum*.

These Shells are said to be obtained by their adhering to the fishermen's lines in deep sea water; and probably their greatly lengthened shape may be peculiar to this locality.

BOTANY AND HORTICULTURE.

SACCHARINE PLANTS.—The most valuable plants producing Sugar in this country are Beet-Root and Parsnip. The White Beet (*Beta cicla*) is a hardy biennial plant, a native of the Sea Coasts of Spain and Portugal, and introduced into this country in 1570. It was from the roots of this plant that the French and Germans obtained sugar with so much success during the late war, while all their West India colonies were in the hands of the British. The following is the ordinary process of extracting the sugar from this plant:—The roots are reduced to a pulp by pressing them between two rough cylinders; the pulp is then put into bags, and the sap it contains is pressed out. The liquor is then boiled, and the saccharine matter precipitated by quicklime; the liquor is now poured off, and to the residuum is added a solution of sulphuric acid, and again boiled; the lime uniting with the acid, is got rid of by straining; and the liquor is then gently evaporated, or left to granulate slowly, after which it is ready for undergoing the common process of refining raw sugars. The French manufacturers have acquired so much experience in this process, that, from every 100 lbs. of Beet, they extract 12 lbs. of sugar in the short space of twelve hours.

The Parsnip (*Pastinaca sativa*) is next in value to the White Beet as a saccharine root. It is a biennial British plant, common in calcareous soils, and used in England chiefly as a vegetable. One thousand parts of Parsnips contain ninety parts of sugar, nine parts of starch, the rest being water and fibre. An excellent ardent spirit is obtained by distillation from this plant; but the wine manufactured from it, in the opinion of many, possesses a finer flavor, and more nearly approximates to foreign wine than that obtained from any other British produce. The process of manufacturing Parsnip Wine is more clearly and fully described, in an interesting little work entitled " The British Wine-Maker," recently published by Mr W. H. Roberts, than in any other work on the same subject. We may refer generally to Mr Roberts' useful and practical treatise as affording comprehensive and scientific information, while it seems a safe guide in the manufacture of wines from British produce. By the use of the *Saccharometer*—an instrument remarkable for the accuracy of its results—the process, as detailed in the work referred to, is rendered simple and of unfailing success, while, without its aid, wines of uniform quality, from year to year, cannot be otherwise produced. This circumstance arises from the fruits themselves yielding in some years a greater or less proportion of saccharine matter than they do in others, and this difference in quality is accurately determined by the application of the Saccharometer.

RINGING OF WALNUT-TREES.—The Baron de Trehoudi, near Metz, in Lorraine, has successfully introduced into his neighbourhood a practice of ringing Walnut-trees. It is accomplished by abstracting a ring of two inches breadth from the outer bark all around, and then plastering over the part with clay, mixed with moist manure. The Walnut-trees thus treated not only prove more prolific, but the fruit is more early.—*Neill's Horticultural Tour.*

GEOLOGY.

THE FOSSIL ELK OF THE ISLE OF MAN.—In the Royal Museum of the College of Edinburgh there is now the most perfect known specimen of this animal. Its dimensions are given below, but that a more distinct notion may be formed of its great stature, we have placed beside it a human skeleton of six feet, drawn upon the same scale.

	Ft.	In.
Height to the tip of the process of the first dorsal vertebra, which is the highest point of the trunk,	6	1
Height to the anterior superior angle of the scapula,	5	4
Length from the first dorsal vertebra to the tip of the os coccygis,	5	2
Height to the tip of the right horn,	9	7½
Lateral or horizontal diameter of the thorax, at the widest part, that is at the eleventh rib,	2	0½
Depth of the thorax, from the tip of the process of the eighth dorsal vertebra, to the sternum at the junction of the eighth rib,	2	2

This superb fossil was dug up in the parish of Kirk Ralaff, and secured for our University Museum by the late Duke of Atholl. It was found imbedded in loose shell-marl, associated with numerous branches and roots of trees; over the marl was a bed of sand; above the sand a stratum of peat, principally composed of small branches and decayed leaves; and on the surface of all, the common alluvial soil of the country.

FOSSIL BOTANY.—The researches of M. Adolphe Brongniart into fossil organic remains, have in a great measure led to a knowledge of what must have been the appearance and temperature of the earth, when these fossils were vegetating on its surface; and also how far the various epochs of the existence of those plants accord with those remote Zoological epochs established by modern geologists.

The primitive vegetation, according to M. Brongniart, in its periods to the three successive formations of soil, from the earliest times, from the Creation down to that of the great Diluvian Change.

To the first period, which is co-etaneous with the simplest and the oldest formation of the globe, and lasted until the occurrence of deposits of coal strata, belong those vegetable bodies, the structure of which is in the highest degree simple. These organic remains are also remarkable for their rarity and the excessive magnitude of their dimensions. With respect to their rarity, compared to those of an analogous character inhabiting the present world, it is asserted, that of the former there are only six distinct families known; whereas of the latter, at least *two hundred* families exist; and with regard to the magnitude of their dimensions, it will be sufficient to instance the Fern trees, which in the actual world, and under the most favorable circumstances, grow to the height of from 20 to 25 feet only, while the same trees rose, in the primitive world, as high as 40 and even 50 feet. Brongniart thinks that the great coal formation which appeared at the termination of this *vegetable* period is due to the destruction of the plants in question. He arrives at the conclusion à priori, as well as from the inspection of the strata in which such plants are found, that life on the surface of the globe began with the vegetable kingdom; that the animals without a vertebrated spine succeeded next; and that probably the oceans contained no fish at the time.

To the second period of Antediluvian vegetation, corresponding to the geological

FIG. 2.5. Wood engraving and hand-colored steel engraving juxtaposed. Illustrations from the *Edinburgh Journal of Natural History* 15 (August 1836), depicting (a) specimens of marine gastropod and fossil elk using wood engraving, and (b) specimens of owl using steel engraving. Images from the Biodiversity Heritage Library, www.biodiversitylibrary.org, contributed by Ernst Mayr Library, Museum of Comparative Zoology, Harvard University.

were or will be published: these often do harm instead of good & are more likely to make "knowers of species" than sound botanists. . . . Figures with dissections have a very high value; superior even to a dried specimen, if faithfully executed; but then they are so expensively got up that very few can afford to purchase them, compared with the number who would buy woodcuts or Lithographs, such as might be depended upon.[70]

As Anne Secord has shown, Hooker and other botanists shared Wilson's concerns about the constraints that a desire to satisfy aesthetically motivated book purchasers placed on botanical illustration. Hooker was a talented botanical artist who in 1827 had become the editor and sole draftsman of the *Botanical Magazine*, with its hand-colored intaglio plates. Yet most copies of his *Botanical Miscellany* (1830–33) were sold with uncolored plates, though not until 1840 (in the successor *Journal of Botany*, 1834–42) were these lithographed rather than engraved.[71] Joseph Paxton likewise sometimes considered that readers might learn more from the wood engravings in his *Magazine of Botany* than from the elaborate color plates.[72]

The complexity of decision making in relation to illustration might be especially evident in natural history periodicals, but other scientific periodicals had similar issues to contend with as they negotiated the new technologies. Some of the new scientific and technological magazines that appeared in the 1830s, such as the *Railway Magazine* (1835–77) and the *Civil Engineer* (1837–68), adopted wood engravings. Others, such as the *Annals of Electricity, Magnetism, and Chemistry* (1836–43), maintained the established practice of using separate plates, but substituted lithography for copper. Only gradually over the next two decades did wood engravings become the standard illustrations in the more arcane scientific periodicals, and in the new "proceedings" that scientific societies began to produce alongside their transactions. In all of this, decisions involved balancing considerations relating to producing a product that was appropriately priced to develop the desired community of readers, while offering printed images that were considered informationally and aesthetically suitable for their purposes.

CONCLUSION

Parallel to the mechanization of book production that took place in the early decades of the nineteenth century, the transformation of illustrative technologies had profound consequences for the sciences. The rise of wood engraving and lithography, in particular, radically altered the possibilities for scientific illustration. Most obviously, the new techniques opened up a new graphic repertoire. The vividness of wood engravings and the quasi-tactile qualities of lithographs extended the palette of scientific artists, offering new effects that were of value for particular purposes. Arguably more radical, however, was the effect that the new technologies had on the economics of scientific illustration. While much remains to be done to understand the financial implications, the new techniques were clearly significantly cheaper than the intaglio technologies that had long dominated scientific illustration. Moreover, given the high relative cost of illustration in the overall budget of publication, such savings had the potential to produce an economic transformation. Nowhere was this more evident than in relation to periodicals. Within the learned societies lithography, and to a lesser extent wood engraving, played a significant role in reducing spiraling costs. At the same time, wood engraving began to be used to offer scientific illustrations within the new cheaper commercial periodicals, often with a twin emphasis on instruction and amusement. Such illustrations formed a key element in the establishment of diverse kinds of scientific periodicals, most notably the types of gardening and natural history magazines developed by John Loudon. For Loudon, the use of wood-engraved illustrations was a core part of his mission to expand the communities of informed and actively engaged gardeners and naturalists.

As other chapters in this book show, the role of periodicals in expanding communities of scientific practitioners was often controversial, but the role of illustrations within that was likewise a matter of dispute. The provision of cheaper illustrations in order to engage and inform new periodical readers might have any number of deleterious consequences. As Anne Secord has shown, the use of imagery in teaching readers about nature was inherently suspect in an age when it was considered that the pleasure it generated might as easily degenerate into sensuality as engender rationality.[73] More than that, there is evidence to suggest that the graphic qualities of the new, cheaper technologies—especially wood engraving—were considered by some to be prone to mislead. However, while this chapter has begun to uncover some of the considerations that came into play in choosing between the growing range of available technologies, much remains to be done. It was certainly far from being the case that the older and more expensive intaglio technologies, often combined with hand coloring, were necessarily to be preferred graphically for scientific purposes. On the contrary, it seems that they were sometimes preferred because of a sense of their beauty or their social exclusivity, to the detriment of their scientific utility. In this regard, the account in this chapter raises as many questions as it answers; in so doing, however, it opens a rich vein for further research in the history of nineteenth-century scientific illustration. Moreover, it is a history that extends far beyond the period explored here, since the rapid rate of change in illustrative technology in the first part of the century, if anything, only accelerated in the decades that followed.

NOTES

1. "Biographic Sketches: Thomas Bewick," *Chambers's Edinburgh Journal*, 2 January 1836, 388–89, on p. 388.

2. *First Report from the Select Committee on Postage; Together with the Minutes of Evidence, and Appendix*, House of Commons Parliamentary Papers, Session 1837–38, 33: 1–516, on p. 319.

3. See, for example, Jonathan R. Topham, 'Scientific Publishing and the Reading of Science in Nineteenth-Century Britain: A Historiographical Survey and Guide to Sources,' *Studies in History and Philosophy of Science* 31 (2000): 559–612; James A. Secord, *Victorian Sensation: The Extraordinary Publication, Reception and Secret Authorship of "Vestiges of the Natural History of Creation"* (Chicago: University of Chicago Press, 2000), esp. ch. 2; and Aileen Fyfe, *Science and Salvation: Evangelicals and Popular Science Publishing in Victorian Britain* (Chicago: University of Chicago Press, 2004).

4. See Michael Twyman, "The Illustration Revolution," in *Cambridge History of the Book in Britain*, vol. 6, *1830–1914*, ed. David McKitterick (Cambridge: Cambridge University Press, 2009), 117–43; Paul Goldman, "The History of Illustration and Its Technologies," in *The Book: A Global History*, ed. Michael F. Suarez and H. R. Woudhuysen (Oxford, UK: Oxford University Press, 2013), 231–44; and Bamber Gascoigne, *How to Identify Prints: A Complete Guide to Manual and Mechanical Processes from Woodcut to Jet Ink* (London: Thames and Hudson, 1986).

5. See, for example, Patricia Anderson, *The Printed Image and the Transformation of Popular Culture, 1790–1860* (Oxford, UK: Clarendon Press, 1991); and Celina Fox, *Graphic Journalism in England during the 1830s and 1840s* (New York and London: Garland, 1988).

6. See, for example, Martin Rudwick, "The Emergence of a Visual Language for Geological Science 1760–1840," *History of Science* 14 (1976): 149–95; Lorraine J. Daston and Peter Galison, *Objectivity* (New York: Zone Books, 2007); and Anne Secord, "Botany on a Plate: Pleasure and the Power of Pictures in Promoting Early Nineteenth-Century Scientific Knowledge," *Isis* 93 (2002): 28–57.

7. Anonymous, "[Review of *History of British Birds*, by Thomas Bewick]," *Annual Review* 3 (1804): 729–37, on p. 729.

8. The distinction between "woodcut" and "wood engraving" was not used consistently in the period, and usage can often prove confusing to the historian.

9. See James Mosley, "The Technologies of Printing," and Tim Clayton, "Book Illustration and the World of Prints" in *Cambridge History of the Book in Britain*, vol. 5, *1695–1830*, ed. Michael F. Suarez and Michael L. Turner (Cambridge: Cambridge University Press, 2009), 163–99, on pp. 182–84, and 230–34, on p. 240.

10. On Bewick, see *Oxford Dictionary of Natural Biography* (*ODNB*) and Jenny Uglow, *Nature's Engraver: A Life of Thomas Bewick* (London: Faber and Faber, 2006), esp. ch. 13–14.

11. Anonymous, "[Review of *General History of Quadrupeds*, by Thomas Bewick]," *Critical Review* 70 (1790): 414–18, on p. 414–15; anonymous, "[Review of *General History of Quadrupeds*, by Thomas Bewick]," *General Magazine* 4 (1790): 540–41, on p. 541.

12. Anonymous, "[Review of *History of British Birds*, by Thomas Bewick]," *Annual Review* 3 (1804): 729–37, on p. 729.

13. Anonymous, "Wood-cuts," in *Supplement to the Third Edition of the Encyclopædia Britannica; or, A Dictionary of Arts, Sciences, and Miscellaneous Literature*, 2 vols., ed. George Gleig (Edinburgh: Thomas Bonar, 1801), 2: 809–12, on p. 811; [W. A. Chatto], *A Treatise on Wood Engraving* (London: Charles Knight, 1839), esp. 568–69, 683, and 710; and "Wood-Engraving," in *Penny Cyclopaedia*, 27 vols., ed. George Long (London: Charles Knight, 1833–43), 27: 522–26, on p. 526.

14. N. M., "Remarks on Engraving on Wood," *Monthly Magazine* 5 (1798): 111–12, on p. 112; the letter was reproduced in the *Edinburgh Magazine*.

15. Anonymous, "Wood-cuts," in *Supplement to the Third Edition of the Encyclopædia Britannica; or, A Dictionary of Arts, Sciences, and Miscellaneous Literature*, 2 vols., ed. George Gleig (Edinburgh: Thomas Bonar, 1801), 2: 809–12, on pp. 811–12.

16. *An Abridgement of Dr. Goldsmith's Natural History of Beasts and Birds: Interspersed with a Variety of Interesting Anecdotes, and Illustrated by Nearly Two Hundred Engravings on Wood, in the Manner of Bewick* (London: Scatcherd and Letterman, and Langley and Belch, 1807).

17. N. M., "Remarks," 111; Chatto, *Treatise on Wood Engraving*, 640. See also Twyman, "The Illustration Revolution," 122.

18. Twyman, "The Illustration Revolution," 124–25; "The Commercial History of a Penny Magazine," *Penny Magazine* 2 (1833): 377–84, 417–24, 465–72, 505–11, on p. 420; John Buchanan-Brown, *Early Victorian Illustrated Books: Britain, France and Germany, 1820–1860* (London: British Library, 2005), 21.

19. William Andrew Chatto, "Wood-Engraving: Its History and Practice," *Illustrated London News* 4 (1844): 251–54, 257–59, 273–74, 293–94, 309–10, 325–26, 357–58, 405, 417, 425, on p. 405.

20. Chatto, *Treatise on Wood Engraving*, 692–93.

21. "Commercial History of a Penny Magazine," 420.

22. Michael Twyman, *Lithography, 1800–1850: The Techniques of Drawing on Stone in England and France and Their Application in Works of Topography* (London: Oxford University Press, 1970), esp. ch. 3; C. Hullmandel, *The Art of Drawing on Stone, Giving a Full Explanation of the Various Styles, of the Different Methods to be Employed to Ensure Success, and of the Modes of Correcting, as Well as of the Several Causes of Failure* (London: C. Hullmandel and R. Ackermann, [1824]).

23. John Ford, *Ackermann, 1783–1983: The Business of Art* (London: Ackermann, 1983), 61–64; anonymous, "Some Account of the Art of Lithography, or Process for Taking Fac-Simile Impressions of Drawings from Stone," *Repository of Arts*, 2d ser. 3 (1817): 222–25, 284–86, and 343–44; 4 (1817): 33–34.

24. Michael Twyman, ed., *Henry Bankes's Treatise on Lithography* (London: Printing Historical Society, 1976 [1813]), 10; *Monthly Magazine* 44 (1817): 355; anonymous, "Some Account," 285; Hullmandel, *The Art of Drawing on Stone*, xv.

25. Hullmandel, *The Art of Drawing on Stone*, 1n; Michael Twyman, ed., *John Phillips's Lithographic Notebook, Reproduced in a Facsimile from the Original at Oxford University Museum of Natural History* (London: Printing Historical Society, 2016), 16, 19–20; James Hamilton, "Artists, Scientists, and Events," in *Fields of Influence: Conjunctions of Artists and Scientists, 1815–1860,* ed. James Hamilton (Birmingham, UK: University of Birmingham Press, 2001), 1–30, on pp. 11–12.

26. Twyman, *Phillips's Lithographic Notebook*; Jack Morrell, *John Phillips and the Business of Victorian Science* (Aldershot, UK: Ashgate, 2005), 25–27.

27. John C. Thackray, "Separately-Published Prints of Fossils in Nineteenth-Century Britain," *Archives of Natural History* 12 (1985): 175–99.

28. Quoted in Thackray, "Separately-Published Prints," 182.

29. Gordon L. Herries Davies, *Whatever Is under the Earth: The Geological Society of London, 1807 to 2007* (London: Geological Society, 2007), 36–39, on p. 37; council minutes, 15 February 1811, Geological Society of London, GSL/CM1/1/21.

30. Council minutes, 21 December 1821, GSL/CM1/1/213.

31. Council minutes, 19 June 1822, GSL/CM1/1/237.

32. Council minutes, 15 February and 21 June 1822, GSL/CM1/1/222 and 242.

33. Council minutes, 8 March 1822, GSL/CM1/1/224.

34. Council minutes, 21 June 1822, GSL/CM1/1/243; 2 February 1826, GSL/CM1/2/70.

35. [Charles Lyell], "[Review of *Transactions of the Geological Society*]," *Quarterly Review* 34 (1826): 507–40, on p. 525.

36. Hullmandel, *The Art of Drawing on Stone*, 52 and plate 6. On Hullmandel, see *ODNB*. On Scharf as a "mechanic," see Caroline Arscott, "George Scharf and the Archaeology of the Modern," in *George Scharf: From the Regency Street to the Modern Metropolis* (London: John Soane Museum, 2009), 26–41, on p. 41.

37. Hullmandel, *The Art of Drawing on Stone*, 79.

38. Anonymous, "A View of the Present State of Lithography in England," *Library of the Fine Arts*, 1 (1831): 201–16, on pp. 211–12.

39. The high proportion of society expenditure devoted to the *Transactions* is detailed in the annual accounts in the council minutes. See especially the review of expenditure in "Council Minutes" 2 (1826–43), Linnean Society, pp. 304–20. See also Andrew Thomas Gage and William Thomas Stearn, *A Bicentenary History of the Linnean Society of London* (London: Academic Press, for the Linnean Society of London, 1988), esp. ch. 20.

40. On the lithographed resolutions (24 March 1823), see "Correspondence Relating to Formation of the Zoological Club," Linnean Society, MSS MISC SP621–34, f. 21.

41. Harold R. Fletcher, *The Story of the Royal Horticultural Society, 1804–1968* (London: Oxford University Press, 1969), 114, 122, and 158.

42. A few exceptions are noted in Wilfred Blunt and William T. Stearn, *The Art of Botanical Illustration,* new edition (Woodbridge, UK: Antique Collectors' Club in association with the Royal Botanic Gardens, Kew 1994), 252. Other notable examples of the use of lithography in natural history in the 1830s include John Gould's *Century of Birds Hitherto Unfigured from the Himalaya Mountains* (1830–32) and Charles Darwin's *Zoology of the Voyage of H.M.S. Beagle* (1839–43).

43. Jan Lewis, *Walter Hood Fitch: A Celebration* (London: HMSO, 1992), 10.

44. Chatto, "Wood Engraving," 405.

45. Jonathan R. Topham, "John Limbird, Thomas Byerley, and the Production of Cheap Periodicals in Regency Britain," *Book History* 8 (2005): 75–106.

46. *Mirror of Literature* 1 (1822–23): 113; Jonathan R. Topham, "The 'Mirror of Literature, Amusement and Instruction' and Cheap Miscellanies in Early Nineteenth-Century Britain," in *Reading the Magazine of Nature: Science in the Nineteenth-Century Periodical*, ed. Geoffrey Cantor et al. (Cambridge: Cambridge University Press, 2004), 37–66, on pp. 52–62.

47. *Mechanics' Magazine* 1 (1823–24): 16 (emphasis added).

48. Anonymous, "[Review of *Mechanics' Magazine*]," *Imperial Magazine* 10 (1828): 562–63, on p. 563.

49. *Mechanics' Magazine*, 1 (1823–24): 1, 8, 11; W. J. Linton, *The Masters of Wood Engraving* (New Haven and London: Author, 1889), 201.

50. [Henry Peter Brougham], *The Objects, Advantages and Pleasures of Science* (London: Baldwin, Cradock, and Joy, 1827), 6; Secord, "Botany on a Plate," 30–31.

51. Anonymous, "Preface," *Penny Mechanic* 1 (1837): iii–iv, on p. iv. The source of the maxim is unknown.

52. Quoted in [Charles Knight], "Education of the People," *London Magazine,* 3rd ser. 1 (1828): 1–13, on p. 8.

53. Charles Knight, *Passages of a Working Life during Half a Century, with a Prelude of Early Reminiscences*, 3 vols. (London: Bradbury and Evans, 1864–65), 2:114–17.

54. See Anderson, *Printed Image*, ch. 2.

55. Valerie Gray, *Charles Knight: Educator, Publisher, Writer* (Aldershot, UK: Ashgate, 2006); Knight, *Passages*, 2:115–16; "Commercial History of a Penny Magazine," 420–21; Chatto, *Treatise on Wood Engraving* 696, 707.

56. John Claudius Loudon, *An Encyclopædia of Gardening: Comprising the Theory and Practice of Horticulture, Floriculture, Arboriculture, and Landscape-Gardening, Including All the Latest Improvements; a General History of Gardening in All Countries; and a Statistical View of Its Present State, with Suggestions for Its Future Progress, in the British Isles* (London: Longman, Hurst, Rees, Orme, and Brown, 1822), iv; Sarah Dewis, *The Loudons and the Gardening Press: A Victorian Cultural Industry* (Farnham, UK, and Burlington, VT: Ashgate, 2014), ch. 1.

57. *Literary Gazette*, 15 May 1824, p. 282.

58. [John Claudius Loudon], [*Prospectus for the "Gardener's Magazine"*] ([London: Longman, Hurst, Rees, Orme, Brown and Green, 1826]), i–ii. The first issue was priced at two shillings and six pence (not five shillings, as sometimes stated), and the price rose to three shillings, six pence after the second issue was published.

59. [John Claudius Loudon], "[Review of *Transactions of the Horticultural Society*]," *Gardener's Magazine* 2 (1827): 332–33, 414–44, on pp. 438–39.

60. *Literary Gazette*, 11 February 1826, p. 473.

61. Dewis, *The Loudons*, 73–74.

62. [John Claudius Loudon], [*Prospectus for the "Magazine of Natural History"*], ([London: Longman, Rees, Orme, Brown & Green, 1828], ii; Susan Sheets-Pyenson, "A Measure of

Success: The Publication of Natural History Journals in Early Victorian Britain," *Publishing History* 9 (1981): 21–36, on p. 22.

63. On Branston, Sowerby, Harvey, and Strutt, see *ODNB* and Chatto, *Treatise on Wood Engraving*, esp. pp. 628–30. Reference to Strutt's involvement appears in a Longman advertisement in the John Johnson Collection, Bodleian Library, Oxford.

64. [John Claudius Loudon], "[Review of *The Animal Kingdom*, by Edward Griffith]," *Magazine of Natural History* 1 (1829): 275–77, on p. 276.

65. Chatto, *Treatise on Wood Engraving*, 425.

66. Secord, "Botany on a Plate"; Ray Desmond, "Loudon and Nineteenth-Century Horticultural Journalism," in *John Claudius Loudon and the Early Nineteenth Century in Great Britain,* ed. Elisabeth B. MacDougall (Washington: Dumbarton Oaks Trustees for Harvard University, 1980), 77–97.

67. Frank Broomhead, *The Book Illustrations of Orland Jewitt* (Pinner, UK: Private Libraries Association, 1995), 22–23.

68. Susan Sheets-Pyenson, "A Measure of Success," 24. See also David E. Allen, "The Struggle for Specialist Journals: Natural History in the British Periodicals Market in the First Half of the Nineteenth Century," *Archives of Natural History* 23 (1996): 107–23; and Sheets-Pyenson, "From the North to Red Lion Court: The Creation and Early Years of the 'Annals of Natural History,'" *Archives of Natural History* 10 (1981): 221–49.

69. "Address to the Public," *Edinburgh Journal of Natural History* 1 (1835): 1.

70. Quoted in Secord, "Botany on a Plate," 35.

71. Hooker discussed the audiences and costs of his new work at length with his publisher John Murray. See Murray Papers, MS. 40574, National Library of Scotland.

72. *Magazine of Botany* 8 (1841): 265–66.

73. Secord, "Botany on a Plate," 30–32.

CHAPTER 3

Proceedings and the Public: How a Commercial Genre Transformed Science

Alex Csiszar

Proceedings are ubiquitous in nineteenth-century scientific life, but the format and genre has been all but invisible in the history of science. Perhaps this is because it is easy to suppose that the word is simply a synonym for "transactions" or "scholarly journal," and that learned societies have always published them. Or maybe the problem is that "proceedings" can refer to a range of distinct objects. On the one hand, "proceedings of learned societies" was a genre that made up a particular section of many independent periodicals. On the other, Proceedings were a periodical format unto themselves under the control of a society or academy.[1] Moreover, these latter might be made up of schematic descriptions of the events at meetings, sets of abstracts of papers read, or simply collections of full papers. Finally, it might be that ubiquity itself is the problem. Like Poe's place names that stretch across the length of the map, they "escape observation by dint of being excessively obvious."[2]

But academic proceedings have a distinctive history, one that is anything but straightforward. As a format, such publications began to emerge in the 1820s, inspired by a genre of journalism that had become a touchstone of civil society, first in politics and then in other cultural spheres including natural philosophy. Moreover, proceedings represent a crucial development in scientific publishing, and the slippages between commercial journal and official report, between popular summary and expert précis, between abstract and paper, are crucial to what makes them so important. Indeed, without understanding the role of proceedings, it is impossible to understand the intimate connection between the commercial periodical press and the consolidation of specialized scientific publishing during the nineteenth century.

The advent of Proceedings publications was not limited to Britain; it swept across the academic landscape in Europe and beyond in the 1830s, giving rise to publications with titles such as *Comptes rendus*, *Bulletin*, and *Sitzungs-berichte*. But this development followed a distinctive path in Britain, in part because of the character of the commercial periodicals that were most active in printing proceedings of societies. In France, for example, literary journals and daily political journals often led the way, crowding academic meetings with journalists (often young savants) paid to take notes at meetings. In contrast, in Britain the key role was played by specialized periodicals that were, paradoxically, more focused on turning a profit than were their French counterparts. This meant that journal publishers and editors often developed close working relationships with those societies. Nowhere was this intimate connection more evident than in the work of the London printer, publisher, and editor Richard Taylor, who not only published two of the longest-running scientific journals in Britain but was also printer to a great number of societies. This chapter will focus especially, though not exclusively, on the group of London societies connected at various times to Taylor's firm. These included the Royal (who employed Taylor between 1828 and 1877), the Geological (1822–), the Astronomical (1822–28), the Linnean (1791–), the Zoological (1826–), and the Chemical (1841–48).[3]

Publishing habits and expectations varied a great deal across these different branches of knowledge, depending, for example, on the value accorded to timely publishing and on the demand for large-format illustrations; and a great deal may be learned by focusing on these individual cases. But tracing the more general evolution of the format reveals a struggle over just what kinds of genres scientific practitioners took to be most appropriate for making authoritative natural knowledge public and, conversely, over what kinds of publics were imagined as the legitimate audiences for natural knowledge claims.

My use of the term "public" in this chapter departs from its most common uses in the historiography of Victorian science, and indeed it departs from its use in many of the other chapters in this volume. The history of science's publics is not simply a history of the social groups engaged in scientific exchanges, but also of beliefs about the nature of the groups—and the forms of judgment appropriate to those groups—that are authorized to make legitimate knowledge claims.[4] The history of scientific publics and scientific communities is as much a discursive history as a social one. Following Proceedings from their origins in attempts to diffuse science to new audiences to their transformation into a preeminent form for specialized publishing helps to show that the rise of popular genres for communicating science has been important not simply because it allowed new social groups to participate in knowledge. These genres have sometimes been incorporated into elite science itself, reshaping elite institutions in the image of the publics they represent.

MEETING IN PUBLIC

In 1821, John Herschel spied a dilemma. He and a group of friends had recently founded a scientific society for astronomers, and one of their aims was to foster the publication of astronomical knowledge.[5] The form they imagined for such publications were what they called "memoirs," polished and complete contributions to knowledge. But the new society's meetings had immediately become a clearinghouse for intelligence on eclipses and all manner of observations. As Herschel explained to Francis Baily, "however desirable it may be to have read to the Society any matter interesting to Astronomy which may occur in the correspondence of the foreign Secretary, it does not follow that we are bound, or authorized to *print* it." Perhaps they might print letters that had "the air of a formal communication," but otherwise they would be printing "garbled extracts" of letters. Alternatively, if they produced summaries "as intelligence," then "our Memoirs become an Astronomical Journal." It was obvious to Herschel that producing a journal would be a bad look for a new society looking to establish credibility.[6]

Until the 1830s, learned societies tended to publish what they called "transactions," a noun whose generic meaning as a collection of polished memoirs had been fixed in the second half of the eighteenth century. When the Royal Society of London took official control of the *Philosophical Transactions* in 1752, the character of that publication changed in nearly every way. It was issued much less frequently than previously, as a committee now took its time to vote on each paper to be published. The papers progressively became longer, and far fewer were published. Eventually the periodical's dimensions were increased to that of a large quarto, and Joseph Banks, the Royal Society's president, hired a new printer, trained in the most extravagant techniques in Paris, to produce it. As new specialized societies and regional philosophical societies came into being, they normally chose to publish quarto transactions on this model. "Transactions" became the British answer to the lavish collections of memoirs published by continental academies, exemplified by the Parisian Royal Academy of Sciences' *Histoire et mémoires*. Their expense, size, and overall character made them stand apart from journals or magazines. The latter were usually published as octavos, with tighter margins and cheaper paper, and generally promised readers a synoptic account of progress in philosophy or one of its branches.[7] Likewise, the term "memoir" was used to distinguish the contents of such collections from the "articles" that might make up a journal or a newspaper (granted, a memoir printed in a volume of transactions might be reprinted in whole or in part in a journal, and thus be called an article as well).

Published proceedings of meetings were a rarity among learned societies at the turn of the nineteenth century, but they had by then become central to the political press. Once "a practice which seemed to the most liberal statesman of the old school full of danger to the great safeguards of public liberty," publishing proceedings of legislative meetings had become, Thomas Babington Macaulay noted in 1828, "a safeguard, tantamount, and more than tantamount, to all the rest together." The House of Commons had all but given up barring such reports in the 1770s, and since that time the journalists in the gallery had become central personages in British political life. By the 1830s, an observer noted that without such reporting, "our representative legislature would almost cease to be regarded as an institution of high worth or significance."[8]

The practice nevertheless remained controversial. The link between an unrestrained press and the anarchy into which revolutionary France had descended was notorious. William Windham warned the legislature in 1799 against allowing "newspapers to detail their proceedings," noting "how those who wrote for newspapers in general had contributed to the overthrow of the different Governments of the world."[9] Proceedings of meetings were indeed a central component of several periodicals with radical connections, including William Cobbett's *Political Register* (a hugely successful two-penny weekly); the *Black Dwarf*, run by Thomas Jonathan Wooler; and the *Republican*, by Richard Carlile.[10]

New press laws in 1819 targeted the radical press and dramatically increased the stamp tax on political journals, putting many out of business. In their place emerged a cheap press modeled on these publications and created by the same printers and publishers trained in the offices of the radical press.[11] By avoiding political discussions, these publications avoided the tax; but their success gave evidence of a massive reading audience and became part of arguments for increasing educational opportunities for the working classes and for the extension of political representation. While they included some original material written by paid contributors, these miscellanies aimed to bring to readers a digest of the most useful and entertaining cultural information otherwise available only in more expensive periodicals.

Some of the most successful of these new weekly miscellanies focused on bringing science, technology, and medicine to "the people." The *Mechanics'*

Magazine was founded in 1823 by Joseph Clinton Robertson and Thomas Hodgskin. Robertson was a patent agent with radical leanings, and Hodgskin had worked under radical publishers and also as a parliamentary reporter for the *Morning Chronicle*. The *Mechanics' Magazine* was to "comprehend a digested selection from all the periodical publications of the day," but with a focus on useful science and technology.[12] In 1824, Hodgskin split with Robertson and launched the *Chemist*. Both papers brought not only the cramped two-column format of the radical periodicals to science, but also the radical suspicion of elite secrecy. Hodgskin began his new scientific periodical by critiquing what he saw as the scientific aristocracy, who "keep [science] in a manner inaccessible to the profaning touch of the vulgar."

> In fact, there is some reason to believe that Royal Societies of every description partake of the opinions and apprehensions of their patrons, and, like them, are not forward to encourage that species of instruction which tends to make the great mass of mankind the accurate judges of their merits rather than submissive scholars.[13]

This "sort of royal science" seemed designed to set up barriers to participation, and this was exemplified by their publications. Likewise, the *Gardener's Magazine*, founded in 1826, called the *Transactions* of the Horticultural Society a "sealed book to country practitioners; quarto paper, large print, and extensive margin little suit our pocket." Its editor commented sardonically that the one advantage of transactions was that "from their high price, they are not likely to be read by practical men; and, therefore, if it were possible that such a thing as an error should creep into them, it would do little harm."[14] This was an argument for diffusion that was not simply about public instruction but one which linked the legitimacy of such societies to the audience for their publications. The restricted readership of transactions meant that knowledge claims could not receive the public scrutiny necessary to produce legitimate knowledge.

These journals were part of a marked expansion in print coverage of science and technology that also threatened to alter the market for more expensive journals of science.[15] Both the *Chemist* and the *Mechanics' Magazine* included a regular column called "Analysis of Scientific Journals," which their editors used to mount regular critiques of the way "journalists of a higher class" covered science. The *Chemist* approvingly quoted another journalist who had argued that such journals "should be considered as the links that connect the learned with the industrious—the strainers and digesters through which the truths of philosophy must pass." At least, Hodgskin remarked, the successes of Robertson and himself had "compelled other editors to set about improvement."[16] There was truth in this. Imitators emerged looking to capture some part of this emerging market. One of these was the *Mechanic's Oracle and Artisan's Laboratory and Workshop* (1824), which mimicked the *Mechanics' Magazine*'s layout and content. It was founded by Alexander Tilloch, the proprietor of the *Philosophical Magazine*, the longest-running scientific journal in Britain. As it happened, Tilloch passed away within a year of its founding, but his attempt to compete on this new terrain suggests how keenly aware the more established scientific press was of the challenge posed by new journals written and priced for a larger imagined public.

While the *Oracle* did not survive the death of its founder, the *Philosophical Magazine* did, and it gradually changed the way it covered natural philosophy. In 1822, Tilloch had brought on as coeditor and co-proprietor his printer Richard Taylor, who took full charge of the publication after his death. Although associated with the upper crust of scientific publishing, Taylor's politics shared a great deal with Robertson and Hodgskin. The son of a dissenter, Taylor—a Unitarian—had been apprenticed to the Chancery Lane printer Jonas Davis in 1798. He also carried forward many of his father's political convictions. In 1824 he collaborated with the radical reformer Francis Place (a mentor to Hodgskin) in pushing for the legalization of trade unions, and he also fought for the repeal of the Test and Corporation Acts, which restricted religious freedom.[17] But Taylor was also making himself a crucial figure among London's scientific aristocracy. When he took over Davis's printing business at the beginning of the new century, he inherited the Linnean Society as a key client, as well as Tilloch's *Philosophical Magazine*. Over the next decades he began to corner the market on the printing of elite science in London. He joined the Linnean Society early on, and even became its undersecretary in 1810. In 1822 both the Geological Society and the Astronomical Society of London hired Taylor to print their transactions. By 1828, the Royal Society and the Zoological Society were also employing him as their printer.

Though both the Astronomical and Geological Societies were ostensibly among those "Royal Societies" that Hodgskin and Robertson disdained, both had differentiated themselves from Joseph Banks's Royal Society by making appeals to other constituencies and mores. The Geological Society became known for allowing discussion and debate at its meetings, while the Astronomical Society's leaders were key figures in efforts to push the Royal Society of London past the aristocratic legacy of Banks.[18] Positioned between these societies and the commercial scientific press, Taylor offered them a means of broadening their public face. In 1823 the Astronomical Society resolved to allow Taylor to copy and publish its minutes in the *Philosophical Magazine*. The next year, the society appointed the prolific author and mathematician Olinthus Gregory as its secretary, and the latter worked tirelessly to produce readable summaries of its memoirs for Taylor's *Magazine*, even correcting the proofs himself.[19] Not long after, the Geological Society entered into a similar arrangement with Taylor, so that by 1825 the *Philosophical Magazine* contained full and regular accounts of the papers that had been read at these two societies.

The *Philosophical Magazine* and other scientific journals had long included accounts of the proceedings of scientific societies, but these were irregular and usually very short. The "proceedings of learned societies" section of such journals, when it existed at all, was a fluctuating mélange of intelligence that happened to be on hand, often consisting of annual public meetings of a wide variety of societies and academies, lists of papers presented, translations of foreign accounts of meeting (such as the brief *comptes rendus* of the Académie des Sciences of Paris that appeared in the *Annales de chimie* beginning in 1816), and other miscellaneous news. In most cases, meetings of learned societies were not considered matter for regular public consumption. When the *Annals of Philosophy* was launched in 1813, its editor, the chemist Thomas Thomson, did pledge to give reports of the most important London meetings by attending them and writing what he heard, though in practice his accounts were almost always brief.[20] The Royal Society tended to be particularly circumspect about such public reports; Thomson complained that it forbade him taking notes during meetings, so that he had to write his accounts from memory. But in the mid-1820s those meetings also began to receive more extended publicity, if only because William T. Brande, one of its secretaries, was also the editor of the *Quarterly Journal of Science*.[21]

The primary reasoning editors and publishers initially deployed for making such reports was to keep a broader imagined public informed about learned societies' activities, to be "the strainers and digesters" of which Hodgskin wrote. Some even took the sardonic attitude of Robertson and Hodgskin to heart. When the *Quarterly Journal of Science* announced a new series in 1827, its editor promised to continue to be "of service to the public" by "stripping the valuable facts of science of the verbiage and prolixity in which they are sometimes enveloped."[22]

It turned out, however, that fellows of societies themselves also found value in these reports. Taylor's close relationship with the Astronomical and Geological Societies made the *Philosophical Magazine* a de facto bulletin of these societies' activities. When he began a new series in 1827 (having bought the

Annals of Philosophy), he offered to print these proceedings as separate copies for the use of fellows in both societies.[23] This was a cheap proposition for the societies because, just as with separate copies of papers produced for authors, the cost was only for paper and presswork, and most of it could be produced from the standing type used from the *Magazine*.[24] But even if these began as separate copies extracted from the journals, they very much resembled something like the issues of a small periodical. Eventually both societies decided to treat them as such, taking public responsibility for their contents (fig. 3.1). Soon they were not simply distributed to members but sent abroad to other societies and academies, and these quasi-periodicals were eventually provided with a title page and table of contents.[25] This is the origin of the Geological Society's *Proceedings* and the Astronomical Society's *Monthly Notices*.

Taylor continued to enlarge his sphere of influence.[26] In early 1828, after he successfully bid to become the Royal Society's printer, the society made

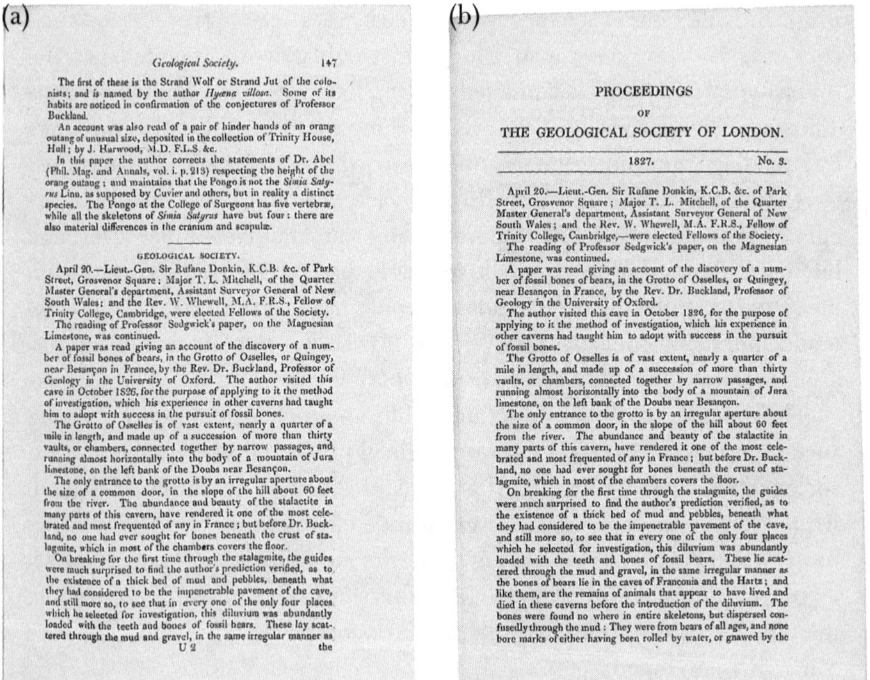
(a)

Geological Society. 147

The first of these is the Strand Wolf or Strand Jut of the colonists; and is named by the author *Hyæna villosa*. Some of its habits are noticed in confirmation of the conjectures of Professor Buckland.

An account was also read of a pair of hinder hands of an orang outang of unusual size, deposited in the collection of Trinity House, Hull; by J. Harwood, M.D. F.L.S. &c.

In this paper the author corrects the statements of Dr. Abel (Phil. Mag. and Annals, vol. i. p. 213) respecting the height of the orang outang; and maintains that the Pongo is not the *Simia Satyrus* Linn. as supposed by Cuvier and others, but in reality a distinct species. The Pongo at the College of Surgeons has five vertebræ, while all the skeletons of *Simia Satyrus* have but four: there are also material differences in the cranium and scapulæ.

GEOLOGICAL SOCIETY.

April 20.—Lieut.-Gen. Sir Rufane Donkin, K.C.B. &c. of Park Street, Grosvenor Square; Major T. L. Mitchell, of the Quarter Master General's department, Assistant Surveyor General of New South Wales; and the Rev. W. Whewell, M.A. F.R.S., Fellow of Trinity College, Cambridge, were elected Fellows of the Society.

The reading of Professor Sedgwick's paper, on the Magnesian Limestone, was continued.

A paper was read giving an account of the discovery of a number of fossil bones of bears, in the Grotto of Osselles, or Quingey, near Besançon in France, by the Rev. Dr. Buckland, Professor of Geology in the University of Oxford. The author visited this cave in October 1826, for the purpose of applying to it the method of investigation, which his experience in other caverns had taught him to adopt with success in the pursuit of fossil bones.

The Grotto of Osselles is of vast extent, nearly a quarter of a mile in length, and made up of a succession of more than thirty vaults, or chambers, connected together by narrow passages, and running almost horizontally into the body of a mountain of Jura limestone, on the left bank of the Doubs near Besançon.

The only entrance to the grotto is by an irregular aperture about the size of a common door, in the slope of the hill about 60 feet from the river. The abundance and beauty of the stalactite in many parts of this cavern, have rendered it one of the most celebrated and most frequented of any in France; but before Dr. Buckland, no one had ever sought for bones beneath the crust of stalagmite, which in most of the chambers covers the floor.

On breaking for the first time through the stalagmite, the guides were much surprised to find the author's prediction verified, as to the existence of a thick bed of mud and pebbles, beneath what they had considered to be the impenetrable pavement of the cave, and still more so, to see that in every one of the only four places which he selected for investigation, this diluvium was abundantly loaded with the teeth and bones of fossil bears. These lay scattered through the mud and gravel, in the same irregular manner as

U 2 the

(b)

PROCEEDINGS

OF

THE GEOLOGICAL SOCIETY OF LONDON.

1827. No. 3.

April 20.—Lieut.-Gen. Sir Rufane Donkin, K.C.B. &c. of Park Street, Grosvenor Square; Major T. L. Mitchell, of the Quarter Master General's department, Assistant Surveyor General of New South Wales; and the Rev. W. Whewell, M.A. F.R.S., Fellow of Trinity College, Cambridge,—were elected Fellows of the Society.

The reading of Professor Sedgwick's paper, on the Magnesian Limestone, was continued.

A paper was read giving an account of the discovery of a number of fossil bones of bears, in the Grotto of Osselles, or Quingey, near Besançon in France, by the Rev. Dr. Buckland, Professor of Geology in the University of Oxford.

The author visited this cave in October 1826, for the purpose of applying to it the method of investigation, which his experience in other caverns had taught him to adopt with success in the pursuit of fossil bones.

The Grotto of Osselles is of vast extent, nearly a quarter of a mile in length, and made up of a succession of more than thirty vaults, or chambers, connected together by narrow passages, and running almost horizontally into the body of a mountain of Jura limestone, on the left bank of the Doubs near Besançon.

The only entrance to the grotto is by an irregular aperture about the size of a common door, in the slope of the hill about 60 feet from the river. The abundance and beauty of the stalactite in many parts of this cavern, have rendered it one of the most celebrated and most frequented of any in France; but before Dr. Buckland, no one had ever sought for bones beneath the crust of stalagmite, which in most of the chambers covers the floor.

On breaking for the first time through the stalagmite, the guides were much surprised to find the author's prediction verified, as to the existence of a thick bed of mud and pebbles, beneath what they had considered to be the impenetrable pavement of the cave, and still more so, to see that in every one of the only four places which he selected for investigation, this diluvium was abundantly loaded with the teeth and bones of fossil bears. These lie scattered through the mud and gravel, in the same irregular manner as the bones of bears lie in the caves of Franconia and the Hartz; and like them, are the remains of animals that appear to have lived and died in these caverns before the introduction of the diluvium. The bones were found no where in entire skeletons, but dispersed confusedly through the mud: They were from bears of all ages, and none bore marks of either having been rolled by water, or gnawed by the

FIG. 3.1. Excerpt from the proceedings of *Geological Society of London*, meeting of 20 April 1827, produced by Richard Taylor (a) as it appeared in *Philosophical Magazine* 2 (August 1827): 147; and (b) as it appeared in the *Proceedings of the Geological Society* 3.

arrangements to take more control of the publication of reports of meetings in the press.[27] At the same time it authorized Taylor to print abstracts from the manuscript minutes in the *Philosophical Magazine*.[28] In 1831 the society began to issue the abstracts from the *Philosophical Magazine* as its own *Proceedings*.[29] The arrangement followed the established pattern, with Taylor promising that the "price of the composition of such part of the Proceedings as can be used in the Philosophical Magazine & Annals—not to be charged, but only the presswork & paper."[30]

PUBLICITY AND CONTROL

Societies that chose to publish their Proceedings were often quick to congratulate themselves for having opened their meetings to public discussion. The Council of the Astronomical explained in its annual report (itself printed in the *Monthly Notices*) that the "public is hereby brought more immediately into contact with the Society":

> The labours of its contributors are canvassed and discussed, while the interest of the author in his subject is yet warm, and when the interchange of ideas respecting it is most beneficial, not only to the public, but to the author himself, whose views may, and probably in many instances will, be enlarged or corrected by such intercourse.[31]

Of course, these attempts to open the society to a broader public owed as much to the pressure from commercial journals as they did to the initiative of these societies. It was, after all, Richard Taylor's distinctive role as both publisher of a journal and printer to the societies that paved the way for their emulating a journalistic genre. Moreover, some authors were becoming increasingly willing to forego whatever honor might come with publishing in the transactions of a society to publish more quickly and easily in the press. As the Geological Society's president noted with regret in 1830, the growing presence of independent journals had begun to change expectations, leaving the society unable to meet "the wishes of those authors especially who have most original matter to communicate."[32]

The implication that this shift toward public engagement was forced by broader cultural obligations was also noted by J. C. Robertson in the *Mechanics' Magazine* itself, even as he bestowed uncharacteristic praise on the Royal Society for publishing its *Proceedings*: "It is by no means an unpleasing spectacle to see so stiff and unbending a Society losing, either by choice or on compulsion, a good portion of its haughty spirit of exclusiveness." Given that "the public have a right to take some interest in the proceedings of the chief scientific body of the nation," Robertson was happy to find that "its members begin to feel that their labours stand a little in need of being more popularised than they have ever hitherto been."[33] By 1838, the *Mechanics' Magazine* was still extolling the society's embrace of a more open stance. Robertson noted that the move to publish more cheaply and regularly had "been absolutely called for by the voice of a scientific public, *rather* more extensive in its numbers, and more impatient in its demands, than the public of the unlocomotive age in which the Society began to flourish."[34]

There was room for other kinds of cynicism about the motives of elite institutions who co-opted a commercial genre to their own ends. Across the channel in Paris, the republican activist, editor, and naturalist François-Vincent Raspail interpreted the founding of the *Comptes rendus* by the Academy of Sciences in 1835 as an illegitimate attempt "to seize a monopoly" on scientific journalism.[35] Raspail had a strong warrant for this interpretation. The academy had announced its new weekly journal in a highly charged political atmosphere, banning several journalists from its meetings at the moment it did so. In Britain, however, the situation was less straightforward. The emergence of authorized Proceedings publications was not usually undertaken as a direct means of curtailing the "proceedings of societies" in other journals. Indeed, in the case of Taylor's publications the two phenomena went hand in hand. The proceedings section of the *Philosophical Magazine* ballooned over the 1830s, and was joined by an equally extensive section in the *Annals of Natural History* in 1838, which took in the Zoological and the Linnean proceedings. Other periodicals, including not only specialized journals but also weeklies such as the *Athenaeum* and *Literary Gazette*, took advantage of Proceedings publications to print excerpts from them when they could get access in time. The *British and Foreign Medical Review*, for example, informed its readers in 1836 of the "small fugitive publication" now available from the Royal Society, which included "perfectly authentic" abstracts of all papers read at its meetings. It promised "to avail ourselves of the facilities afforded by them . . . and we shall, in general, transcribe the very words of the original, without any comment of our own."[36]

While it is tempting to interpret the rise of Proceedings as a natural result of what Robertson called a "scientific public, *rather* more extensive in its numbers," we should not take for granted that the "scientific public" was a stable concept whose relevance to elite science was accepted by all. Rather, just as the growing importance of reporting on political meetings represented shifting conceptions of political legitimacy in England, at stake here was the question of what kind of public could legitimately judge the validity of knowledge claims. This was less a matter of quantifying or even identifying any particular group of readers than of imagining modes of collective organization.

Extending such publicity was entirely compatible with attempting to keep it under tight control. When the Royal Society authorized the printing of proceedings in Taylor's *Philosophical Magazine*, it explicitly prohibited access to its manuscript minutes for the purpose of reporting on them without permission.[37] Its council sometimes objected when unauthorized summaries of papers read at meetings came to its attention, although it was very willing to distribute its new printed *Proceedings* to journal editors on the understanding that these would be excerpted.[38] The Geological Society was particularly active in making sure that its official proceedings were diffused and reprinted widely. After agreeing to a request from the *Athenaeum* for its printed Proceedings in 1831, its council decided to send copies regularly to the *Literary Gazette*, the *Monthly Review*, and the *Metropolitan Magazine* as well.[39] The council even lodged a complaint with one editor in 1833 when its proceedings were *not* inserted in that journal as fully as expected.[40] Their interest in publicizing their meetings was genuine, but they wanted the publicity on their own terms. When editors of the *Constitutional* daily paper applied for permission for a reporter to attend Geological Society meetings in 1836, the request was denied.[41]

That societies were generally open to their Proceedings being reprinted or excerpted shows that these octavo publications were not yet a product from which they expected any direct financial return. They were normally given to fellows gratis, or nearly so, and were exchanged with societies and academies; but little effort was made to sell them to the public. What the societies accomplished instead was to promote a standard format and account of their meetings for public consumption. Most obviously, in the case of the Geological Society, where discussions at meetings were allowed after papers, standardizing Proceedings was a strategy for keeping those discussions from being publicized.[42] More generally, it helped keep anything that might happen at a meeting—including impromptu discussions, audible reactions in the audience, and so on—from being the subject of reporting. To compare, the Paris Academy of Sciences did publish reports of discussions in its *Comptes rendus*, at least in the early years, and journalists continued to be allowed to print eyewitness accounts of meetings, as long as they did so politely. Indeed, by 1843, it was possible for a critic in *Blackwood's Magazine* to scold the Royal Society for keeping its meetings relatively secret affairs, despite the nominal publication of its Proceedings. That critic was William Robert Grove, a fellow of the society who was elected to its council a few years later. He set about improving the situation, putting forward a motion to have abstracts of papers put into type and printed in proof as soon as possible, allowing authors and editors to reprint those abstracts elsewhere in a more timely fashion. Even then, however, the form of its *Proceedings* remained largely constant.[43]

GENRE TROUBLE

The balance between proceedings as a genre making up independent journals and as a discrete periodical format controlled by societies remained uncertain throughout this period. Once societies took public responsibility for printing Proceedings, it was inevitable that the nature of their contents would change. Some of this was straightforward. The proceedings that appeared in the *Philosophical Magazine* and other independent journals usually consisted entirely of summaries and abstracts of scientific papers read, only occasionally including other nonscientific business, since the latter was unlikely to interest most readers. The manuscript minutes kept by the secretary, upon which these were often based, included other information about society business, including nonscientific correspondence, library acquisitions, and gifts of books and specimens. Since the printed Proceedings normally began as a service to fellows, many societies chose to include this business as well, having Taylor make the necessary additions to the standing type.

More complicated was the status of the abstracts and summaries that made them up. Who was responsible for their contents? What was their relationship to the longer memoirs that appeared in quarto publications? What imagined audience were they written for? As long as they appeared in a commercial journal, the question of who was responsible for them was usually ambiguous to readers. They might be based on a summary transcript of what was heard by an onlooker, on the secretary's official manuscript report of the minutes, or even on a writer's own account sent directly to the editor. Even if the latter were the case, there was little warrant for assuming that individual summaries were the sorts of things that could be assigned definitively to authors. But the situation was different when the same summaries appeared in official publications. In some cases it was clear that they were taken verbatim from authors' memoirs or letters, but even in cases where they were written as reported speech, it was possible to suppose that individual authors had authorized those summaries. It was a short step from this to treating such texts not simply as summary reports but as independent papers, especially since many did not ultimately correspond to any longer memoirs published elsewhere.

When a society published an abstract of a paper but chose not to publish the fuller memoir on which it was based, that memoir might eventually appear in some other periodical, or even as part of an independent book. But often the memoir was reworked or scrapped altogether, the author deciding that the abstract satisfied his itch to publish those results.

We should keep in mind that the term "abstract" did not quite have its later connotation of a short standard-length summary. Abstracts of communications varied massively in their size and style, and in the kinds of detail they included. For example, while the median length of the abstracts the Royal Society published during the 1830s was about 350 words, it published dozens of abstracts that topped 1,000 words in length during that decade, often including specific experimental details and data, and sometimes including tables. The one component almost never present in proceedings journals, however, was illustrations. Still, many of these full abstracts could easily be interpreted by readers as communications sufficient unto themselves.

In fact, many readers even considered abstracts of memoirs as *preferable* to the fuller versions on which they were based. Such a judgment was commonly rendered by referees tasked by societies to make recommendations about what they should publish. Good evidence comes from the Geological Society, whose complex system for vetting papers submitted—starting from a formal decision about what got read at meetings (and thus abstracted in the *Proceedings*), and followed by a referee report on whether to publish in the *Transactions*—meant that an abstract was often already available in print by the time a referee rendered their verdict about the longer memoir. Inspection of these reports shows that among the most frequent criticisms of memoirs was that they were too long. Suggestions for curtailing the length of a memoir could easily slide into rejecting it altogether by referencing the abstract already published as being sufficient or superior to the original. Here is an early typical example: "The abstract which has been read from the minutes of the Society, containing all the information which is to be found in the body of the paper, would by most readers be preferred I believe to the original."[44]

In 1834 Henry de la Beche testified that the quality of the abstracts published by the Geological Society was coming to be perceived as a problem. The utility of the Geological Society's *Proceedings* had been proven "by other societies putting forth similar publications," De la Beche wrote, and yet some were complaining that they were becoming "too good." The problem was that "for the most part they contain the cream of the papers read, many of which afterwards appear in the 'Transactions,' and that therefore they hurt the sale of the latter."[45] Already some were calling for dispensing with the *Transactions* altogether and replacing it with an octavo publication like the *Proceedings* but with fuller papers. De la Beche thought this a bad move, since profitability was not the only consideration, and one of the main functions of societies such as theirs was to put out works that would otherwise be lost for want of a market; chief among these were those requiring "extensive illustration," as appeared in the quarto transactions.[46]

Thus, what may have begun as an abstract could very easily take the place of a memoir as the definitive record of a discovery claim. In the case of shorter communications without illustrations, the distinction between abstract and paper disappeared altogether. Indeed, the Astronomical Society began to use its Proceedings journal as a venue for publishing shorter communications almost immediately (this may be why they chose to call it *Monthly Notices* instead of *Proceedings*). Such short notices—of new phenomena noted by observers, or of ephemerides—became a regular part of the publication, and the sense that it was primarily a venue for abstracts of longer memoirs quickly receded. It was just these sorts of communications that John Herschel had worried about publishing in 1821, but by 1828 the Council celebrated the fact that it now had a venue by which to publish "matter of merely temporary interest."[47] That astronomical memoirs rarely required illustrations made this all the more natural. By the 1840s, papers that appeared in any form in the *Monthly Notices* were generally to be considered "sufficiently published," and did not appear in the *Memoirs* at all.[48] The secretary of the society was tasked with producing a wholly distinct set of short abstracts to be dispatched to the journals instead.[49]

Societies that were founded or began publishing during the 1830s were confronted with a serious choice: whether to publish transactions, or only Proceedings, or both. When the Zoological Society of London inaugurated its Committee of Science and Correspondence to foster zoological research in 1830, they confronted just this question. The council decided that it ought to begin with "a monthly publication in the cheapest form, under the name of proceedings," and simply put off the question of publishing transactions to a later time. This option turned out to be especially cheap, for it had already hired Richard Taylor as its printer, and Taylor used a great deal of the contents of its *Proceedings* in the *Philosophical Magazine* (shifting them to the *Annals of Natural History* in 1838).[50] Eventually, in 1835, the council began to put out *Transactions* as well, but a pattern had already been set. Authors submitting papers took publication in the *Proceedings* to be a form of definitive publication, even if those individual contributions were subsumed into the regular textual flow of meeting minutes.[51]

When the Royal Geographical Society was formed in 1830 the council and various committees likewise went back and forth on the question of the "form and distribution of the Society's Transactions." Several committee and council meetings ended in disagreement. One of the fellows involved in these debates was the publisher John Murray, who "strongly dissuaded the Society from publishing 4to Transactions." Quartos, Murray explained, were "a form which was at present almost unsaleable." Murray suggested they simply publish a "Journal in 8vo of Geographical Science," to be published about monthly.[52] By publishing a journal, he meant not simply that it should be cheaper and more frequent than quarto transactions, but that it should also include analytical notes on geographical works both British and foreign, along with other geographical intelligence. If they agreed to do this, he even offered to publish it at his own risk. Not everyone was willing to go so far, however, and what emerged was a hybrid. The society called its periodical a "journal," and it was indeed an octavo, including analyses and summaries of other works, but its main contents were original memoirs. Worse, it was published as an annual volume. Murray was unimpressed with what he saw as an unmarketable product, and he refused to publish it at his own risk after all.[53]

Another key question was just what kinds of readers the abstracts and papers in Proceedings were for. That the primary impetus for editors to publish proceedings of meetings was the desire to bring knowledge to wider audiences suggested that such accounts ought to be as readable and nontechnical as possible. But as abstracts became substitutes for longer memoirs, it was easy to view them as written primarily for the most informed of readers. This was particularly true when referees had a hand in the work of compression, for they were likely to call for the removal of anything in the paper that was synthetic rather than strictly original, assuming expert readers like themselves who would be able to supply the crucial historical context that made sense of those original results. De la Beche pointed out, comparing memoirs with abstracts, that "the larger are not always the heaviest, nor the shorter the most pleasant reading or the most instructive."[54] Across the channel, Jean-Baptiste Biot warned of a related problem in the academy's embrace of short notices in the *Comptes rendus*: If you constrain writers to present the results of their work "as so many aphorisms," he warned, ". . . their value will be but imperfectly understood, and only the authority of the author's name can provide confidence in the result—always an extremely dangerous situation in science."[55]

Questions of trust and authority were particularly acute, for when scientific societies began to publish on a schedule resembling those of speedier independent publications, conundrums arose about just what kind of responsibility they had for the work they chose to publicize. One reason why the Geographical Society had not followed through on its plan to publish a monthly journal, for example, was that its council took it to be its duty to act as a collective judge and filter of the papers it received. But because such formal vetting introduced delays, it appeared impractical to publish as quickly as Murray advised.

Conflicts over editorial authority arose in several societies. At the Zoological Society, where authors were in the habit of submitting short notes to have them appear in the *Proceedings*, controversies could arise over just what should be inserted at what length, and how much power the editor ought to have. In the first half of 1838, a series of unpleasant skirmishes between John Edward Gray and the editors of the society's *Proceedings* led the Publications Committee to make a change. Abandoning the fiction that the *Proceedings* simply contained summaries of meetings, they resolved that decisions about what would go into them were now to depend on referees, just as they did for the *Transactions*. Moreover, nothing that was presented orally at a meeting would even be considered unless their authors submitted a corresponding written text within ten days.[56] Although the *Proceedings* looked virtually unchanged, its character was thus transformed into a periodical consisting of formal submissions that went through a lengthier review process. Gray himself essentially stopped contributing altogether, sending his short papers instead to Taylor's new *Annals of Natural History*, where he could be sure of more efficient publication.

The same question of authority arose at the Astronomical Society in 1847, where some on the council objected to "the great discretion which is necessarily left to the editor" in what got published in the *Monthly Notices*.[57] But here timeliness was taken to be almost nonnegotiable. While astronomy was not the only branch of science that depended on the routine publication of observational facts, the utility of those facts—for example, in the case of ephemerides—could be highly time-sensitive. Richard Sheepshanks, who directed the *Monthly Notices* for many years, explained that the council's wish to issue the *Monthly Notices* promptly each month clashed with the society's perceived responsibility as a collective judge. "All that can be done," he noted, "is to entrust the care of the *Monthly Notices* to a discreet person."[58] The problem was that if printing in the *Notices* generally implied not printing in the quarto *Memoirs*, then "the Editor of the MN must, to a very great extent, decide on the publication in the Memoirs" as well. Sheepshanks claimed that in practice, cases where there was any real doubt about what to do with a paper were quite rare, and in those instances things might be taken more slowly. (This view is belied by his private correspondence on these matters, where he often aired complaints about the judgment of his fellow council members.)[59] In private he thought that often "by far the best judge" of what to print in the *Memoirs* was the person responsible for editing the abstracts, rather than the council.[60]

The council, agreeing that the efficiency of their monthly was crucial to its utility, acquiesced to Sheepshanks's argument for individual control.[61] But it inserted a note on the verso of the title page to the *Monthly Notices* to the effect that the publication was entrusted to an editor responsible only for arrangement and compression and who exercised no "systematic control" of the contents. Readers ought only to judge articles "by their intrinsic merits, or by the reputation of their respective authors."[62] While British societies had long printed disclaimers—arguably disingenuous—that authors alone were responsible for their memoirs, this statement went further by disavowing any claim to having filtered the contents at all.

The central role played by a trustworthy individual editor and the emphasis on economy and speed made the *Monthly Notices* very much resemble a commercial journal. And yet the society was still keen to insist on the distinction, pointing out that only original material submitted to the society appeared in its pages. No other intelligence, translations, or extracts were admissible: "Compilation of this kind would scarcely be in accordance with the position and character of the Society."[63] It also did not want to bother selling individual issues (though it did encourage anyone interested in receiving them to become a fellow and thus receive them for free). But if in 1848 the society's council thought that it was still important to maintain these distinctions between the mandate of a scientific society and that of a journal, elsewhere these last barriers were already beginning to fray.

PROCEEDINGS AS A COMMODITY

As Proceedings gradually became venues for independent papers, they took on some of the roles that transactions played in the publishing life of societies, while their cheaper octavo format resembled those of independent journals. Still, some maintained the look of a printed account of meeting minutes, and as a rule they rarely printed illustrations. Nevertheless, as Proceedings grew in size and expense, they led societies to confront a financial conundrum: Proceedings threatened the sale of a society's transactions, but as they were often provided to fellows for free and much of their content could be found in independent journals, they were not themselves a significant source of income.

Confronted with this dilemma, the Geological Society was among the first to take active steps to make its *Proceedings* more closely resemble an independent journal, thus turning it into a saleable product. While the *Proceedings* had included de facto whole papers from an early date, it began officially to solicit entire papers for its octavo publication in 1844. An indication of its new ambitions was that illustrations were now to be a key part of the publication. This also meant that the society would need to charge fellows for it.[64]

But the society quickly went further and looked for a publisher to take on the *Proceedings* as a commercial concern. Longman and Company agreed to publish it at their own risk, and the *Proceedings* was renamed the *Quarterly Journal*. It would be focused on original papers, but it would also, like other commercial journals, include miscellaneous geological intelligence from other sources, and thus provide readers with a statement of the "progress of geological inquiry" in general. The stipulations were that the *Journal*'s content would be controlled by the society's officers, but would provide papers and editorial services at no cost, save for work on the miscellaneous (intelligence) portions. Longman would keep any profit; advertisements were allowed, but there were to be none on the cover. Longman made sure to advertise the *Journal* itself extensively alongside its many other wares (fig. 3.2).[65]

The society expected that the majority of papers received would be considered for the new *Journal* rather than the *Transactions*, but it was keen to insist that this did not spell the end of its quarto publication.[66] The Geological had a long-standing referee system, and it adopted it to apply to the *Journal*. It would be part of a referee's job to decide whether a paper could be published in an octavo form with illustrations (if required), or whether there were illustrations that required the larger quarto form. The form provided to referees asked explicitly whether the paper might be shortened so as to be appropriate for an octavo journal, thus institutionalizing the long-standing custom of compressing papers for the *Proceedings*. Officially, it remained the job of the vice secretary, rather than of authors, to produce these compressed versions.

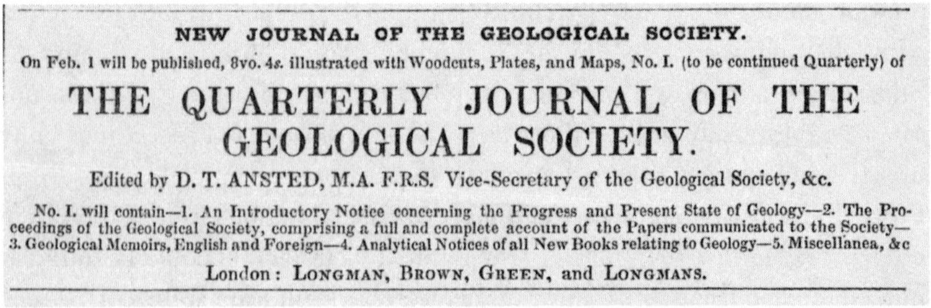

NEW JOURNAL OF THE GEOLOGICAL SOCIETY.

On Feb. 1 will be published, 8vo. 4s. illustrated with Woodcuts, Plates, and Maps, No. I. (to be continued Quarterly) of

THE QUARTERLY JOURNAL OF THE GEOLOGICAL SOCIETY.

Edited by D. T. ANSTED, M.A. F.R.S. Vice-Secretary of the Geological Society, &c.

No. I. will contain—1. An Introductory Notice concerning the Progress and Present State of Geology—2. The Proceedings of the Geological Society, comprising a full and complete account of the Papers communicated to the Society— 3. Geological Memoirs, English and Foreign—4. Analytical Notices of all New Books relating to Geology—5. Miscellanea, &c

London: LONGMAN, BROWN, GREEN, and LONGMANS.

FIGURE 3.2. Advertisement by Longman for the new *Quarterly Journal of the Geological Society*, published in the *Athenaeum* on 18 January 1845, 58.

It is indicative of the controversial nature of this move that the council was not quite willing to trust the society's reputation wholly to a "journal." It made sure that some copies would still be printed as the *Proceedings of the Geological Society*, excluding the miscellaneous section, in part so that it could send these as presents to other societies. The council deemed it inappropriate to give a journal as a gift; the members thought that "the Council cannot present the Journal, but only the Proceedings, to the Royal Society." The next year, however, they abandoned such delicacy and just sent the *Journal*.[67]

The Geological Society's embrace of a journal gave rise to a trend. In 1845, the Horticultural Society of London gave up on its *Transactions*, finding it too expensive, and switched to an octavo journal with proceedings—though, like the Geographical Society's *Journal*, this was only issued annually.[68] In 1847, the Chemical Society also turned its publication into a quarterly journal, intending to publish more quickly and with true periodicity, but also to include summaries of chemical papers published elsewhere. The latter intention remained more an aspiration than reality, but the journal did publish more content, first as a quarterly and then as a monthly beginning in 1862.[69]

The Zoological Society considered making the change at the same time. A committee recommended "the publication of a Quarterly Illustrated Journal in 8vo of an intermediate character of a more popular form than the present formal & scientific publications of the Society." On inquiring with the other societies that had already started publishing journals, it decided that, transactions being "expensive and exclusively scientific" and Proceedings being rather formal, they were "neither of them adapted to the requirements of modern literature."[70] They did not follow through entirely on the plan to found a journal, perhaps because the council was not convinced that "in such a journal a popular form is by no means incompatible with a strictly scientific character."[71] But they did make crucial changes that turned the *Proceedings* into a saleable publication focused on original content. Starting in 1847, each abstract or paper was clearly separated out, numbered, and given an unambiguous title and author in capital letters (fig. 3.3). Most important, illustrated plates were now allowed. The society hired Longman to sell the periodical. It produced versions with and without the illustrations, providing fellows with the unillustrated version at no cost. It began to publish far more content, with longer papers, so that the *Proceedings* doubled in average size within a decade.[72]

At the Astronomical Society, the *Monthly Notices* were already largely independent of the *Memoirs*, but in 1847 the society began to sell them to the public. In 1852 the society reversed its earlier policy by deciding to open its monthly to astronomical intelligence from elsewhere, including extracts and

(a)

25

March 13th, 1838.

William Yarrell, Esq., in the Chair.

Mr. Ogilby read a letter from Mr. V. der Hoeven, in which the writer expresses his belief that the large Salamander preserved in a living state at Leyden ought to be regarded as a species of Harlan's genus *Menopoma*; its specific characters consisting in the absence of the branchial apertures, which are present in the species upon which Harlan founded his genus. M. V. der Hoeven thinks it probable that the branchial apertures were present in the Leyden Salamander in the young state, and he proposes to adopt the generic term *Cryptobranchus* in preference to that of *Menopoma*, and to give it the specific name of *Japonicus*. He further states that his observations upon this singular reptile will shortly be published in a Dutch Journal.

Mr. Owen observed, with reference to the opinion of M. V. der Hoeven respecting the relations of the Gigantic Salamander of Japan to the *Menopome* of the Alleghany Mountains, that the persistence of branchial apertures was a structure so likely to influence not only the habits of an amphibious reptile, but also the structural modifications of the osseous and vascular parts of the respiratory organs, as to render it highly improbable that the *Menopome* should be related generically to a species having no trace of those apertures. He thought, therefore, that the question of the *Menopome* and gigantic Japanese Salamander being different species of the same genus, could be entertained only on the supposition, that the branchial apertures were a transitional structure in the former reptile as they are in the latter. That this was the case he considered as highly improbable; for, besides the ossified state of the hyoid apparatus, there was evidence in the Hunterian Collection that both the male and female generative organs in the *Menopome* have arrived at maturity without any change having taken place in the condition of the branchial apparatus usually considered as characteristic of the *Menopome*. He therefore considered it to be undoubtedly generically distinct from the gigantic Salamander of Japan, the true affinities of which could only be determined satisfactorily after a complete anatomical investigation, especially of its sanguiferous, respiratory, and osseous systems.

Mr. Ogilby exhibited a drawing, made by Major Mitchell, of a Marsupial animal found by that officer on the banks of the river Murray, during his late journey in the interior of New South Wales. Mr. Ogilby stated his original belief that the animal in question belonged to the *Perameles*, under which impression he had proposed to name it *Per. ecaudatus*, from its entire want of tail, a character.

No. LXIII.—PROCEEDINGS OF THE ZOOLOGICAL SOCIETY.

(b)

16

January 25, 1848.

Dr. Gamble in the Chair.

The following papers were read :—

1. NOTE ON THE CAPTURE OF THE AUROCHS (*Bos Urus*, Bodd). BY M. DIMITRI DE DOLMATOFF, MASTER OF THE IMPERIAL FORESTS IN THE GOVERNMENT OF GRODNO.

(Communicated by Sir Roderick Murchison.)

Après avoir été nommé en 1842, maître des forêts du Gouvernement de Grodno, je me suis empressé, autant par devoir que par goût pour ma vacation, de porter une attention particulière sur la forêt de Bialowieza, ce dernier asile du Bison de l'Europe, et j'ai fait la description de cette forêt primitive et de son hôte intéressant, dignes tous les deux d'être cités au nombre des curiosités, qu'offre notre belle et immense patrie. Mon ouvrage fut accueilli favorablement par notre Gouvernement, mais depuis cinq années d'observations et de recherches assidus m'ont convaincu que cet ouvrage est incomplet ; et ont fait naître en moi le désir de rédiger un traité sur le Bison ; car mes propres expériences, renferment des faits curieux, et exempt de toute erreur.

Je me suis attaché particulièrement à combattre par des expériences l'opinion erronée, accréditée par tous les écrivains qui ont traité cette matière, nommément comme quoi le veau du bison ne pouvait être

FIG. 3.3. Pages from the *Proceedings of the Zoological Society of London*: (a) vol. 6 (1838): 25, three communications as part of a running summary of a meeting; and (b) vol. 16 (1848): 16, the beginning of a communication with a distinct title and author, along with an illustration.

translations from other publications such as the *Comptes rendus* and the *Astronomische Nachrichten*.[73] The question also arose occasionally of whether to amalgamate the two periodical publications. Merging made sense financially, but the society proved reluctant to give up either quarto or octavo forms.[74] The concern here was not illustrations, but prestige. On the one hand, doing away with the prestigious quartos traditionally associated with scientific societies meant that authors might take their most substantial astronomical contributions elsewhere. But losing the octavo brought the converse risk: "that the trivial matters as the contributors call them which are sent for printing in octavo, would not be sent for printing in a form apparently more elaborate and associated with graver communications."[75] Another consideration was modern reading practice. There was a great deal of disagreement on the question of the best format for reading; while some preferred the larger format, it turned out that "some much prefer reading such matters in the octavo form."[76] With no agreement in sight, an awkward compromise was worked out in 1859 whereby the *Monthly Notices* were kept up in octavo, but also reimposed in two columns so that a quarto version could be printed and bound with the annual volume of *Memoirs*.[77] Some objected that this was rather unsightly, and that the society should simply concede to making "the 4to publication subsidiary to the 8vo as formerly the 8vo had been subsidiary to the 4to."[78]

The transition happened at the Linnean Society a few years later in 1855. At first, the plan was simply to expand the Proceedings by allowing illustrations and more original papers, splitting them into botanical and zoological parts.[79] A committee decided to go further, changing the title of the periodical to *Journal of the Proceedings of the Linnean Society*, a mouthful that corresponded to a reorganization of content that gave pride of place to papers and abstracts by pulling them out from the flow of meeting minutes. The actual "proceedings of meetings," virtually emptied of substantive content, were shunted to a specially paginated section.[80] The latter were still sent to fellows gratis, but they ceased to be included in the journal after 1864, and the society stopped producing them altogether in 1876.[81]

At the Royal Society, the official shift to publishing full papers in the *Proceedings* took place more slowly. For two decades its *Proceedings* remained more or less focused on abstracts and other miscellaneous matter (in the beginning it also served as a venue for public referee reports, but this experiment flagged).[82] While the secretaries occasionally sent papers to the *Proceedings* or published abstracts in place of memoirs, the longest articles in the *Proceedings* usually corresponded to memoirs that were subsequently published in the *Transactions*.[83] That situation changed drastically in the 1850s, in part for economic reasons. In 1852 a finance committee reported that the size and cost of producing the *Transactions* had ballooned, and recommended that referees exercise "greater strictness in selection." One way to compensate was to print fuller abstracts of papers in the *Proceedings*. Median length of articles in the *Proceedings* reached one thousand words by 1860, and many of the longest now corresponded to no memoir in the *Transactions*.[84] But it was only in 1863 that referees were systematically encouraged to recommend "papers of merit" for printing *in extenso* in the *Proceedings*. Again, the transition was signaled by the introduction of more extensive illustrations, including lithographic plates.[85] In 1886, the society considered fully equalizing their two periodicals by insisting that "the selection of the 4to. or 8vo. form for the publication of any paper be not made in any way dependent on the scientific character of the paper."[86] They did not entirely follow through on this idea, even though for many of the London specialized societies this had long been standard practice, if their transactions survived at all.

As Proceedings came increasingly to resemble independent journals, their relationship to commercial publications changed. For one, societies began to invest more resources into marketing their periodicals. Besides advertising extensively in other papers, at the Geological Society, for example, much effort was put into convincing fellows to buy the *Journal*, going so far as enlisting other fellows to engage in letter-writing campaigns to their peers to subscribe on the grounds that "it is the duty of each to promote its honour and interest."[87] Even at the Chemical and Astronomical Societies, whose journals were either free to fellows or offered at a steep discount, it was becoming customary to think of fellows as essentially nothing more than subscribers to the periodical. A review of Astronomical Society finances in 1862 commented very approvingly of having gained fifty-six fellows in the past year, and expected even greater increases "if the Monthly Notices can be made more generally interesting by the addition of instrumental and descriptive matter."[88]

All of this meant that societies had entered the marketplace of the scientific press, and were thus competitors with the journals published by firms such as Taylor and Francis. It is thus not surprising that the early 1840s represents the apex of the "proceedings of societies" sections in those journals. Until then, Taylor had printed increasingly extensive excerpts from the proceedings of the Royal, Geological, Zoological, and Linnean Societies.[89] When the Geological and Linnean Societies founded their journals, their proceedings disappeared completely from the *Philosophical Magazine* and the *Annals*.[90] The Linnean Society also refused to give permission to editors who wished to have access to its proceedings or journal for reprinting.[91] The Geological Society, on the other hand, created a new condensed version of its proceedings with "the title of the papers read and their general object only" for distribution to the *Athenaeum* and other periodicals.[92]

Up until 1847, Taylor had continued to print nearly the whole of the *Proceedings of the Zoological Society* verbatim in the *Annals*. But as that society's own *Proceedings* grew, the proportion that made it into the *Annals* began to diminish. Normally, excerpting was accomplished not by compressing the summaries, but by selecting which abstracts and papers to include. This approach seems to have led to dissension among the Zoological Society's contributors, who began to complain to the council about the reprints in the *Annals* in 1865. William Francis, now editor and proprietor, was already giving less space to the proceedings, and he agreed to stop printing them altogether.[93]

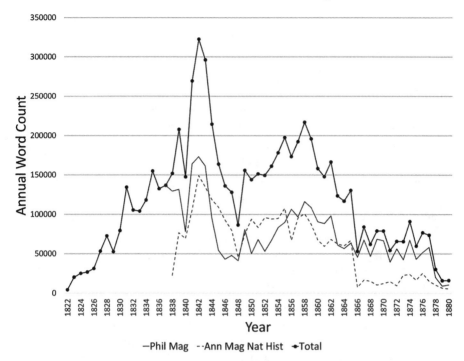

FIG. 3.4. The annual word count of the "Proceedings of Learned Societies" sections in Richard Taylor's journals, the *Philosophical Magazine* and *Annals and Magazine of Natural History*, between 1822 and 1880.

By the 1870s proceedings of societies thus made up a very small portion of both the *Annals* and the *Philosophical Magazine* (fig. 3.4). The one society that continued to be written up regularly late in the century was the Geological, which had again begun sending its shortened proceedings to the *Philosophical Magazine* in 1855. But these proceedings were now explicitly a special gathering of papers or abstracts, with individual author and title listed in the table of contents.

Proceedings did not by any means disappear from the press. The editors of many of the new journals founded in the 1860s who focused on reporting scientific news continued to take it as part of their duty to produce reports on the meetings of societies.[94] Periodicals such as the *Intellectual Observer*, the *Reader*, and the *Quarterly Journal of Science* all made sure to provide accounts of a variety of societies. While some societies rebuffed them, others were more accommodating. The Royal Society's council received many requests for permission to report on meetings, and while it did not wholly forbid reporters, it tended to direct them to the society's own printed accounts, which it was willing to provide in proof at the end of each meeting.[95]

CONCLUSION

Most of the new scientific journals of the 1860s did not survive, but proceedings of societies continued to be included in a variety of scientific publications.[96] They also found a relatively stable home in *Nature*, which made sense as it quickly became a periodical aimed largely at professional practitioners. In contrast, when Richard Proctor founded *Knowledge* in 1881, he departed from tradition by not even attempting to provide readers with proceedings of scientific meetings. While some complained of the omission,[97] one reader—who had himself worked as a scientific correspondent for a newspaper—responded by pointing out that he had found the Royal Society, for example, to be hostile to his attempts to report on their meetings. He finally gave up, citing both the difficulty involved and futility of doing so, for even the titles of most scientific papers were "about as readable as the Post-Office Directory, and practically worthless as a record of scientific progress." Of the abstracts published by societies such as the Chemical, which ran to thousands of pages per year, he had this to say: "To the professional chemist these are invaluable; to the general public they are mere waste paper."[98]

This marks a convenient end to the narrative arc of the "Proceedings" as a distinctive periodical format, at least in Britain. While many such publications had explicitly become "journals," most that continued to include "Proceedings" in their title were largely containers for individual specialized papers.[99] This transition corresponds roughly to the rise of genuine abstract journals and services late in the century, but it would be misleading to situate Proceedings journals as their precursors. Abstracting services were geared toward collecting information from a diverse range of mostly periodical sources, an impulse shared more with the intelligence-gathering role of commercial journals than with the proceedings sections of those journals. When societies such as the Chemical and Physical became interested in abstracting services, it was largely this other role of independent journals that they were appropriating. But they differed from such "miscellaneous intelligence" sections in that they took as their quarry a specific format, the scientific paper, rather than scientific news in general. That this should be seen as a pressing need depended on the consolidation of the scientific paper as a format that deserved archiving in itself. And this happened largely via the domestication by societies and academies of their Proceedings, turning "abstracts" into a model for what a "scientific paper" ought to be.

Why does this arcana of formats and meeting minutes matter? Most obviously, the rise and transformation of Proceedings journals had a profound effect on the categories of periodical formats involved in publishing science. As societies took control of their proceedings and transformed them into journals, old distinctions between the periodicals that voluntary societies published and the products of commercial publishers broke down. By competing with the latter, societies inadvertently helped to legitimate those commercial products as potential equals in the publication of scientific knowledge. By late in the century, there was enough fluidity between them that it became unremarkable both for a society to absorb a commercial journal (as the Royal Geographical Society did to the *Geographical Magazine* in 1878), and for a commercial journal to overtake a specialized society's publication as the journal of record in that field (as the *Geological Magazine* eventually did to the Geological Society of London's *Journal* late in the century).[100]

But perhaps even more important than format were the consequences for the genre of the "scientific paper." During the twentieth century, it was very common (as it largely remains in the early twenty-first century) to suppose that the "scientific paper" was a genre that in many ways stood apart from commercial considerations. If greedy corporations had begun to corner the market on their distribution, scientific papers—and the scientific journals that contained them—had long ago been invented by learned societies and academies to serve the needs of scientific practitioners, and had remained remarkably stable ever since. In the words of the physicist John Ziman, "The general form of a scientific paper has changed less, in nearly 300 years, than any other class of literature except the bedroom farce."[101] But the history of proceedings shows that the "scientific paper" is not only less stable as a genre than has often been supposed, but that it was shaped precisely through negotiating the boundaries between—and respective responsibilities of—the editors of commercial publications and voluntary societies.

One way to understand the appropriation of the journal format by bodies of elite science would be as evidence for the growth of public interest in the activities of these groups. Indeed, there is no doubt that the audience for the most successful journals published by societies was far larger and more socially diverse than that for even the most successful transactions volumes. But this reading sells short the significance of this transformation in at least two ways. On the one hand, when elite societies began to appropriate features of commercial journals for their own publications, they were tacitly acquiescing to the argument from the radical press that their legitimacy depended on a different kind of imagined public. This was a structural transformation in the politics of knowledge that had far-reaching consequences for the nature of scientific expertise. Conversely, the uptake of journals by elite science wrought equally profound transformations on the nature of scientific journals themselves. As societies transformed their Proceedings publications into journals, they entered more fully the marketplace of the press, while at the same moment working hard to tame the emerging journal format by establishing formal routines by which to decide what to publish, by setting up strict generic expectations of papers regarding originality and even style, and by encouraging the notion that certain journals—usually those associated with scientific societies—were the only legitimate venues for original scientific claims.[102] Paradoxically, this appropriation and gradual redefinition of the scientific journal as a format focused on expert science took place even as the earlier claim that scientific claims should be subject to a wide reading public remained relatively intact.[103]

Gradually, the term "proceedings" lost its specific meaning and became a conventional label usually applied to collections of papers connected with some scientific organization.[104] But this label's persistence captures a crucial ambiguity that still defines the scientific literature today. Proceedings began as a public-facing genre; they were a means by which elite collectives gave an account of themselves to an imagined public. In that sense, they played a crucial role in the public legitimacy of such expert collectives. But this same format became a container for the most elite form of knowledge expression. Scientific papers, especially when they were shortened and stripped of the introductory materials, historical framing, and detailed methodological information that often accompanied longer memoirs, were of use only to the most informed inner circle of readers, and were an enigma to everyone else. Here, in the curious history of a format, we begin to discern the emergence of an enduring problem of science and democracy, one that continues to haunt debates about science and its publics today.

NOTES

1. For clarity, I capitalize "Proceedings" when the term refers unambiguously to periodicals published by learned societies, but otherwise I leave it uncapitalized. I use the term "format" to refer to periodical publications, and "genre" when proceedings are one component of a periodical publication.

2. For an exception to the neglect of Proceedings in the history of scientific publishing, see

James A. Secord, "Science, Technology and Mathematics," in *The Cambridge History of the Book in Britain*, vol. 6: 1830–1914, ed. David McKitterick (Cambridge: Cambridge University Press, 2009), 443–74, on 453–54. Poe's observation was made in "The Purloined Letter."

3. Dates in parentheses refer to the period when Taylor's firm was employed by each society. Taylor was also printer of the annual reports of the British Association for the Advancement of Science. Although the evolution of these reports is connected to the rise of Proceedings journals, that story is distinctive enough that I have chosen to set it aside here.

4. On this discursive sense of "public," see Harold Mah, "Phantasies of the Public Sphere: Rethinking the Habermas of Historians," *Journal of Modern History* 72 (2000): 153–82; Keith Michael Baker, *Inventing the French Revolution: Essays on French Political Culture in the Eighteenth Century* (Cambridge: Cambridge UP, 1990); and Thomas Broman, "The Habermasian Public sphere and 'Science in the Enlightenment,'" *History of Science* 36 (1998): 123–49.

5. This section and part of the next are based on my book *The Scientific Journal: Authorship and the Politics of Knowledge in the Nineteenth Century* (Chicago: University of Chicago Press, 2018).

6. Herschel to Francis Baily, 2 June 1821, Bodleian Library, MS. Autogr. d. 14, f. 6–7. The Council of the Geological Society of London also flirted with publishing an octavo journal before choosing to go with its quarto *Transactions*. See Leonard Horner to George Greenough, 4 April 1809, Greenough Papers, UCL Add. 7918/5/823.

7. See Jonathan R. Topham, "Anthologizing the Book of Nature: The Circulation of Knowledge and the Origins of the Scientific Journal in Late Georgian Britain," in Bernard Lightman, Gordon McOuat, and Larry Stewart, eds., *The Circulation of Knowledge between Britain, India, and China: The Early-Modern World to the Twentieth Century* (Boston: Brill, 2013), 119–52; and Jonathan Topham, "The Scientific, the Literary and the Popular: Commerce and the Reimagining of the Scientific Journal in Britain, 1813–1825," *Notes and Records of the Royal Society* 5 (2016): 305–24.

8. "Parliamentary Reporting," *The Companion to the Newspaper*, 1 April 1833, 17–20, on 20; Thomas Babington Macaulay, [review of Henry Hallam, *The Constitutional History of England*] *Edinburgh Review* 48 (1828): 96–169, on 165. On parliamentary reporting in London, see A. Aspinall, "The Reporting and Publishing of the House of Commons' Debates, 1771–1834," in *Essays Presented to Sir Lewis Namier*, ed. R. Pares and A. J. P. Taylor (London, 1956), 227–57; Dror Wahrman, "Virtual Representation: Parliamentary Reporting and Languages of Class in the 1790s," *Past & Present* 136 (1992): 83–113; Andrew Sparrow, *Obscure Scribblers: A History of Parliamentary Journalism* (London: Politico's, 2003).

9. William Windham speech recorded in *The Parliamentary Register* 7 (London, 1799), 475.

10. On radical British journalism in the 1810s and 1820s, see for example Kevin Gilmartin, *Print Politics: The Press and Radical Opposition in Early Nineteenth-Century England* (Cambridge: Cambridge University Press, 1996).

11. On the radical origins of the cheap periodicals of the 1820s, see Jonathan Topham, "John Limbird, Thomas Byerley, and the Production of Cheap Periodicals in the 1820s," *Book History* 8 (2005): 75–106.

12. "To the Mechanics of the British Empire," *Mechanics' Magazine* 1 (1823): 16.

13. "Apology and Preface," *The Chemist* 1 (1824), vii.

14. W. R. Y., "Remarks on the Disappointments Incident to Orchardists, and on Describing and Characterising Fruit Trees," *Gardener's Magazine* 3 (1827): 32; "Retrospective Criticism," *Gardener's Magazine* 5 (1826): 105. Both of these are quoted in Brent Elliott, *The Royal Horticultural Society: A History 1804–2004* (Chichester, UK: Phillimore, 2004), 182.

15. See James Secord, "Progress in Print," in *Books and the Sciences in History*, ed. M. Frasca-Spada and N. Jardine (Cambridge: Cambridge University Press, 2000), 369–89.

16. *Chemist* 1 (1824): 275. The journal quoted is *The Scotsman*.

17. This précis of Taylor's early life is based largely on W. H. Brock and A. J. Meadows, *The Lamp of Learning: Two Centuries of Publishing at Taylor & Francis* (London: Taylor & Francis, 1998), 19–63.

18. On the influence of the early Geological and Astronomical Societies on the Royal Society, see David Philip Miller, "Method and the 'Micropolitics' of Science: The Early Years of the Geological and Astronomical Societies of London," in *The Politics and Rhetoric of Scientific Method: Historical Studies*, ed. J. Schuster and R. Yeo (Dordrecht, Netherlands: Reidel, 1986), 227–57.

19. See Astronomical Society Council Minutes, 21 February 1823, Royal Astronomical Society Library, Council Minutes (hereafter RAS/CM) vol. 1, for the decision to give the *Philosophical Magazine* access to the minute books. See RAS Letters, 1820–29, for several letters from Gregory in 1824 on his progress in this regard. And RAS/Papers/17/1 for manuscripts and proof sheets from the *Philosophical Magazine*.

20. "Advertisement," *Annals of Philosophy* 1 (1813).

21. For Thomson's complaint, see Walter Crum, "Sketch of the Life and Labours of Dr. Thomas Thomson," *Proceedings of the Royal Philosophical Society of Glasgow* 3 (1855): 250–64, on 256. It is notable that there was a brief period in 1802 when Thomas Young was given permission to print proceedings of the Royal Society for the new journal run by the offices of the Royal Institution. That periodical went under almost immediately, however, and when it was revived at the end of the 1820s its editor inquired whether this permission still held; but this was denied. Edmund Robert Daniell to Royal Society, 15 February 1830, Royal Society of London Library (hereafter RSL)/MC/1. Reply in RSL Council Minutes, 11 March 1830, RSL/CMO/11.

22. "To our Readers and Correspondents," *Quarterly Journal of Science* 22 (1827), front matter.

23. See Taylor to the Astronomical Society in Council Minutes, 9 March 1827, RAS/CM/1, and similarly to the Geological Society in Council Minutes 19 March 1827, Geological Society Library (hereafter GSL)/CM/2.

24. Taylor's account of charges to these societies, extant at St. Bride Library, makes clear that he used standing type from the *Philosophical Magazine* for both the Geological's and the Astronomical's Proceedings, and charged accordingly. See the volume labelled *Journal 1823*, 31–32. On the printing of separate copies, see Csiszar, *The Scientific Journal*, 54–56.

25. The Astronomical Society, in an attempt to save money, switched printers in late 1828. They hired the firm of Priestley and Weale to act as printer and publisher to the society, which in turn employed J. Moyes as printer, who agreed to provide copies of the *Monthly Notices* to the society at no cost. In 1831, when it ended the first volume of the *Monthly Notices*, the society had the first several numbers reprinted by Moyes to create a uniform volume. RAS Council Minutes 12 December 1828 and 10 June 1831, RAS/CM/2.

26. Save for the setback that he lost the Astronomical Society account in late 1828.

27. Taylor was confirmed as the society's printer of the *Transactions* on 14 February 1828, and on 28 February the decision was made that "the Secretaries be authorized to communicate to the public, at their discretion, accounts of the proceedings of the Society." For Taylor's winning bid, see RSL/DM/1, f. 95–96.

28. Many of these abstracts also began to appear in the *Literary Gazette* in April 1828. Although I have found no evidence of formal permission granted for these, it is likely that one of the society's secretaries gave informal authorization. The arrangement seems to have lasted less than a year.

29. Although the individual numbers were labeled *Proceedings* and this is how they were generally referred to at the time, when it bound the first volume the society chose to call it volume 3 of the *Abstracts of the Papers Printed in the Philosophical Transactions*. Confusingly, this was done to imply continuity with the two volumes of abstracts published by the society of papers from the *Philosophical Transactions* for 1800 to 1830, even though the latter was actually a distinct publishing project initiated *after* the society began printing its Proceedings.

30. R. Taylor to P. M. Roget, 1831, RSL/DM/1/97. See also Taylor's *Journal 1830–1840* volume (St. Bride Library), 1. The decision to print abstracts of papers as proceedings is recorded in the Council Minutes of 16 December 1830, RSL/CMO/11.

31. "Report of the Council of the Society to the Eighth Annual General Meeting," *Monthly Notices* 1 (8 February 1828): 49–56. For the parallel discussion at the Geological Society, see the report of the council at the annual general meeting of 15 February 1828, in *Proceedings of the Geological Society of London* 1, 48.

32. Adam Sedgwick, "President's Address at the Annual General Meeting (19 February 1830)," *Philosophical Magazine* 7 (1830): 291.

33. "Abstracts of the Philosophical Transactions," *Mechanics' Magazine* 21 (1834): 204–7, on 205.

34. "Abstracts of the Philosophical Transactions," *Mechanics' Magazine* 28 (1838): 246–49, on 246.

35. On Raspail and the academy, see Bruno Belhoste, "Arago, les journalistes et l'Académie des sciences dans les années 1830," in Harismendy, ed., *La France des années 1830 et l'esprit de réforme* (Rennes, France: Presses universitaires de Rennes, 2006), 253–66; and Csiszar, *The Scientific Journal*, 92–117.

36. Review of Philosophical Transactions and Proceedings of the Royal Society, *British and Foreign Medical Review* 2 (1836): 212–18.

37. Council Minutes, 28 February 1828, RSL/CMO/10. Permission to access the papers read to the society was denied to the *Quarterly Journal of Science* on 11 March 1830, RSL/CMO/11.

38. See Royal Society to William Ritchie, 4 February 1832, objecting to an abstract appearing in the proceedings section of the *Literary Gazette*, RSL/MC/2/18.

39. GSL Council Minutes, 8 June 1831, GSL/CM/6.

40. GSL Council Minutes, 1 May 1833. GSL/CM/6.

41. GSL Council Minutes, 2 November 1836, GSL/CM/4.

42. On this point, see John C. Thackray, introduction to *To See the Fellows Fight: Eye Witness Accounts of Meetings of the Geological Society of London and Its Club, 1822–1868* (Faringdon, UK: British Society for the History of Science, 2003).

43. William Robert Grove, "Physical Science in England," *Blackwood's Edinburgh Magazine* 54 (1843): 514–25. For the details of Grove's motion and the resulting resolution, see RSL Council Minutes, 20 May 1847, RSL/CMP/2. On Grove's perspective on reform, see I. R. Morus, "Correlation and Control: William Robert Grove and the Construction of a New Philosophy of Scientific Reform," *Studies in History and Philosophy of Science* 22 (1991): 589–621.

44. Undated and unsigned report in GSL/COM/P/4/1. There are several more examples of this reasoning in extant reports for 1830–41 in GSL/COM/P/4/2. Charles Darwin, for example, often rejected papers on this basis. Similar judgments that a short version in the *Proceedings* would be sufficient were made by referees at the Royal Society (see, for example, RSL/RR/1/27, 34, and 66), though in this case they usually had not yet seen the abstract.

45. De la Beche to George Greenough, 10 November 1834, Greenough Papers, UCL Add. 7918/1/44.

46. Ibid.

47. "Report of the Council of the Society to the Eighth Annual General Meeting," 49.

48. "Explanatory Notice" with front matter for volume 8 (1848).

49. "General Duties of the Assistant Secretary," in RAS Council Minutes, 9 March 1838, RAS/CM/3.

50. Zoological Society of London Council Minutes, 15 December 1830, Zoological Society of London Library (hereafter ZSL)/CM/2.

51. The society's *Proceedings* were, as Gowan Dawson has shown, a key venue by which natural historical authors practiced serialized publishing, or "publication in parts." See "Paleontology in Parts: Richard Owen, William John Broderip, and the Serialization of Science in Early Victorian Britain," *Isis* 103 (2012): 637–67.

52. Royal Geographical Society Council Minutes, 18 December and 29 December 1830, Royal Geographical Society Library (hereafter RGS)/CM/1.

53. RGS Council Minutes, 18 February 1832, RGS/CM/1.

54. De la Beche to Greenough, 10 November 1834.

55. Jean-Baptiste Biot, [review of the *Comptes rendus hebdomadaires*], *Journal des savants*, November 1842, 641–61, on 655.

56. Distinct controversies involving Gray and editorial matters were discussed at Publications Committee meetings on 9 January 1838, 13 March 1838, and 13 June 1838. At the last, where it was reported that Gray had been pressuring John Barlow, the secretary, to insert a long paper in the *Proceedings*, Barlow put forward the new resolution. This passed on 10 July, and the resolution prohibiting purely oral notes from being reported in the *Proceedings* passed 24 July 1838. That the society followed through on the new rule is confirmed in the Publications Committee minutes, where papers for the *Proceedings* are after this date referred. ZSL/LCA/2.

57. Sheepshanks. "Report to Council on Monthly Notices," 14 May 1847, RAS/45.

58. Ibid.

59. See, for example, Sheepshanks to De Morgan, 2 February 1847 and 4 March 1847, RAS/ MSS DE MORGAN/1.

60. Sheepshanks to De Morgan, 4 March 1847.

61. That the issue of speedy publication should arise in early 1847 at the Astronomical Society is no accident. British astronomers were still dealing with the fallout of the scandal over

the discovery of the planet Neptune, with the French declaring speedy print publication to be crucial to any priority claim.

62. Note on verso of title page to *Monthly Notices*, starting with volume 9.

63. "Explanatory Notice" with front matter for volume 8 (1848).

64. GSL Council Minutes, 10 May 1843, GSL/CM/6.

65. The agreement with Longman is reproduced in GSL Council Minutes, 8 January 1845, GSL/CM/6. Longman pulled out after one year, citing losses incurred, but was kept on to sell the volumes on commission on behalf of the society.

66. "Introductory Notice," *Quarterly Journal of the Geological Society of London* 1 (1845): 1.

67. GSL Council Minutes, 2 April 1845 (and 18 February 1846, for the decision to send the journal), GSL/CM/6.

68. Elliott, *The Royal Horticultural Society*, 182.

69. For the decision to establish the *Quarterly Journal*, see Council Minutes of the Chemical Society on 3 May and 17 May 1847 (Library of Chemical Society of London, Council Minutes, vol. 1). For the expansion of the journal into a monthly, see the minutes on 5 December 1861 (Council Minutes, vol. 3).

70. "Report of the Scientific Committee appointed 17 June 1847," ZSL/GB/0814/GABQ.

71. Ibid.

72. "Report of a Committee to Enquire into the Means of Invigorating the Society in All Its Branches" [1848], ZSL/GB/0814/GACS. Recommendation passed by council, 16 February 1848, ZSL/CM/9.

73. The new plan to include miscellaneous matter involved leaving out ephemerides and other information that was already handled better by the *Astronomische Nachrichten*. It began with volume 13 and was described extensively by Richard Sheepshanks in "Proposal for Remedying the Monthly Notices," sent to De Morgan, RAS/MSS DE MORGAN/4.

74. See, for example, Printing Committee Report in Council Minutes, 14 January 1848, RAS/CM/5.

75. RAS Council Minutes, 8 January 1858, RAS/CM/6.

76. Ibid.

77. RAS Council Minutes, 20 November 1858, RAS/CM/6.

78. Ibid.

79. Linnean Society Council Minutes, 1 May 1855, Linnean Society Library (hereafter LSL)/CM/3.

80. LSL Council Minutes, 18 December 1855, LSL/CM/3.

81. LSL Council Minutes, 1 December 1875, LSL/CM/5.

82. Csiszar, *The Scientific Journal*, 138–58.

83. This is based on an analysis of all articles in volumes 3–5 of the *Proceedings*, covering 1831 to 1850.

84. RSL Council Minutes, 19 February 1852, RSL/CMP/2.

85. RSL Council Minutes, 5 November 1863 and 17 March 1864, RSL/CMP/3.

86. Report of the Publication Committee, 18 March 1886, RSL/CMP/6.

87. Report of Committee of Quarterly Journal, 28 April 1858, GSL/COM/SP/3/1.

88. Report of Finance Committee in Council Minutes, 9 May 1862, RAS/CM/6.

89. It must be noted, however, that Taylor's reports of proceedings were often far from timely, regularly six months after the fact or more. In cases where the society publication was more timely, this is rather surprising. William Jardine, one of Taylor's coeditors of the *Annals*, was similarly perplexed by these delays. Jardine to Richard Taylor, 10 April 1843, Authors letters D–K, Taylor & Francis Papers, St. Bride Library.

90. Both continued to use Taylor & Francis as their printer, however.

91. See the denial of the request for abstracts from the *Quarterly Journal of Science*, LSL Council Minutes, 14 January 1864, LSL/CM/4.

92. GSL Council Minutes, 29 November 1843, GSL/CM/6. Some other societies also continued separately to provide abstracts of their papers to journals. This was, for example, one of the jobs of the secretary of the Chemical Society of London.

93. Publications Committee Minutes, 14 November 1865, ZSL/LCA/4. Meeting with William Francis reported on 9 January 1866.

94. For an overview of these, see Ruth Barton, "Just before 'Nature': the Purposes of Science and the Purposes of Popularization in Some English Popular Science Journals of the 1860s," *Annals of Science* 55 (1998): 1–33.

95. William Sharpey to F. C. Mathieson, 22 March 1864; Walter White to William Crookes (editor of the *Quarterly Journal of Science*), 2 December 1864; G. G. Stokes to Thomas Bendyshe (new proprietor of *The Reader*), 8 December 1865; Walter White to editor of *China Telegraph*, 27 January 1866, all in RSL/MS/426.

96. For reports on the Royal Astronomical Society in the *Astronomical Register* and the *Observatory*, see chapter 8 in this volume.

97. "Reports of Societies," *Knowledge* 1 (1881): 116.

98. W. Mattieu Williams, "'Knowledge' and the Scientific Societies," *Knowledge* 1 (1881): 143–44.

99. This idea was confirmed also by the Royal Society's *Catalogue of Scientific Papers*, which indexed Proceedings publications extensively. See Alex Csiszar, "How Lives Became Lists and Scientific Papers Became Data: Cataloguing Authorship during the Nineteenth Century," *British Journal for the History of Science* 50 (2017): 23–60.

100. For the Royal Geographical Society absorbing the *Geographical Magazine*, see RGS Council Minutes, 25 November 1878. On the competition between the Geological Society and the Geological Magazine, see chapter 4 in this volume.

101. John Ziman, *Public Knowledge: An Essay Concerning the Social Dimension of Science* (Cambridge: Cambridge University Press, 1968), 105.

102. For more on the consolidation of the new expert-oriented journal format, see Csiszar, *The Scientific Journal*.

103. On this point, see Broman, "Habermasian Public Sphere," 142–44; and Steven Shapin and Simon Schaffer, *Leviathan and the Air-Pump: Hobbes, Boyle, and the Experimental Life* (Princeton, NJ: Princeton University Press, 1985), 343.

104. *Conference* proceedings, however, followed a different trajectory, such that the label maintained a more specific significance, and these continue to play a distinctive and central role in certain fields such as computer science.

* 2 *

Defining the Communities of Science

CHAPTER 4

"An Independent Publication for Geologists":
The Geological Society, Commercial Journals, and the Remaking of Nineteenth-Century Geology

Gowan Dawson

In September 1862, Charles Lyell heard a disturbing rumor at the Geological Society of London. As he told Thomas Henry Huxley,

> Horner tells me that Prestwich was thinking it possible that Mackie might succeed R. Jones as editor of our journal! Whether as a man of business, or a writer in an inflated style, & for some other reasons he is just about the worst man we could have. . . . I am sorry for him & his wife (who would make the better assist. secry. of the two) but this would be a most mistaken charity.

Lyell's informant, Leonard Horner, was too ill to fulfil his own responsibilities as the Geological Society's president, relying on Huxley, as secretary, to undertake them for him. But Lyell was more concerned about the subordinate post of assistant secretary, which entailed editing the *Quarterly Journal of the Geological Society* and, unlike the fixed-term presidency, was open-ended. As he reflected anxiously: "If we got a 3rd rate man it would be for ever."[1] While editors of the transactions published by scientific societies had traditionally remained discreetly invisible, choosing an appropriate editor of "our journal" was, for Lyell, the Geological Society's most pressing consideration.[2]

The allegedly "3rd rate man" who prompted Lyell's anxiety was Samuel Joseph Mackie, who had been elected a fellow of the Geological Society in 1851, with the personal support of such luminaries as Roderick Murchison, and whose advocates for now taking over from Thomas Rupert Jones as editor of the *Quarterly Journal* went well beyond Joseph Prestwich.[3] Indeed, two months later, in November 1862, the *Medical Times and Gazette* reported: "Mr. Samuel J. Mackie, F.G.S., is a candidate for the office of Assistant Secretary to the Geological Society . . . We see in the list of his supporters the names of Professor [Richard] Owen, the Astronomer Royal [George Biddell Airy] . . . and many others who are pre-eminent in science."[4] More significantly, Mackie already had considerable experience of editing geological journals, having founded the *Geologist* in 1858, which, Mackie claimed two years later, "now . . . stands . . . unrivalled in circulation by any British scientific periodical."[5] Notwithstanding this vaunted success in attracting readers, the *Geologist*'s—and thus Mackie's—financial situation remained extremely parlous, as Lyell uncharitably hinted. Although this, as William Brock has shown, was the fate of almost all commercial science journals in the nineteenth century, especially those devoted to just a single subject, Lyell's animadversions were shared by other members of the Geological Society's council.[6] In February 1863 it appointed as editor of the *Quarterly Journal* Henry Michael Jenkins, a twenty-one-year-old novice without a fellowship of the society (although he was expediently elected to one soon after), whose severe "asthmatical affection" made it difficult for him to work "in unfavourable conditions of weather."[7] Evidently anyone—including his wife, in Lyell's opinion—was preferable to Mackie.[8]

Lyell's visceral response to Horner's hearsay regarding Mackie and the *Quarterly Journal* is symptomatic of the tensions between official society publications and independent commercial journals that, while common across many sciences in nineteenth-century Britain, were more pronounced and consequential in geology than in any other area. When the Geological Society's *Quarterly Journal* was established in 1845 as a response to long-standing difficulties with the society's expensive and irregular *Transactions*, it was explicitly modeled on the material format of commercial periodicals, with its octavo size, regular publication schedule, and a commercial publisher, Longman, that agreed to defray the costs in return for a share of the profits. Even the innovation of having a single, named editor, with David Thomas Ansted the first to take on the role, emulated the practices of commercial science journals that, as with *Nicholson's Journal*, were sometimes named after their high-profile editors.[9] The commercial scientific press had proliferated, if not always profitably, since the close of the eighteenth century; and reductions in taxes and other production costs, as well as the introduction of the penny post, meant that by the 1840s it was becoming possible to sustain a specialist journal on a single, discrete field of science like geology.[10]

Such a "special-class scientific periodical," as Mackie himself acknowledged, could never achieve a "higher circulation than from 800 to 1000."[11] There were, after all, only "so many geologists" who could be induced to become "permanent purchasers," regardless of occasional spurts of wider interest that "might increase the casual sale" for a time. These "special scientific periodicals" nevertheless could, with prudent management, be "maintained remunerative" even on such a limited circulation, and, unlike the Geological Society's generously subsidized but ponderous official publications, they afforded cheap, efficient, and expeditious formats for disseminating new knowledge, debating controversial issues, and establishing priority claims.[12] The *Quarterly Journal* was meant to bring the Geological Society into line with these developments in early nineteenth-century scientific publishing but, despite its conscious emulation of many of the formal innovations pioneered by periodicals that had to turn a profit, it remained aloof from other aspects of the commercial marketplace. As Lyell's barbed comments to Huxley made clear, neither Mackie's financial risk taking nor his overwrought, populist prose style would have been fitting for the Geological Society's flagship publication.

The Geological Society's elitism and centralization of intellectual authority, which had been evident since its foundation in 1807, ensured that it remained at odds with commercial journals that sought to foster different, more marketable approaches to the earth sciences. The financial imperatives of the literary marketplace necessitated that these journals, beginning in the 1840s and continuing through the rest of the century, endeavored to cultivate a much larger readership than the Geological Society's official publications, which for the most part addressed the needs and interests of what Martin Rudwick has termed the elite group of "gentlemanly specialists" who, albeit small in number, dominated the earth sciences in the early and mid-nineteenth century.[13] This same disregard for what might attract a larger, paying audience, including factors such as theoretical speculation, controversy, and imagination, continued with the ostensibly innovative *Quarterly Journal*—as Longman, hemorrhaging money because of poor sales, soon discovered to its cost. The pecuniary difficulties imposed by a limited circulation, as will be seen shortly, had significant implications for the intellectual and scientific value of the Geological Society's periodicals.

At the same time, those journals, including Mackie's *Geologist*, that had to at least break even envisaged their broader audiences, in part for nakedly pragmatic reasons, as active participants in geology, able to come to their own conclusions about the most contentious disputes and to provide original observations and information that would be valuable to the specialist community, both of which could be accomplished by continuing to purchase the periodical. These new communities of practitioners, drawn from the various graduated zones of geological competence identified by Rudwick below the ranks of the gentlemanly specialists, had been excluded and ignored by the Geological Society.[14] Commercial journals, which instead viewed them as essential to their continued existence, utilized the potential of these new communities to forge a less exclusive and more egalitarian conception of the earth sciences, directly challenging the hierarchical exclusivity of the Geological Society. Significantly, the gentlemanly specialists, seemingly including even the disdainful Lyell, would eventually feel compelled to take advantage of the new opportunities afforded by the more inclusive approach of journals such as Mackie's *Geologist*. In fact, the format of Mackie's problematically populist periodical would, by the end of the nineteenth century, provide the vehicle for a new community of professional geologists to finally supplant the intellectual authority of the Geological Society and its gentlemanly specialists.

THE GREAT EVIL OF DELAY: *TRANSACTIONS* (1811-56) AND *PROCEEDINGS* (1827-45) OF THE GEOLOGICAL SOCIETY

At the beginning of the nineteenth century, the fledgling science of geology was both tainted by controversy, especially in those approaches that contradicted religious orthodoxies by insisting on the vast age of the earth, and riven by bitter disputes. Among many other intellectual ruptures, geology was most profoundly split between the speculative cosmological theoreticians who had first initiated the science in the late eighteenth century and a new school of inductive empiricists who now sought to redefine it at the start of the next century.[15] In none of these competing approaches to the study of the earth, however, was publication a particular problem.

The genre of imaginative geotheory was suited to lengthy, sumptuous treatises such as the three quarto volumes of James Hutton's *Theory of the Earth* (1795). In other areas, though, more emphasis was put on the collecting and exhibition of specimens, and geology remained a science in which oral communication regularly took precedence over publication.[16] As Roy Porter has suggested, the "lack of pressure to publish" was a distinct feature of British geology until well into the nineteenth century.[17] Even the founders of the Geological Society, who took the lead in redefining geology as a strongly empirical science rather than a theoretical and speculative one, did not initially prioritize print. George Greenough, the society's first president, published little himself, and instead disseminated knowledge by inviting guests to visit his mansion, with its extensive geological collections, in fashionable Regent's Park. In France, by contrast, the state funding of science meant that geologists such as Alexandre Brongniart were duty-bound to announce their results in official publications.[18] Back across the Channel, even as late as 1846 Charles Darwin still considered that "geology is at present very oral," and that "geologists never read each other's works."[19] Similar observations about the failure of geologists to read (and, more pertinently, to purchase) the specialist journals in their field continued to be made throughout the rest of the century, with Horace Woodward complaining in 1907 that certain of the Geological Society's "Fellows were more concerned in appending F.G.S. to their names than in adding the *Quarterly Journal* to their bookshelves."[20] This lingering ambiguity toward print, which persisted with practical collectors as much as with dilettante F.G.S.s, was one of the factors that made it so difficult for publishers to make a profit from geological periodicals (fig. 4.1).

In 1834 William Swainson reflected on the recent emergence of several new "societies for the promotion of science," noting that the "most important object of these associations is the publication of essays . . . read at their meetings." Although he was generally critical of these new "London scientific bodies," whose primary purpose was in producing an "annual volume" to be "*distributed* to members, and sold to the public," Swainson made a single exception, avowing that the "Geological Society is unquestionably the most active, and most popular," and might even "be looked upon as a model for all the others."[21] Following its foundation in 1807, however, Swainson's model scientific society had initially published only a short pamphlet, *Geological Inquiries* (1808), which was intended to stimulate observers in the field while also ensuring that they adhered to the new empirical methods enjoined by the Geological Society's founders. It was not until February 1810 that there was any discussion within the society's council of a serial publication (then to be called *Memoirs*), and, after a formal decision at a special general meeting in March, the first volume of what was now named the *Transactions* was published, with a print run of 750, in the summer of the following year.[22]

It was only with the publication of the first volume of the *Transactions* in 1811 that the Geological Society, as Rudwick has proposed, determined precisely what kind of organization it was to become.[23] While many early members had envisaged the society as either a hub for a utilitarian collection of mineral resources—hence the *Geological Inquiries* pamphlet—or merely a convivial dining club, the appearance of the *Transactions*, printed, under council orders, with "ink . . . of the best quality" in an opulent quarto

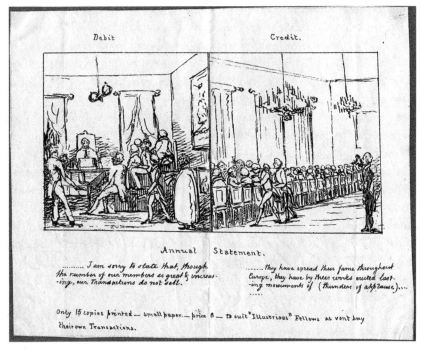

FIG. 4.1. Cartoon by Henry De La Beche from the early 1830s, satirizing the fellows of the Geological Society for being reluctant to buy copies of the *Transactions*, yet willing to laud the achievements of its geologist contributors. Image at left, labeled "Debit," depicts the private annual financial statement of the society's council. Image at right, labeled "Credit," depicts the grand annual address of the society's president. The cartoon was pasted into a scrapbook belonging to Roderick Murchison, with a mocking note: "Only 15 copies printed—small paper—price 0—to suit 'Illustrious' Fellows as vont buy their own Transactions." GSM/DR/Mu/S/3, Geological Survey Archives, Keyworth. Reproduced by permission of the British Geological Survey, CP18/075.

format that resembled the Royal Society's prestigious *Philosophical Transactions*, consolidated the position of those members who wanted it to attain the status of a metropolitan learned society.[24] The Royal Society's president, Joseph Banks, feared that this model of the Geological Society would herald the fragmentation of the natural sciences, with the broad inclusivity of the *Philosophical Transactions* being supplanted by an array of specialist society publications.[25]

The preface to the first volume of the Geological Society's *Transactions*, written in June 1811 by the newly founded Papers Committee, proclaimed that "every latitude has been allowed to authors, with regard to their theoretical inferences from the observations which they record; it being understood . . . that the writers alone are responsible for the facts and opinions, which their papers may contain."[26] The same volume's title page nevertheless carried a Latin epigraph from Francis Bacon's *Novum Organum* (1620) that, with its insistence on producing "*certo et ostensive scire* [certain and demonstrable knowledge]" over "*belle et probabiliter opinari* [attractive and plausible conjectures]," adumbrated the society's corporate policy of stringent Baconian empiricism.[27] Rather than theoretical latitude, the papers in the initial volumes of the *Transactions* instead exhibited, as William Henry Fitton observed in 1817, "a character of strict experiment or observation, at the expense of all hypothesis, and even of moderate theoretical speculation."[28] With its embargo on hypothetical theories, the *Transactions* was the Geological Society's principal vehicle in its campaign for, as Ralph O'Connor has put it, "rebranding 'geology'" as a rigorously inductive science rather than, as it had been in the eighteenth century, merely a speculative a priori one.[29]

Such scientific rebranding, however, risked denuding geology of the very things that had made it intellectually appealing, for, as Fitton noted, "speculations . . . are much more seducing than . . . dry determinations . . . and the student is glad to shelter himself in generalities from the revolting details."[30] In fact, as Rachel Laudan has argued, the early members of the Geological Society "overreacted" in repudiating so completely the more philosophical approaches to geological method common both in Scotland and on the Continent, and, as a consequence, the initial volumes of the *Transactions* were "filled with articles that were narrow in scope and unimpressive in quality."[31] Those papers that surpassed the general standard, such as John MacCulloch's influential account from 1817 of the role of temporary lakes in the formation of the Parallel Roads of Glen Roy, only gestured at "keep[ing] clear of all speculations purely hypothetical." As MacCulloch acknowledged, there were certain natural phenomena, like the long-vanished lakes, that must inevitably "stimulate us to seek for a theory . . . beyond the limits of our immediate observation."[32] While the *Transactions*'s title page retained its Baconian epigraph into the 1850s, the prohibition of theory that it enjoined had already been partially abandoned by the journal's second series, which began in 1822.

With this second series of the *Transactions*, the Geological Society dispensed with the services of William Phillips, who had printed the initial volume in 1811 and then published the following four, retaining ownership of the copyright and pocketing whatever small profits might be gleaned from their sale (although none had yet sold out their full print runs).[33] The printing of the second series, in a reduced print run of five hundred and with a new, more modern typeface, was instead delegated to the firm of Richard Taylor, whose brother John was, conveniently, the Geological Society's treasurer.[34] More significantly, the society itself assumed the role of publisher, with the financial risks underwritten by wealthy members. This was ostensibly for the "purpose of reducing the price, and thereby promoting the diffusion of Geological information," but, no longer beholden to an independent publisher needing to recoup their investment by maintaining a remunerative circulation, the society, and particularly its presiding council, could now take an even greater control over the tone and contents of the *Transactions*, ensuring that it remained restricted to the needs and interests of the small elite of gentlemanly specialists who dominated the society.[35] In fact, while the initial volume of the first series had been sold at the same price, one pound and twelve shillings, to both society members and the book trade, the first volume of the new more economical second series was priced differentially, at one pound and five shillings to members and one pound, eleven shillings, and six pence to the public, further restricting its circulation beyond the society's confines.[36]

If the ban on theory had been relaxed by the 1820s, especially once abstract stratigraphic methods became a leading concern within the society, the *Transactions* still persisted in eschewing the imaginative and romantic presentation of the earth's history that, in other areas of early-nineteenth-century print culture, had made geology both hugely popular and profitable. Books and periodicals relating sublime visions of the ancient past, as O'Connor has shown, captivated the same audiences who also paid to view panoramic history paintings and attend large-scale theatrical entertainments.[37] Such alluring rhetoric was self-consciously resisted in the *Transactions*, with, for instance, John Davy, in an 1821 account of the geology of Ceylon, remarking: "A poet's pen could hardly do justice to its beautiful and sublime scenery. I should wander from my subject were I to indulge my feelings, and . . . I must confine myself to the skeleton . . . [of] minute details."[38] Not all contributors were so self-denying, particularly after the discovery in the following decades of new and increasingly bizarre fossil creatures. These compelled William Buckland, in 1835, to resort to literary analogies in comparing the pterodactyl to both "Milton's fiend" and the "dragons of romance and heraldry," while also imaginatively recreating "flocks of such-like creatures flying in the air and shoals of no less monstrous Ichthyosauri and Plesiosauri swarming in the ocean."[39] This vivid panoramic scene, invested with a frisson of religious controversy by its multitudes of demonic monsters, was a notable exception from the customary, more restrained tone of the *Transactions*, which, as Adelene Buckland has argued, deliberately kept the romantic sublime and even the more placid pleasures of narrative "at arm's length."[40] Even while geohistory became both more extensive and ever more remarkable, the *Transactions* continued, with very few exceptions, to expound on the structure rather than the story of the earth. It maintained a deliberately dry descriptiveness that was intended in part to render geology safe and gentlemanly, and to temper the controversial and dangerous reputation that still clung to the earth sciences throughout the first half of the nineteenth century.

The picturesque did occasionally enter the *Transactions* in its extensive illustrations, which were more plentiful than in any other scientific periodical of this period and regularly featured landscape views of geologically significant localities. Its lavish allowance of illustrations enabled the *Transactions*, as Rudwick has proposed, to develop a "standardized visual language" that was crucial for the development of geological communication among elite practitioners; but it came, literally, at a heavy price.[41] The costly copper-plate engravings used in the first series, as Phillips had warned the council, made the *Transactions*'s break-even point half the print run (publishers usually anticipated that one-third of a print run would cover their costs). Even when, with the second series, the Geological Society switched to the new and cheaper technique of lithography, the inclusion in each volume of more than twenty often hand-colored quarto plates kept the price of printing the *Transactions* exorbitantly high. At the same time, moreover, the self-conscious exclusion of textual elements that might have appealed to a larger, paying readership, whether theoretical hypothesis or imaginative narrative, made the high cost of the *Transactions*'s numerous illustrations still more problematic. From 1817 the *Transactions* initially appeared in separate parts, with two or three then bound together into a volume, although the Geological Society's finances still required a period of recovery following the publication of each part, during which the funds for publishing could be replenished.[42] This necessary interlude of fiscal renewal meant that the parts of the *Transactions* appeared only very slowly and irregularly, with, for instance, a seven-year hiatus between the initial two volumes of the second series, which were published in 1822 and 1829, and then the next three volumes coming out in quick succession between 1835 and 1836.

These pecuniary problems and the consequent delays in publication they imposed had hugely significant implications for the scientific value of the *Transactions*, especially as, by the mid-nineteenth century, claims to priority in scientific discoveries were increasingly being both staked and adjudicated in serialized print, rather than, as previously, in oral presentations or informal correspondence.[43] As Leonard Horner, then in his healthier first term as the

Geological Society's president, acknowledged ruefully of "our 'Transactions'" in 1846, the "great delay in . . . publication" was "robbing authors in some instances of the honour of priority in discovery." Indeed, the "uncertainty when a paper that had been read would be published," Horner conceded, "very materially diminished the usefulness of the Society, and . . . cooled the zeal of many of our Members, and forced them to send their memoirs elsewhere."[44] The alternative of choice for such dispirited society members was the *Philosophical Magazine*, a successful commercial journal owned by the *Transactions*'s printer, Richard Taylor, that had the considerable advantage of appearing regularly on the traditional "magazine day" at the end of each month. This rapid production schedule, as well as the commercial imperative for such periodicals to carry the very latest news of scientific developments, also meant that the *Philosophical Magazine* did not require the same complicated and often circuitous refereeing procedures as the *Transactions*, where a vote of approval from the Geological Society's council was necessary before the already lengthy publication process could even begin.[45] What Horner termed the "great evil of delay" had compelled the council to supplement the *Transactions* with a new publication, the *Proceedings*, which from 1827 printed abstracts of papers delivered at the society's meetings "as promptly as may be" in a more expedient octavo format that was distributed gratis to fellows of the society resident in London.[46] The short summaries in the *Proceedings* were printed by Richard Taylor and also appeared, in the very same format and with the express approval of the Geological Society, in his *Philosophical Magazine*.[47] They were therefore an effective means of promulgating the society's geological agenda, though, like the lengthier articles published in the *Philosophical Magazine*, they were denuded of the illustrations that were such an integral, if problematically expensive, component of the papers in the *Transactions*.

William Buckland observed in 1820 that the "transactions of the Geological Society . . . are quoted as standard authority, wherever this science has been admitted," and, as the fledgling science became increasingly established, the *Transactions* remained the most prestigious and authoritative site for publishing geological research in Britain, albeit addressing only a small circle of gentlemanly specialists.[48] By the early 1840s, however, the perpetual delays in its publication ensured that, as Archibald Geikie reflected, "few but professed geologists ever thought of buying or reading . . . the ponderous quarto Transactions . . . to the detriment alike of the science and the authors of the papers."[49] With such a restricted readership, which invariably fell far short of the print run of five hundred, both the Geological Society, whose finances were at breaking point by 1843 after an attempt to augment the *Proceedings* with illustrations proved a financial disaster, and the earth sciences more generally had reached what Gordon Herries Davies has called a "publication crisis."[50] At precisely this moment the commercial press afforded a new model of what a geological periodical might look like, and a new sense of the audiences it might reach.

THE UNFETTERED EXPRESSION OF INDEPENDENT OPINIONS: *GEOLOGIST* (1842-43), *QUARTERLY JOURNAL OF THE GEOLOGICAL SOCIETY* (1845-1971) AND *LONDON GEOLOGICAL JOURNAL* (1846-47)

In 1827 the Geological Society, uniquely among London's now burgeoning scientific organizations, decided to allow informal debates on the papers that were read at its meetings. These "oral discussions of the geologists," as Swainson observed, were soon "proverbial in the scientific world" for their "high intellectual gratification."[51] However, the society's most distinctive and fruitful innovation, which Roderick Murchison lauded as the "true safeguard of our scientific reputation," had no tangible impact on either of its official periodicals.[52] Notwithstanding their manifest scientific value, the often vigorous discussions that followed a paper, contesting particular points and relating them to larger theoretical disputes, were excluded from the printed record.[53] This gave the erroneous impression that the published papers were merely indubitable statements of objective fact, on which the entire society, as befitted its gentlemanly origins, was in unanimous accord. While this was a convenient facade at a time when the society's claim to exclusive authority in geological matters was being challenged by both scriptural literalists and transmutationists, it had deleterious consequences for the quality and interest of the *Transactions* and *Proceedings*, particularly for those outside the small elite of gentlemanly specialists who did not have the opportunity to attend the meetings.[54] The published papers, as Geikie opined, were "embalmed in the Society's printed publications" and denuded of the "fire and humour" with which they were originally read and received in the society's "dingy room in Somerset House" in central London.[55] Sometimes even the actual paper that had been read at Somerset House was censored in the editorial process, as happened to Adam Sedgwick's 1844 abstract in the *Proceedings* on the putatively Cambrian rocks of Wales, to remove any elements that had provoked subsequent discussion and controversy.[56]

The society's strict embargo on reporting its discussions extended to the commercial press, where vibrant controversies had long been a means of drawing in and retaining readers. In 1830 the metropolitan weekly the *Athenæum* noted with disappointment that "it has been intimated to us that the Council do not wish the proceedings of their meetings published," expressing "surprise at this . . . apparently strange determination."[57] Throughout the 1830s the *Athenæum* continued to abide by what it called the "laws of etiquette in force at the Geological Society" and would "only intimate, in general terms" the nature of the "debates which arise at the assemblies of that body."[58] By the following decade, however, other commercial journals, with a more pressing interest in geology, were less willing to remain so obligingly discreet.

In 1841 a precocious twenty-one year old, Charles Moxon, wrote to the Geological Society's council informing them that he would soon initiate an unprecedented innovation in scientific publishing: an independent commercial "periodical exclusively devoted to the Science of Geology, in all its bearings."[59] Although the society's secretary, William John Hamilton, declined Moxon's request for official support, he advised that "it would not do for the Society to throw cold water upon any plans which in proper hands might do good to the science."[60] Even such tentative patronage, though, did not last for long, as it quickly became apparent that Moxon's vision for this new journal, which first appeared in January 1842, was entirely at odds with the society's own publishing practices with the *Transactions* and *Proceedings*. The *Geologist*, whose title foregrounded the role of individuals rather than institutions, specialized in, among other things, the "most recent information on theoretical questions," on which the Geological Society's official periodicals, even after the relaxing of the erstwhile ban on theory, were still noticeably reticent.[61] With the assistance of the Anglo-French publisher Hippolyte Baillière, Moxon's octavo journal was published every month and cost just a shilling, enabling it to target a readership of not only "those persons thoroughly conversant" with geology, but also the less erudite "student" of the subject; and it was, as Moxon professed, "our anxious wish to please both parties."[62] The *Transactions*, even in its cheaper second series, still cost an exorbitant one pound, eleven shillings, and six pence to the public; and the *Geologist*, with its considerably more affordable cover price, endeavored to attract an entirely new audience for geology periodicals and, by doing so, create a new community for the discussion and even the adjudication of geological issues.

In his editorial "Monthly Notice" for May 1842, Moxon acknowledged that the Geological Society was an "institution, which, from its position and influence, should rank as the tribunal before which all geological questions" are "weighed and decided upon"—but, he protested, it had abnegated these responsibilities by the failure of its official periodicals to record the "interesting and highly valuable discussion[s]" that followed papers at its meetings.[63] Instead, the *Transactions* and *Proceedings* simply printed what were often much-contested interventions in long-standing controversies, such as the famously testy "Cambrian-Silurian" dispute between Murchison and Sedgwick, as if they had been accepted, with the society's authoritative endorsement, without any disagreement at all.[64] As Moxon pointed out, this polite pretence would inevitably result in the paradoxical situation of "two or more communications opposed in their statements" being "published virtually under the sanction of the same Society," often in the very same number of one of its periodicals.[65]

Printing such antagonistic statements alongside each other without any hint of the disputes of which they were part, Moxon warned, would "cast rather obliquity than perspicuity upon the subjects reported." Still more provokingly, he labeled the heavily subsidized *Proceedings* a "very sinecure publication" in which "so many observations" from the vibrant oral discussions were profligately "lost to the world." The commercial imperatives of the *Geologist*, by contrast, would ensure that such valuable copy was properly used, and Moxon magnanimously offered his journal's services: "Gladly would we ourselves, as the only periodical devoted to the science of geology, give immediate insertion to the abstract of the *proceedings in full* . . . for the benefit of the geological public."[66] This putative "geological public," comprising those excluded from the Geological Society's narrow elite of gentlemanly specialists, would thereby be able to come to their own conclusions on even the most contentious questions, with Moxon observing that "through the medium of a periodical . . . everybody may read and judge for himself."[67] Aghast at this avowedly egalitarian agenda, the society's secretary responded, with haughty formality, that the "council have seen with extreme regret, in . . . numbers of 'the Geologist' . . . that you have allowed allusions to . . . the discussions which take place at their evening meetings . . . to be inserted."[68] Rescinding his initial tentative support for the *Geologist*, Hamilton threw cold water on Moxon's impertinent plans.

Moxon insisted that "*words said in public are public property*," and, for him, this ostensibly trifling contretemps over intellectual property was actually vital for the integrity and future success of the earth sciences.[69] He avowed that "in justice to Geologists, contemporaries, and scientific men generally, we shall, when occasion requires, continue to notice" the Geological Society's prohibited discussions, insisting that "if the Society will not itself act in a spirit of fairness, we . . . must do our duty in the matter."[70] At issue was the question of what kind of journal would most benefit the progress of geology. While the *Transactions* and *Proceedings* afforded merely an austere record of the papers delivered at Somerset House that could induce only confusion and "apathy,"

the *Geologist*'s mix of pungent editorial opinion and the airing of current controversies would generate discussion and debate, thereby facilitating a "real
spirit of enquiry" among its community of readers.[71] Indeed, the Manchester
Geological Society, which was founded in 1838 following the emergence of
similar regional societies in Cornwall (1814), Newcastle (1829), Edinburgh
(1834) and Yorkshire (1837), often held discussions based on articles "given in
the 'Geologist'" rather than papers read at its meetings.[72] By 1843, Moxon was
forecasting that his journal would become just as important as the activities of
"societies" for the "active geologists" into whose "hands" it "will fall."[73] The
only problem was that, in the highly competitive mid-nineteenth-century literary marketplace, the *Geologist* did not fall into enough of those hands to make
it remunerative; and the first attempt at an independent geology periodical
folded at the end of 1843 after only eighteen numbers.

If Moxon's experiment was a failure, the Geological Society, engulfed by a
crisis in financing its periodicals, recognized that the same commercial route
might be a means of resolving its own publishing problems. In May 1843 the
society appointed a committee of publication to survey the financial wreckage
of the *Transactions* and *Proceedings*, the latter of which alone cost £144, one
shilling, and four pence to produce each year, offset by sales worth only a
meager eight pounds and nineteen shillings.[74] Having discarded William Phillips from the *Transactions* back in 1822, the society determined to once again
work with a commercial publisher, this time on a new periodical that, as noted
earlier, would embrace many of the formal and technological innovations of
early-nineteenth-century scientific publishing.

The *Quarterly Journal*, published from February 1845 as a "collaborative
effort" between the society and the firm of Longman, subsumed the *Proceedings*'s erstwhile function of providing short abstracts of the papers read at
Somerset House.[75] At the same time, it also published the longer versions of
the same papers previously carried in the *Transactions*'s ponderous "quarto
form of publication," now printed in an octavo format "attended with more
expedition, and more commensurate with the funds of the Society" (fig. 4.2).[76]
An advance prospectus in the *Publishers' Circular* even announced that the
"regular periodical publication of the Journal will ultimately ensure the great
advantage of a certain and immediate notice, within three months at most, of
the date of the communication," though Horner, a year later in 1846, had to
temper such precipitous promises by clarifying that "memoirs will in general appear within six months of their having been read at the meetings, and
sometimes even more speedily."[77] As well as speed, the *Quarterly Journal*,
unlike the *Proceedings*, could also accommodate numerous illustrations in its

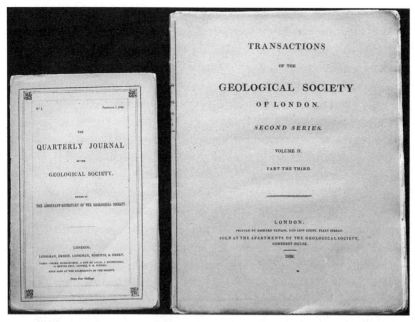

F I G .   4 . 2 . Wrappers of the octavo *Quarterly Journal* and quarto *Transactions* of the Geological Society, with the former, liveried in a sprightly green with decorative borders, giving prominence to information regarding price, date of publication, the unnamed editor, and its British and foreign publishers, all in the style of a commercial journal. Courtesy of Jim Secord.

octavo format at a relatively inexpensive cost, with Horner pronouncing that the "great improvements of late years make it possible to have distinct and accurate illustrations by woodcuts, lithographs, and zincographs, upon a page of that size."[78] By combining the rapidity of the *Proceedings* with the higher production values of the *Transactions*, the *Quarterly Journal* quickly rendered them both obsolete (though one further part of the irregular *Transactions* came out in 1856), and in 1851 the Geological Society disposed of its remaining stock of unsold volumes to a London bookseller for a knockdown price.[79]

In a marked departure from its two predecessors, the *Quarterly Journal* also included "information of a mixed character, on Geological subjects" of the kind that appeared regularly in commercial journals such as the *Philosophical Magazine* or even Moxon's *Geologist*.[80] It was, as Horner acknowledged, a "departure from a sound principle, for a Society to publish, under its authority, anything beyond that which properly belongs to it as an integral part of its own proceedings." This pioneering "miscellaneous part of the Journal" comprised such "choice exotic fruits" as translations of articles from foreign geological periodicals and notices of new books in the field.[81] While, on the surface, the contents of the *Quarterly Journal* increasingly resembled those of its commercial rivals, it nevertheless retained the Geological Society's long-standing aversion to any hint of controversy, continuing the strict policy of not recording the informal discussions at the society's meetings (a ban that was only lifted in 1868). More problematically, the *Quarterly Journal* even extended this prohibition on disagreement to its book reviews.

While general periodicals such as the *Athenæum* had long cultivated large readerships with their self-consciously acerbic and opinionated style of reviewing, the *Quarterly Journal*'s reviews of recent "Geological publications" were instead "confined to an analysis of their contents" and did "not express any opinion on the general merits of the works analysed." In fact, reviewers were instructed that "if any omission or inaccuracy on the part of an Author should be noticed, such notice will be strictly limited to the particular matter adverted to, and will be expressed in guarded and courteous tones."[82] This was still the tone of gentlemanly restraint that had dominated the *Transactions*, but now transposed onto a format that, with book reviewing usually predicated on the kind of vigorous debate that yielded a profit in the literary marketplace, was unsuited to such genteel values. While the gentlemanly specialists of the Geological Society recognized the financial necessity of attracting a broader readership, they still found the inevitable compromises that this entailed, as with Lyell's dismay at the prospect of Mackie's editorship discussed earlier, egregiously distasteful. As Longman soon recognized, the *Quarterly Journal* fell between the two stools of official society transactions and independent commercial journals.

Not that the *Quarterly Journal* was itself without controversy behind the scenes. It quickly descended into what Herries Davies has called a "fiasco" as a "venture in commercial publishing," and Longman, hemorrhaging money, refused to renew its contract with the Geological Society after only one year (though it remained as publisher, with the society assuming the financial risk).[83] Soon after, in January 1847, the society's council censured the *Quarterly Journal*'s founding editor, David Thomas Ansted, for a "lack of zeal and diligence," prompting him to resign.[84] Ansted was replaced by James Nichol, whose own brief stint as assistant secretary lasted only until 1849, but even once Thomas Rupert Jones established greater editorial continuity, remaining in the post from 1849 to 1863, his actual control of the journal was much more circumscribed than it was for editors in the commercial press. In 1852 the Geological Society's president, William Hopkins, ensured that the council consented to another of Sedgwick's peevish interventions in his long-standing dispute with Murchison over stratigraphic nomenclature appearing in the *Quarterly Journal*. Once it was published, though, it became apparent that Sedgwick's "manner and tone," as Horner anxiously observed, went "far beyond what is allowable in scientific controversy" and his paper should "never have been printed in our journal."[85] The council, furious at Hopkins's manipulation of their publication processes, attempted to retract the already published paper—a "recall" that, according to James Secord, was "unique in the annals of Victorian science," as well as ultimately unrealizable.[86] Public expressions of controversy had to be expunged from the *Quarterly Journal*, even if this meant potentially wrenching its pages from readers' hands.

For the third geological periodical founded in the 1840s, Edward Charlesworth's *London Geological Journal*, even the most rancorous controversy was something to be cultivated rather than suppressed. Charlesworth had already honed a distinctively abrasive editorial style at the *Magazine of Natural History* from 1838 to 1840, and he also became embroiled in a furious dispute with Lyell over crag formations that compelled the Geological Society, in 1842, to peremptorily refuse his application for the post of curator of the society's museum. The council censured his "intemperate conduct," while Darwin observed to Lyell that "poor Charlesworth is of an unhappy discontented disposition."[87] This disposition imbued the tone and conduct of the *London Geological Journal*, which began in September 1846, though for Charlesworth the "unfettered expression of independent opinions" and righteous "opposition to *authority*" were necessary components of the "influence exercised on the progress of science by the publication of periodical records."[88] Like Moxon's *Geologist*, the *London Geological Journal* endeavored to create a new, broader community for the adjudication of geological issues, bringing otherwise undisclosed disputes and disagreements before its readers. Charlesworth warned that "our opposition will be directed" in particular against the "men-worship" that "holds sway in what is known as the 'geological world,'" and the *London Geological Journal* persistently impugned the expertise and even the integrity of such fashionable gentlemanly specialists as Buckland and Richard Owen.[89] The latter, for instance, was subjected to a very "public rectification of . . . an error" he had made regarding belemnite fossils during a discussion at the Geological Society, which had been left unchallenged in both the *Proceedings* and the new *Quarterly Journal*. This stinging rebuke to Owen's authority was heralded by Charlesworth as "due to the establishment of an independent Journal" and "proof . . . of the service likely to be rendered to Palæontology by . . . the 'London Geological Journal.'" For Charlesworth, even more than it was for Moxon, the public airing of controversy, no matter how acrimonious, enabled periodicals to ensure "honesty and good faith in the pursuit of science"—attributes that the Geological Society, with its conceited exclusivity and pandering to prestige, was sadly lacking.[90]

The *London Geological Journal*'s assertively metropolitan title was in fact a misnomer, as Charlesworth edited it from provincial Yorkshire, where, as he conceded, the need to have "consulted authorities and works of reference not accessible to me at York" sometimes resulted in the postponement of his contributions.[91] Although the Yorkshire Philosophical Society, which had appointed Charlesworth curator of its museum following his rejection by the Geological Society, assisted with the cost of the lithographic plates, and these were often contributed by unpaid artists from the York School of Design, the *London Geological Journal* was notably expensive, with each issue costing three shillings and six pence.[92] Charlesworth claimed to have "about 200 sub-scribers," but the original plan to "issue six numbers in the course of twelve months" was abandoned after only the first number, and instead subsequent parts appeared at irregular intervals before ceasing altogether after May 1847.[93] The initial wave of commercial geological periodicals that emerged in the mid-1840s had quickly proven financially unviable, with only the *Quarterly Journal*, albeit reliant on the resources of the Geological Society, surviving beyond the decade. They had nevertheless established a precedent that would soon be returned to with more lasting success.

A VALUABLE MEDIUM OF COMMUNICATION:
GEOLOGIST (1858-64) AND *PROCEEDINGS OF THE GEOLOGISTS' ASSOCIATION* (1859-)

The Geological Society's initial emphasis on observation and fact gathering had helped geology to become a popular pastime with men and women across the social spectrum, though their interests and aptitudes, especially in local geological matters, had never been catered to in any of the society's official periodicals. By the 1850s, the earth sciences were crucial to the endeavor to include the working classes in the improving agenda of science in the wake of the Chartist uprisings of the previous decade, and for the first time there was a potential readership sufficiently literate, interested, and affluent to sustain a commercial journal dedicated only to geology. Such a "popular organ of a Science which has of late years advanced with gigantic strides, and which is daily attracting an increasing share of attention from all classes of society" was begun in January 1858 by Mackie, who sought to remedy what he perceived as a profound "deficiency in the scientific periodical literature."[94] Mackie's new journal, which resuscitated Moxon's practitioner-focused title the *Geologist*, would, he proposed, be "not only a welcome monthly visitor, but a cherished friend," with its "pages . . . contain[ing] such things as a beginner can comprehend."[95] While the same explicitly popular mode of address had earlier been used to widen the audiences for the initial generation of commercial science journals, Mackie was the first to adopt it in a geological periodical.[96] As the *Morning Chronicle* observed approvingly, with the "establishment of a periodical like 'The Geologist' . . . Mr. Mackie is doing a great service to general science in so arduously popularizing that which, until very recently, was a seven-sealed book of mystery."[97] Significantly, Mackie, who, as noted earlier, was a fellow of the Geological Society, was adamant that the *Geologist* was "not attempting to compete with existing magazines" and had "no intention nor wish to anticipate the published records of the Geological Society," even if its monthly schedule afforded a greater "speed for novelties" than the "'Quarterly Journal' of that body."[98] Instead, the *Geologist* would "break a new fallow for ourselves" and generate an entirely new audience by "stir[ring] up a taste for geology" among a previously untapped constituency of participants. In fact, Mackie promised his putative competitors that "by creating more Geologists, we are likely to increase rather than diminish the number of their readers."[99] The *Geologist*, as Mackie presented it, would not impinge upon the interests of more specialist journals, and would merely supplement the Geological Society's *Quarterly Journal* with a becoming deference.

Tellingly, however, one of the readers who would "eagerly look for the GEOLOGIST as the first of each month comes round" and "heartily wish[ed] it success" was the still irascible Charlesworth, who, predictably, nevertheless berated Mackie for allowing "anonymous communications" that impeded forthright scientific interactions.[100] Charlesworth's public endorsement of the *Geologist* indicates that Mackie was never as submissive to the interests of the Geological Society as he initially suggested, and, in reality, his ostensibly popular journal often diverged markedly from the society's agenda, advancing an entirely different conception of how the earth sciences should be practiced. When the subscription rate of the society's *Quarterly Journal* was reduced in the summer of 1858 to twelve shillings, Mackie, notwithstanding his previous abjuration of any rivalry, evidently considered it as commercial competition (after all, the annual cost of the *Geologist* was also twelve shillings), and he protested: "We think the Council will err in reducing the price."[101] Mackie expressed a "feeling of pride" in acknowledging that the *Geologist* followed "two previous attempts by others" to establish independent geological periodicals, thus explicitly aligning himself with both Moxon and Charlesworth as part of what Susan Sheets-Pyenson has called the "important tradition of dissent from publications sanctioned by the Geological Society" (though, notably, Sheets-Pyenson does not herself include Mackie's *Geologist* in this tradition).[102]

At a time when Thomas Henry Huxley, as the Geological Society's president in the late 1860s, was announcing in the *Quarterly Journal* that the "influence of uniformitarian views has been enormous and . . . most beneficial and favourable to the progress of sound geology," and that the newer doctrine of "evolutionism . . . is assuredly present in the minds of most geologists," the *Geologist* was notably ambivalent about both of these central tenets of mid-Victorian geology.[103] In an editorial in 1863, for instance, Mackie renounced standard uniformitarian explanations and instead averred that "catastrophe fissured out the 'narrow stream'" that became the English Channel, which, he insisted, had been formed much more recently than most geologists assumed, even "bring[ing] down the fracture of the Channel to the age of man!"[104] Similarly, in a review of *On the Origin of Species* (1859), Mackie challenged Darwin's proposition that the "changes produced by natural selection *usually* require great ages of time," and was instead "disposed to consider that such changes might . . . be rapidly accomplished, and that in some cases they might even be brought about in . . . a single generation." According to Mackie, this was made feasible by referring the "modifications which species . . . have undergone to the direct will of God."[105] While the *Geologist* aimed to "popularize and to extend the noble science of geology without sacrificing, in any way, its proper dignity," it was not necessarily the same secular and uniformitarian version of geology that now increasingly held sway at the Geological Society.[106]

Although its principal objective was popularization, Mackie nevertheless wanted to render the *Geologist* "something more than a mere mirror of passing events," and it quickly became notorious for challenging established authorities and intervening in some highly contentious issues.[107] When in late 1862 John Evans identified the fishlike skull of the archaeopteryx, thereby threatening to undermine Owen's avian classification of the hybrid feathered fossil, Mackie ran a highly partisan analysis of the discovery which upheld a conclusion entirely at odds with that of Evans, instead suggesting that the "existence of the head . . . goes far to support the admirable inferences of Professor Owen."[108] At precisely this time, Owen, as noted earlier, was supporting Mackie's abortive campaign to become editor of the *Quarterly Journal*, and Mackie repaid the favor in the *Geologist*'s pages.[109] Aware that Mackie was deliberately distorting his discovery, Hugh Falconer urgently cautioned Evans: "Keep your own counsel, and . . . have nothing to say to the *Geologist*."[110]

As with Lyell's contemptuous response to Mackie's ambitions regarding the *Quarterly Journal*, the *Geologist* and its overbearing editor often elicited angry disdain and exasperation from the leading geological specialists.

Yet the *Geologist* also introduced an innovative feature that enabled Mackie to boast, justifiably for once, that "Sir Charles Lyell, and other eminent geologists . . . regard with interest the success of this publication as a valuable medium of communication, and as an important means for the accumulation of new facts."[111] The novelist Charles Kingsley, an eager reader of the *Geologist*'s early numbers, suggested to Mackie, "Why not establish in your magazine a regular geological 'Notes and Queries' department," advising that it would benefit "artizans desirous of self-instruction, and ladies of rank desirous of instructing their children . . . who have no time to work out geological problems for themselves." The erudite replies of a few "kind-hearted sages who will bear with our ignorance" would also, Kingsley shrewdly noted, add "plenty of possible subscribers to your magazine."[112] After all, the success of the literary and antiquarian weekly *Notes and Queries*, founded in 1849 with the subtitle *A Medium of Inter-Communication*, had already shown the commercial potential of such inclusive formats. Mackie soon gave notice that "we now commence a 'Notes and Queries' department," and was glad to acknowledge the significance of the "regular correspondents" who sent in both questions and replies, avowing that "such a combination of workers . . . more than anything else, will tend to keep up the standard of the GEOLOGIST as a work of merit, and to maintain its name in popular favour." He nonetheless conceded that "these, of course, must be voluntary helpers, as our resources would not permit such a staff of paid contributors," and this almost limitless source of free copy was an essential component of the *Geologist*'s business model.[113]

While the Geological Society had, from its foundation but particularly since the 1830s, maintained a hierarchical conception of geology that treated provincials and plebeians as "mere sources of information," in the *Geologist* Mackie instead rekindled an older tradition, exemplified in early-nineteenth-century correspondence networks, in which a broad range of participants from all levels of society contributed to the making of scientific knowledge.[114] As Mackie proposed, the "combination of local workers" whose letters appeared in the *Geologist* would "render good service to all learned societies by bringing together rapidly many coincident facts, which may have a natural bearing upon . . . those papers which practical geologists are preparing to produce in a more scientific form than would be ordinarily adapted for our pages."[115] Mackie's industrialization in the *Geologist*'s steam-printed pages of the erstwhile tradition of the reciprocal correspondence network would bring together, in a mutually beneficial "free intercourse among all classes of geologists," both "local workers" with a detailed knowledge of specific localities and what Mackie proudly termed "our scientific readers and supporters."[116] Even when members of the former group did, as in one instance, have "regular access, through a friend, to the Quarterly Journal of the Geological Society," they were unable to "obtain the slightest clue that would tend to solve the . . . geological phenomena" in which they were interested. In Mackie's "really popular, and, on that account, valuable periodical," on the other hand, they could instead pose their own targeted questions.[117] Metropolitan men of science, meanwhile, were able to pump provincials for local knowledge and specimens that, when collated together, could contribute to broader generalizations, thus bringing these very different communities of practitioners together in a reciprocal interchange of ideas and information.

In fact, just ten months after privately warning Evans in February 1863 to "have nothing to say to the *Geologist*," Falconer himself was willing to speak to the very same journal. In the issue for December 1863 he published a letter requesting those "possessing collections from Grays Thurrock" to "oblige the undersigned, by communicating to him whether they possess good specimens" that might assist with the "settlement of some undecided points connected with the 'mammalian fauna' of the pre-glacial deposit."[118] Never slow to parade such eminent endorsements, Mackie expressed himself "pleased to have these pages made use of, as has been done by Dr. Falconer[,] . . . as a medium of making known requirements of particular material for valuable labours in progress."[119] Notably, the *Geologist* was included in the Royal Society's *Catalogue of Scientific Papers* (1867–1902), which helped to define what in the late nineteenth century were considered scientifically credible periodicals (even if much of the miscellaneous content included by Mackie did not fit the format of a scientific paper).[120] For leading geological specialists such as Falconer, Mackie and his *Geologist* were simultaneously an exasperating populist nuisance and a highly effective means of scientific communication across all regions and social classes.

If the Geological Society had, half a century earlier, been a learned society that only subsequently spawned a periodical with the *Transactions*, Mackie's *Geologist* reversed this trajectory by forging a new society for the study of geology from a proposition initially made in its pages. With the "'Geological Society' . . . too far advanced in the strict course of scientific method" and still prioritizing the needs of gentlemanly specialists, Mackie eagerly took up the proposal, made by a reader in the "Notes and Queries" section of the *Geolo-*

gist's August 1858 number, for a "common means of communication among those who, while not devoting their lives to the pursuit, yet take an active interest in the facts and teachings of Geology."[121] Mackie's urgings led directly to the formation of the Geologists' Association in January 1859, to which, as he assured the *Geologist*'s readers, he felt the "paternal" affection of a "father to his son," as it was "through this journal the Association had birth."[122] The same familial regard was not always reciprocated by the amateur membership of the Geologists' Association, though Mackie served on its press committee and, from 1859 to 1860, printed the association's *Proceedings* on the same presses as the *Geologist*.[123]

By this time, however, the parent journal was struggling to break even, with Mackie pleading with "my readers . . . to extend still more the circulation of the magazine," and it seems likely that the annual sum of twenty-seven pounds and ten pence he received for printing the *Proceedings of the Geologists' Association* was actually subsidizing the *Geologist*.[124] Mackie himself had already been declared bankrupt in 1857, and his personal financial difficulties only intensified when in April 1858 he took over from the printer Frederick Reynolds as the *Geologist*'s proprietor. In November 1860 Mackie was "for a short period . . . a prisoner in [Abraham] Slowman's lock-up house" in Chancery Lane, and it seems that he continued to edit the *Geologist* from within the walls of this infamous debtors' prison.[125] By the end of the following year, Mackie had reluctantly raised the cover price from one shilling to one shilling and six pence, and made unspecified "arrangements" with the publisher Lovell Augustus Reeve to "ensure the permanent commercial stability of the journal," though within three years Reeve, who was now the proprietor, compelled Mackie to allow him to sell the *Geologist* for just twenty-five pounds.[126] As Mackie reflected dejectedly when the journal was "brought unexpectedly to a close" in June 1864:

> A new geological magazine is announced, and having received an intimation from my publishers that to continue "The Geologist" in rivalry with it would be attended with anxiety, and perhaps with loss, I have decided to retire from the field rather than take part in a contest that might prove injurious to both.[127]

This formidable newcomer, which purchased the *Geologist* and incorporated it into its full title, was the *Geological Magazine*, Longman's latest attempt, after its loss-making travails with the Geological Society's *Quarterly Journal*, to tap the market for commercial geology periodicals.

THE ADVANTAGES OF TERSENESS: *GEOLOGICAL MAGAZINE* (1864-), *GEOLOGICAL AND NATURAL HISTORY REPERTORY* (1865-67), *GEOLOGICAL RECORD* (1875-89), AND *ANNALS OF BRITISH GEOLOGY* (1891-94)

With its hostile takeover of the *Geologist*, the *Geological Magazine* could announce itself in July 1864 as the "only public Journal of Geology in Great Britain," pledging to maintain a scrupulously "independent character which will alone cause it to be regarded as an impartial tribune."[128] Just as Mackie had vowed in the opening number of the *Geologist*, the *Geological Magazine* likewise presented itself as merely a "supplement, as far as possible, [to] the authoritative and old-established Journal of the Geological Society," though this submissive assurance was even more specious than Mackie's had been.[129] The *Geological Magazine*'s initial editor, Thomas Rupert Jones, had previously edited the *Quarterly Journal* for more than a decade, and the two periodicals shared a publisher, with Longman continuing to publish the *Quarterly Journal* without assuming any financial risk. The *Geological Magazine*, however, adopted a decisively different approach, especially once Jones's youthful assistant, Henry Woodward, took over as editor in July 1865. In the following year, the Anglo-German publisher Nicholas Trübner acquired the journal from Longman, who, fearing another financial disaster, had "given [it] over for dead."[130] The *Geological Magazine*, Woodward later reflected, regularly published "'Rejected Addresses' that a too conservative element in former Councils of the Geological Society thought it right to discountenance," and he noted wryly that many of the papers deemed unsuitable for the *Quarterly Journal* were "now regarded as a classic."[131] In fact, Woodward located the *Geological Magazine* in the very same tradition as the "early efforts to provide an independent publication for geologists," invoking "Mr. Charles Moxon . . . Mr. Edward Charlesworth . . . [and] Mr. S. J. Mackie" as his direct precursors in this endeavor.[132] No less than either incarnation of the *Geologist* or the *London Geological Journal*, the *Geological Magazine* was a rebuke, albeit a more courteous one, to the authoritarian publishing practices of the Geological Society.

Thomas George Bonney, as president of the Geological Society, reflected sardonically that the "GEOLOGICAL MAGAZINE arose upon the foundation—I had almost said the ruins—of an earlier publication"—by which, of course, he meant Mackie's *Geologist*. Woodward himself was more considerate about the financially stricken precursor that was, as he acknowledged, "merged (by purchase) in the GEOLOGICAL MAGAZINE." Significantly, he even appeared to conflate the two journals when observing in 1886 that, during the "period embraced by the joint volumes of the 'Geologist' and the GEOLOGICAL MAGAZINE combined (*i.e.* 28 years), we have witnessed . . . some of the greatest advances which have ever been made in our science" (fig. 4.3).[133] The foremost of these advances for Woodward was the "Darwinian theory of evolution," and in its unequivocal support for Darwinism, as well as in many other ways, the *Geological Magazine* was very different from the more heterodox *Geologist*, whose erstwhile editor continued to mock the "Darwinian raptures in every transmutationist's heart."[134] Yet even when contributors were at pains to emphasize these differences, there remained significant continuities between Mackie's and Woodward's journals.

When in January 1865 the Geological Society's clerk George Roberts proposed to Woodward that the *Geological Magazine* should begin a "Geological Notes and Queries" section, he noted haughtily: "Although a desultory system of notes and enquiries did obtain during the existence of the respected predecessor of the GEOLOGICAL MAGAZINE, the scheme upon which it was cast differed somewhat from the arrangement I would suggest."[135] In fact, the "form of 'Notes and Queries'" that Woodward swiftly commenced was, in almost every respect, identical to that adopted in the *Geologist*, with both "Students and working Geologists" being promised that their "inquiries will be answered direct."[136] The *Geological Magazine*'s "*Notes and Queries*" also "form[ed] a medium of communication between distant subscribers," including "local observers," whose particular knowledge could be reciprocally tapped "with a few queries" by leading specialists, who might otherwise be "precluded, from a want of local knowledge, from taking . . . part" in a "controversy" such as that over the Old Red Sandstone.[137] Like Mackie, Woodward saw his periodical as a means of harnessing the powers of "our great Geological army" by affording the "scientific world an easily accessible medium of discussion."[138] Nor was the *Geological Magazine* immune from the "periods of financial depression" to which a "Class periodical—appealing only to Geologists" was inevitably prone, and Woodward recognized, no less than his insolvent predecessor, that inclusive formats such as "Notes and Queries" were an effective means of maintaining a remunerative circulation as well as creating new communities of geological practitioners.[139]

The evident continuities between the *Geologist* and the *Geological Magazine* seem to have been noticed by Mackie, who soon began a new weekly rival, the *Geological and Natural History Repertory*, in which he made ominous warnings against "any opposition" that might "be attempted to deprive me of the just reward of my labours and expenditure."[140] He then, more specifically,

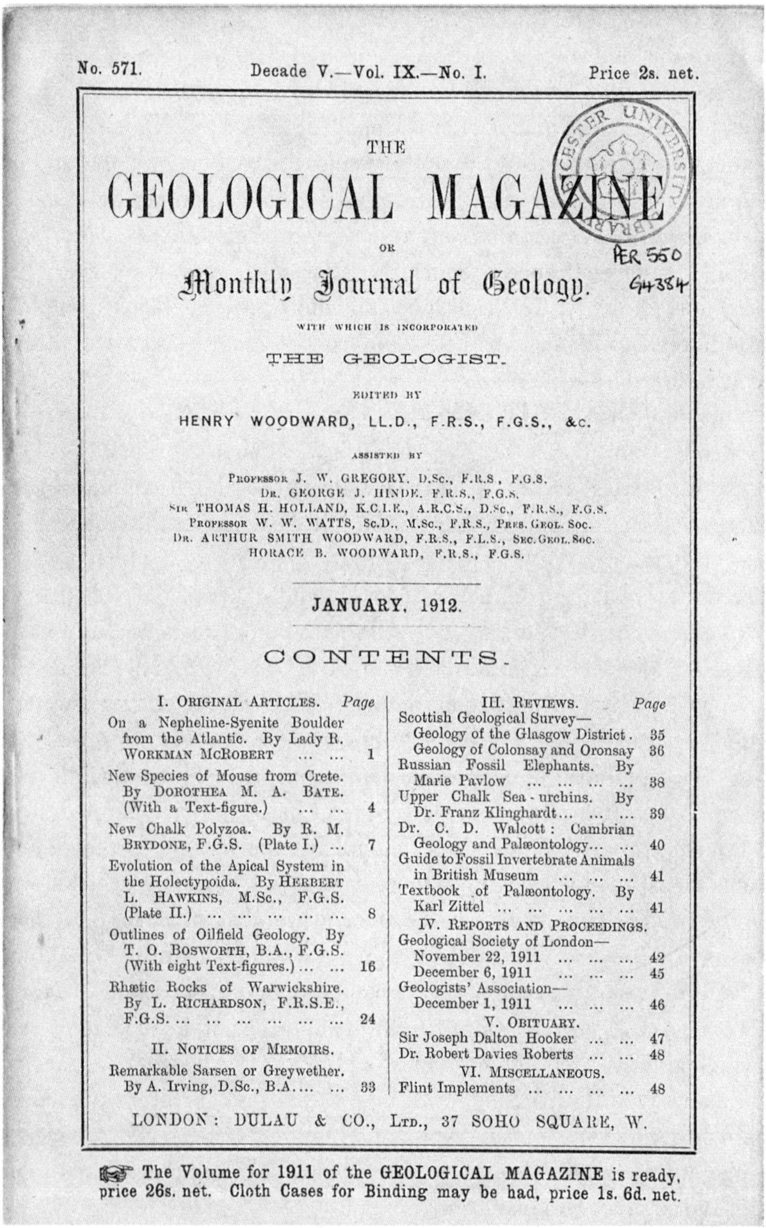

No. 571. Decade V.—Vol. IX.—No. I. Price 2s. net.

THE
GEOLOGICAL MAGAZINE

OR

Monthly Journal of Geology.

PER 550
G4384

WITH WHICH IS INCORPORATED

THE GEOLOGIST.

EDITED BY

HENRY WOODWARD, LL.D., F.R.S., F.G.S., &c.

ASSISTED BY

PROFESSOR J. W. GREGORY, D.Sc., F.R.S., F.G.S.
DR. GEORGE J. HINDE, F.R.S., F.G.S.
SIR THOMAS H. HOLLAND, K.C.I.E., A.R.C.S., D.Sc., F.R.S., F.G.S.
PROFESSOR W. W. WATTS, Sc.D., M.Sc., F.R.S., PRES. GEOL. SOC.
DR. ARTHUR SMITH WOODWARD, F.R.S., F.L.S., SEC. GEOL. SOC.
HORACE B. WOODWARD, F.R.S., F.G.S.

JANUARY, 1912.

CONTENTS.

LONDON : DULAU & CO., LTD., 37 SOHO SQUARE, W.

☞ The Volume for 1911 of the GEOLOGICAL MAGAZINE is ready, price 26s. net. Cloth Cases for Binding may be had, price 1s. 6d. net.

FIG. 4.3. Wrapper of the *Geological Magazine* for January 1912, which, even in the early twentieth century when the *Geological Magazine* was established as the most authoritative specialist journal in the earth sciences, still carried a prominent acknowledgement of its origins in Samuel Joseph Mackie's avowedly populist and heterodox *Geologist*. By permission of the University of Leicester.

berated the Geological Society after reading "in the 'Geological Magazine,' that the Wollaston fund was to be given to its editor."[141] Mackie proclaimed that the acceptance of the fund, the society's most prestigious award for younger geologists, would be an act of "venality," and he angrily demanded to know what "should make Mr. Woodward meritorious of this honour."[142] What likely provoked Mackie's ire was that his new *Geological and Natural History Repertory*, which once more included "under 'Notes and Queries,' jottings from many a widely different source, upon multifarious topics," again proved a financial calamity, folding in 1867 after twenty-nine issues.[143] The *Geological Magazine*, meanwhile, had survived its own early pecuniary difficulties, and, with a format not dissimilar to Mackie's, was becoming increasingly established.

Bonney accorded this success to the *Geological Magazine*'s unique market "position, which, on one hand, has common ground with our [i.e. the Geological Society's] Quarterly Journal; on another . . . with such a publication as 'Nature,'" which was begun by Norman Lockyer in 1869.[144] In fact, with the *Quarterly Journal* now increasingly publishing longer monographic articles (the cost of which eventually compelled Bonney to "impress . . . upon authors the advantages of terseness"), the *Geological Magazine*, whose emphasis on more concise contributions tallied with *Nature*'s new format of short notices, had much more in common with Lockyer's innovative and influential scientific weekly.[145] Like *Nature*, the *Geological Magazine* provided a new generation of university-based professional scientific workers with the most efficient and effective forum for communicating news of their latest findings.[146] With their authoritative processes of peer review, commercial science journals such as *Nature* were also increasingly supplanting learned societies as the central institutions where new forms of specialist expertise were adjudicated and guaranteed, and the *Geological Magazine*, whose contents were closely superintended by Woodward and his long-standing editorial assistants John Morris and Robert Etheridge, itself similarly soon took over from the Geological Society as the most authoritative site for evaluating geological research.[147]

In May 1883 Charles Hitchcock wrote to Woodward on behalf of the "younger American geologists" who were desirous of establishing an "Association for holding meetings with . . . a geol. magazine . . . which was to be the organ of the Association." This plan would come to fruition within three years, with the founding of the Geological Society of America in 1888, but what is striking is that Hitchcock, who was then the president of the Geological Department of the American Association for the Advancement of Science, did not approach the original Geological Society in London about a potential partnership, and instead asked Woodward, "Would it be practicable for us to combine with your mag," with the possibility of the "transference of part of the editorial work to the United States."[148] While Woodward seems to have rebuffed this request for a transatlantic alliance, Hitchcock's proposal shows that by the 1880s the *Geological Magazine* was, for a new generation of geologists, more significant for the development of the earth sciences than either the *Quarterly Journal* or even the Geological Society itself. In the following decade, Henry Neville Hutchinson, lamenting the "ponderous and almost unreadable . . . quarterly journals" issued by learned societies, urged that as an alternative the "Geological Society could undertake to grant an annual subsidy" to the "'Geological Magazine'; that most excellent periodical, which is read with so much pleasure and profit by geologists in all parts of the world."[149] By the early twentieth century, the *Geological Magazine*'s preeminence was such that it was a "common saying that almost every recruit to British geology since 1864 has begun authorship in its pages."[150] For the rising generation of professional geologists on both sides of the Atlantic, the *Geological Magazine* was emphatically the journal of choice.

One of the ways that Woodward attempted to serve the interests of this new generation was by printing "brief abstracts of geological papers," though, with the exponential growth in scientific papers published in the final decades of the nineteenth century, it "soon became obvious that the GEOLOGICAL MAGAZINE was not large enough to embody all the abstracts that were forthcoming."[151] This unprecedented bibliographic proliferation was an urgent problem across the sciences at the end of the nineteenth century, and in the winter of 1873, Woodward convened a committee of leading geologists to consider the matter in relation to the earth sciences. They agreed to establish and personally underwrite an entirely new form of geological periodical: an annual digest of abstracts. While the *Geological Record*, which began in late 1875, soon attracted a "number of Subscribers . . . large enough to cover the cost of publication," its editor, William Whitaker, struggled with the perpetual "increase in the number of papers, &c. noticed," and there were significant delays in processing the backlog for each year, with, for instance, the abstracts for 1880 not being published until 1889.[152] Whitaker jovially adopted a "pet phrase from our palæontological brethren . . . 'The Imperfections of the GEOLOGICAL RECORD'" to acknowledge the shortcomings of his own abstracting, which, as well as "omitted papers" that were not recorded at all, also included recording the "same thing two or three times over."[153] By 1891 the *Geological Record* had been replaced by the *Annals of British Geology*, edited by John Frederick Blake, which attempted to make the task more manageable by excluding foreign publications. This parochial "limitation of the Record to our own country," though, did not reflect the increasing internationalization of the earth sciences, and Blake's *Annals of British Geology* lasted for only three years despite the pressing need for an effective abstracting service.[154]

CONCLUSION

The *Geological Magazine*, by contrast, had always endeavored to be a "medium of intercommunication for English and Foreign geologists," and it was this cosmopolitan outlook that evidently prompted Hitchcock's proposal to Woodward to further "enlarge the sphere of your Geol. Mag." with a "fuller discussion of American geolog. topics."[155] This intrinsic internationalism, even without Hitchcock's proposed contribution, ensured that the *Geological Magazine* was more equipped than any other geology periodical to serve the needs of its specialist readers during what Melinda Baldwin has called an "era of increasing international ties between scientific workers" before the outbreak of the First World War.[156] Under Woodward's long-standing editorship, which continued until 1918, the *Geological Magazine* increasingly shifted from the focus on nationally based communities, whether of gentlemanly specialists or more plebeian practitioners, fostered by previous geology journals, and instead brought together an international community of professional academics.

Such ascendancy within the emergent community of professional geologists was not, of course, accomplished without significant alterations to Woodward's original approach, and the initially enthusiastic Hutchinson, a prominent popularizer of the earth sciences, later cautioned that "with regard to the Geological Magazine I think it would have a far wider circulation if it were not written in such a pedantic style. It is far beyond the reach of the ordinary geologist."[157] With the consolidation of the *Geological Magazine* as the most authoritative and prestigious specialist periodical in the field, however, the tradition of independent commercial journals that had begun in the 1840s with Moxon's *Geologist* and Charlesworth's *London Geological Journal*, and then recommenced in the following decade with Mackie's *Geologist*, finally deposed the *Transactions*, *Proceedings* and *Quarterly Journal* of the Geological Society, the last of which remained as what Herries Davies has called a "stately edifice" whose continued existence was made possible by the subscriptions of fellows with no aspirations to publish in its pages.[158] After more than half a century of often acrimonious disputes, scientific authority in geology was now vested in the public market for serialized print—albeit a small, specialist one—rather than in a privately subsidized institution.

NOTES

1. C. Lyell to T. H. Huxley, 9 September 1862, Thomas Henry Huxley Papers 6.72, Imperial College of Science, Technology, and Medicine Archives, London. Quoted by permission of the Archives Imperial College of Science Technology and Medicine.

2. On the traditional status of editors of society publications, see Steven Shapin, "O Henry," *Isis* 78 (1987): 417–24 (418).

3. On Mackie's election as a fellow of the Geological Society, see Eric F. Freeman, "The Founders of the Geologists' Association II: The Mysterious Mr. Mackie," *Proceedings of the Geologists' Association* 107 (1996): 85–96 (87).

4. "Notes, Queries, and Replies," *Medical Times and Gazette*, 22 November 1862, p. 564.

5. "Preface," *Geologist* 3 (1860): [iii–iv] ([iii]).

6. W. H. Brock, "The Development of Commercial Science Journals in Victorian Britain," in *Development of Science Publishing in Europe*, ed. A. J. Meadows (Amsterdam: Elsevier, 1980), 95–122 (96).

7. J. W. Judd, "The Anniversary Address of the President," *Quarterly Journal of the Geological Society* [hereafter *QJGS*] 43 (1887): 38–82 (45); "Obituary. Henry Michael Jenkins, F.G.S.," *Geological Magazine* [hereafter *GM*] 4 n.s. (1887): 95–96 (96).

8. As Lyell may have intuited, Susan Mackie was a highly resourceful woman who took out several patents on her mechanical inventions. See Freeman, "Founders of the Geologists' Association," 92.

9. *Nicholson's Journal* was the shorthand name regularly given to the *Journal of Natural Philosophy, Chemistry and the Arts*, founded by William Nicholson in 1797.

10. See David E. Allen, "The Struggle for Specialist Journals: Natural History in the British Periodicals Market in the First Half of the Nineteenth Century," *Archives of Natural History* 23 (1996): 107–23.

11. "Notice to Subscribers, Contributors, & Advertisers," *Geological and Natural History Repository* [hereafter *GNHR*] 1 (1865–67): [iii].

12. "Preface," *Geologist* 5 (1862): iii–iv (iii).

13. Martin J. S. Rudwick, *The Great Devonian Controversy: The Shaping of Scientific Knowledge among Gentlemanly Specialists* (Chicago: University of Chicago Press, 1985), 17–18 and passim.

14. See ibid., 418–28. Rudwick identifies *"accomplished* geologists," whose primary expertise was in other sciences, *"amateur"* or *"local* geologists," and finally the "general public" as the graduated zones below the "gentlemanly specialists."

15. See Martin J. S. Rudwick, *Bursting the Limits of Time: The Reconstruction of Geohistory in the Age of Revolution* (Chicago: University of Chicago Press, 2005).

16. See Simon J. Knell, *The Culture of English Geology, 1815–1851: A Science Revealed through Its Collecting* (Aldershot, UK: Ashgate, 2000), 7 and 320–23. On science and orality more generally, see James A. Secord, "How Scientific Conversation Became Shop Talk," in *Science in the Marketplace: Nineteenth-Century Sites and Experiences*, ed. Bernard Lightman and Aileen Fyfe (Chicago: University of Chicago Press, 2007), 23–59.

17. Roy Porter, "Gentlemen and Geology: The Emergence of a Scientific Career, 1660–1920," *Historical Journal* 21 (1978): 809–36 (815).

18. See James A. Secord, "The Geohistorical Revolution," *Metascience* 16 (2007): 375–86 (379).

19. Frederick H. Burkhardt et al., ed., *The Correspondence of Charles Darwin*, 26 vols. (Cambridge: Cambridge University Press, 1985–), 3:338.

20. Horace B. Woodward, *The History of the Geological Society of London* (London: Longman, Green, 1907), 157.

21. William Swainson, *A Preliminary Discourse on the Study of Natural History* (London: Longman, Rees, 1834), 304, 313, and 429.

22. See Woodward, *History of the Geological Society*, 43–44; and Gordon L. Herries Davies, *Whatever Is under the Earth: The Geological Society of London 1807 to 2007* (London: Geological Society, 2007), 36–37.

23. Rudwick, *Great Devonian Controversy*, 20–21.

24. Quoted in Herries Davies, *Whatever Is under the Earth*, 37.

25. See ibid., 39.

26. "Preface," *Transactions of the Geological Society* [hereafter *TGS*] 1 (1811): v–ix (ix).

27. See Ralph O'Connor, "Facts and Fancies: The Geological Society of London and the Wider Public, 1807–1837," in *The Making of the Geological Society of London*, ed. C. L. E. Lewis and S. J. Knell (London: Geological Society, 2009), 331–40 (331–32).

28. [William Henry Fitton], "Transactions of the Geological Society," *Edinburgh Review* 28 (1817): 174–92 (175).

29. O'Connor, "Facts and Fancies," 332.

30. [Fitton], "Transactions," 176.

31. Rachel Laudan, "Ideas and Organizations in British Geology: A Case Study in Institutional History," *Isis* 68 (1977): 527–38 (531 and 533).

32. John MacCulloch, "On the Parallel Roads of Glen Roy," *TGS* 4 (1817): 314–92 (316 and 389).

33. See Woodward, *History of the Geological Society*, 63–64.

34. On the relations between the Taylor brothers, see W. H. Brock and A. J. Meadows, *The Lamp of Learning: Two Centuries of Publishing at Taylor and Francis*, 2nd edition (London: Taylor and Francis, 1998), 45.

35. "Notice," *TGS* 1 n.s. (1822): 5.

36. See Herries Davies, *Whatever Is under the Earth*, 37 and 39.

37. Ralph O'Connor, *The Earth on Show: Fossils and the Poetics of Popular Science, 1802–1856* (Chicago: University of Chicago Press, 2007), 263–323 and passim.

38. John Davy, "On the Geology and Mineralogy of Ceylon," *TGS* 5 (1821): 311–27 (313–14).

39. William Buckland, "On the Discovery of a New Species of Pterodactyle in the Lias at Lyme Regis," *TGS* 3 n.s. (1835): 217–22 (219 and 218).

40. Adelene Buckland, *Novel Science: Fiction and the Invention of Nineteenth-Century Geology* (Chicago: University of Chicago Press, 2013), 88.

41. Martin J. S. Rudwick, "The Emergence of a Visual Language for Geological Science 1760–1840," *History of Science* 14 (1976): 149–95 (181).

42. See Herries Davies, *Whatever Is under the Earth*, 39 and 88.

43. See Alex Csiszar, *The Scientific Journal: Authorship and the Politics of Knowledge in the Nineteenth Century* (Chicago: University of Chicago Press, 2018), 159–97.

44. Leonard Horner, "Anniversary Address of the President," *QJGS* 2 (1846): 145–221 (149).

45. See Rudwick, *Great Devonian Controversy*, 26–27.

46. Horner, "Anniversary Address," 149; "Proceedings of the Geological Society of London," *Proceedings of the Geological Society* 1 (1834): 1.

47. See chapter 3 in this volume.

48. William Buckland, *Vindiciæ Geologicæ* (Oxford: Oxford University Press, 1820), 4.

49. Archibald Geikie, *Life of Sir Roderick I. Murchison*, 2 vols. (London: John Murray, 1875), 2:56.

50. Herries Davies, *Whatever Is under the Earth*, 93.

51. Swainson, *Preliminary Discourse*, 314.

52. Quoted in Rudwick, *Great Devonian Controversy*, 25.

53. On the nature of the discussions, see John C. Thackray, *To See the Fellows Fight: Eyewitness Accounts of the Meetings of the Geological Society of London and Its Club, 1822–1868* (Stanford in the Vale, UK: British Society for the History of Science, 1999).

54. On these challenges, see O'Connor, *Earth on Show*, 201–9.

55. Geikie, *Life of Sir Roderick I. Murchison*, 1:195–96.

56. See James A. Secord, *Controversy in Victorian Geology: The Cambrian-Silurian Dispute* (Princeton, NJ: Princeton University Press, 1986), 167–68.

57. "Geological Society," *Athenæum*, no. 160 (1830): 730.

58. "Our Weekly Gossip," *Athenæum*, no. 578 (1838): 841.

59. "The Geologist," *Geologist* 1 (1842): 1–2 (1).

60. Quoted in John C. Thackray, "Charles Moxon and *The Geologist*: A Geological Iconoclast and His Organ," *Geology Today* 3 (1987): 69–70 (69).

61. "Geologist," 1–2.

62. Ibid., 2.

63. "Monthly Notice," *Geologist* 1 (1842): 129–32 (130 and 131).

64. On the dispute between Murchison and Sedgwick, see Secord, *Controversy in Victorian Geology*.

65. "Monthly Notice," *Geologist* 1 (1842): 161–64 (162).

66. "Monthly Notice," *Geologist* 1 (1842): 195–97 (195 and 196). Perhaps because of their irregularity, Moxon seems not to have considered the Geological Society's *Transactions* or *Proceedings* as proper periodicals.

67. "Monthly Notice," *Geologist* 1 (1842): 227–29 (228).

68. Ibid., 227.

69. Ibid., 228.

70. Ibid., 228–29.

71. "Monthly Notice," 162 and 163.

72. "Proceedings of Societies," *Geologist* 1 (1842): 208–20 (208). On provincial geological societies, see Simon Naylor, *Regionalizing Science: Placing Knowledges in Victorian England* (London: Routledge, 2010), 25–28.

73. "Sketch of the Progress of Geology," *Geologist* 2 (1843): v–xii (vi).

74. See Herries Davies, *Whatever Is under the Earth*, 93.

75. Ibid., 93.

76. "Introductory Notice," *QJGS* 1 (1845): 1–3 (1).

77. "Works in the Press," *Publishers' Circular* 8 (1845): 10–13 (12); Horner, "Anniversary Address," 149.

78. Horner, "Anniversary Address," 149.

79. See Herries Davies, *Whatever Is under the Earth*, 95.

80. "Introductory Notice," 2.

81. Horner, "Anniversary Address," 150.

82. "Introductory Notice," 2–3.

83. Herries Davies, *Whatever Is under the Earth*, 96. See also Martin J. S. Rudwick, "Historical Origins of the Geological Society's *Journal*," in *Milestones in Geology*, ed. M. J. Le Bas (London: Geological Society, 1995), 5–8 (7).

84. Quoted in John C. Thackray, "Ansted, David Thomas (1814–1880)," *Oxford Dictionary of National Biography*, ed. H. C. G. Matthew and Brian Harrison, 60 vols. (Oxford, UK: Oxford University Press, 2004), 2:268–69 (268).

85. Quoted in Secord, *Controversy in Victorian Geology*, 232.

86. Ibid., 233.

87. Quoted in Barbara J. Pyrah, *The History of the Yorkshire Museum* (York, UK: William Sessions, 1988), 55 and 53.

88. Edward Charlesworth, "On the Occurrence of a Species of Mosasaurus in the Chalk of England," *London Geological Journal* [hereafter *LGJ*] 1 (1846–47): 23–32 (28); "The London Geological Journal," *LGJ* 1 (1846–47): 79–85 (79).

89. "London Geological Journal," 85.

90. "The London Geological Journal," *LGJ* 1 (1846–47): 122–29 (127 and 128).

91. "Short Communications, Miscellaneous Intelligence, &c.," *LGJ* 1 (1846–47): 130–32 (132).

92. See Pyrah, *History of the Yorkshire Museum*, 60.

93. "London Geological Journal," 81; "To Readers and Correspondents," *LGJ* 1 (1846–47): [iii].

94. "The Geologist," *Geologist* 1 (1858): 1–5 (1).

95. "Preface," *Geologist* 1 (1858): i; "Geologist," 2.

96. See Jonathan R. Topham, "The Scientific, the Literary and the Popular: Commerce and the Reimaging of the Scientific Journal in Britain, 1813–1825," *Notes and Records* 70 (2016): 305–24 (310–11).

97. "Literature," *Morning Chronicle*, 26 January 1859, p. 6.

98. "Geologist," 2; "Notes and Queries," *Geologist* 1 (1858): 155–62 (155 and 156).

99. "Geologist," 3 and 2.

100. "Notes and Queries," *Geologist* 2 (1859): 37–43 (38).

101. "Notes and Queries," *Geologist* 1 (1858): 400–401 (401).

102. "Preface, *Geologist* 3 (1860): i–ii (i); Susan Sheets-Pyenson, "Geological Communication in the Nineteenth Century: The Ellen S. Woodward Autograph Collection at McGill University," *Bulletin of the British Museum (Natural History)* 10 (1982): 179–226 (184).

103. T. H. Huxley, "Anniversary Address of the President," *QJGS* 25 (1869): xxviii–liii (xli and xlvii).

104. "Thoughts on Dover Cliffs," *Geologist* 6 (1863): 281–93 (292).

105. "Reviews," *Geologist* 3 (1860): 464–72 (471).

106. "Preface" (1858), i.

107. Ibid.

108. "The Aeronauts of the Solenhofen Age," *Geologist* 6 (1863): 1–8 (7).

109. On Mackie's involvement in the dispute between Evans and Owen, see Gowan Dawson, *Show Me the Bone: Reconstructing Prehistoric Monsters in Nineteenth-Century Britain and America* (Chicago: University of Chicago Press, 2016), 310–13.

110. Quoted in Joan Evans, *Time and Chance: The Story of Arthur Evans and His Forebears* (London: Longman, Green, 1943), 116.

111. "Preface," *Geologist* 4 (1861): i–ii (ii). Mackie does not specify in what form he "received flattering encouragement" from Lyell, who did not contribute to the *Geologist*, though correspondents did regularly offer information that, as with examples of post-Pliocene flint implements, they considered "supplies the desideratum alluded to by Sir Charles Lyell" in his published writings. "Correspondence," *Geologist* 7 (1864): 24.

112. "Letter from the Rev. C. Kingsley," *Geologist* 1 (1858): 75–77.

113. Ibid., 77; "Notes and Queries" (1858), 155.

114. James A. Secord, introduction to Charles Lyell, *Principles of Geology* (London: Penguin, 1997), ix–xliii (xv). On the earlier tradition of correspondence networks, see Anne Secord, "Corresponding Interests: Artisans and Gentlemen in Nineteenth-Century Natural History," *British Journal for the History of Science* 27 (1994): 383–408.

115. "Notes and Queries" (1858), 156.

116. "Preface," *Geologist* 7 (1864): iii–iv (iii); "Notes and Queries" (1858), 156. On the industrialization of correspondence, see chapter 6 in this volume.

117. "Notes and Queries," *Geologist* 2 (1859): 91–93 (91).

118. "Correspondence," *Geologist* 6 (1863): 455–59 (457).

119. "Preface," *Geologist* 6 (1863): iii–v (v).

120. See Csiszar, *Scientific Journal*, 223–39.

121. "Notes and Queries" (1859), 39.

122. "Reviews," *Geologist* 5 (1862): 318–20 (319).

123. On Mackie's complicated relationship with the Geologists' Association, see Freeman, "Founders of the Geologists' Association," 85–86.

124. "Preface" (1860), [iii].

125. "Court for Relief of Insolvent Debtors," *London Gazette*, 27 November 1860, p. 4763.

126. "Preface" (1861), iv.

127. "Preface" (1864), iii.

128. "The Past and Present Aspects of Geology," *GM* 1 (1864): 1–4 (4).

129. "Geological Progress," *GM* 3 n.s. (1886): 1–3 (1).

130. "The 'Coming of Age' of the *Geological Magazine*," *GM* 3 n.s. (1886): 45–48 (48).

131. "A Retrospect of Geology in the Last Forty Years," *GM* 1 n.s. (1904): 1–6 (1).

132. "The Completion of Fifty Years of the *Geological Magazine*," *GM* 1 n.s. (1914): 241–44 (244).

133. "Coming of Age," 45 and 47.

134. Ibid., 47; "The Bradfield Dam," *Sheffield Daily Telegraph*, 1 April 1864, p. 3.

135. "Geological 'Notes and Queries,'" *GM* 2 (1865): 89–91 (91 and 89).

136. "To Our Geological Friends and the Scientific Public," *GM* 4 (1867): 1–2 (1).

137. "Correspondence," *GM* 2 (1865): 135–39 (137); "Correspondence," *GM* 4 (1867): 333–36 (333). It was the Scottish geologist James Powrie who desired local information to help make his case in the controversy over the ichthyological fossils of the Old Red Sandstone.

138. "Geological Progress," *GM* 2 (1865): 337–39 (337); "Geological Progress," *GM* 3 (1866): 1–3 (1).

139. "Coming of Age," 48.

140. "The Editor's Introduction of His Work to the World," *GNHR* 1 (1865–67): 5–8 (5–6).

141. "Proceedings of Societies," *GNHR* 1 (1865–67): 225–34 (233).

142. Ibid., 233.

143. "Editor's Introduction," 7.

144. "Coming of Age," 45.

145. T. G. Bonney, "Anniversary Address of the President," *QJGS* 42 (1886): 8–115 (49).

146. On *Nature*'s appeal to a new generation of scientific workers, see Melinda Baldwin, *Making "Nature": The History of a Scientific Journal* (Chicago: University of Chicago Press, 2015), 49–55.

147. On commercial journals as adjudicators of scientific expertise, see Csiszar, *Scientific Journal*, 272–74 and passim.

148. C. H. Hitchcock to H. Woodward, 21 May 1883, Ellen S. Woodward Collection, QL26 W66, vol. 4, McGill University Library and Archives, Montreal, Canada. Quoted by permission of Rare Books and Special Collections, McGill University Library.

149. H. N. Hutchinson, "The Work of a Scientific Society," *Science-Gossip* 2 n.s. (1895): 90–93 (90 and 91).

150. "Obituary. Henry Woodward," *GM* 58 n.s. (1921): 481–84 (482).

151. "Retrospect of Geology," 3.

152. "Preface," *Geological Record* [hereafter *GR*] 1 (1875): iii–iv (iii); "Preface," *GR* 2 (1877): iii–iv (iii).

153. "Preface," *GR* 4 (1880): iii.

154. "Preface," *Annals of British Geology* 1 (1891): v–vii (v).

155. "To Our Geological Friends," 1; Hitchcock to Woodward, 21 May 1883, Woodward Collection, QL26 W66, vol. 4.

156. Baldwin, *Making "Nature,"* 119.

157. H. N. Hutchinson to A. Smith Woodward, 24 May 1917, DF PAL/100/63/201, Palaeontology, Departmental Correspondence, Natural History Museum Archives, London. By permission of the Trustees of the Natural History Museum.

158. Herries Davies, *Whatever Is under the Earth*, 148. Rudwick comments that the *Quarterly Journal*'s "cost was absorbed into the Fellows' annual fee, so that its purchase became in effect a compulsory condition of membership." "Historical Origins," 7–8.

CHAPTER 5

Natural History Periodicals and Changing Conceptions of the Naturalist Community, 1828–65

Geoffrey Belknap

For John Claudius Loudon (1783–1843), natural history periodicals in the 1830s could "be divided into two classes: those which are supported by the voluntary contribution of their readers; and those which are forced into circulation by the hired communications of eminent writers." For Loudon, it was clear that "the first class alone answers the legitimate object of a Journal of Science."[1] Loudon's vision for a natural history periodical—which he expressed by establishing and editing the *Magazine of Natural History* (1828–40)—was an inclusive one where "voluntary contributors" and "readers" were given the opportunity to communicate, debate, record, and develop the scope of knowledge about the natural world.

Thirty years later, a different vision was developed for another self-designated natural history periodical, where the authority to produce natural history knowledge was to be placed in the hands of the "eminent writers" and editors of the periodical. As this chapter will show, over the nineteenth century notions of expertise became increasingly tied to the performance of scholarly authorship, expressed through the contribution of "original articles" to scientific periodicals.[2] This change is typified by the example of the *Natural History Review* (1856–65), a periodical first edited by a group of Irish naturalists, which became from 1861 the mouthpiece for "original articles" written by pro-Darwinian biologists, anatomists, physicians, and naturalists. Mid-nineteenth-century natural history periodicals offered multiple visions of who could produce, participate in the making of, and communicate natural history knowledge. In using the term "visions" here, this chapter is borrowing from James Secord's analysis of the interaction between readers and scientific texts as creating "visions of science." For Secord, nineteenth-century authors like John Herschel, Mary Somerville, and Thomas Carlyle were publishing experimental treatises which "projected a vision of the future" predicated on the "growing . . . hopes of the power of knowledge."[3] For these authors, scientific books were experimental objects, which aimed to change the role of scientific discourse and practice for a broad audience. The print revolution that occurred in the first half of the nineteenth century, in Secord's analysis, allowed for printers, publishers, and authors to shape new concepts of the scope and power of science. For Secord, however, the book was the primary site for the creation of these visions of science. This chapter asks what the role was of the subject-specific periodical in shaping new visions of science for a broad range of audiences.

Loudon's vision for a natural history periodical encouraged experts and nonexperts alike to learn, debate, and participate in the practice of natural history.[4] As the ownership and titles of periodicals changed, so did the editors, readers, and contributors—and with these new communities came new visions for a natural history periodical. The argument of this chapter is that natural history periodicals in the period between 1828 and 1865 evolved over time in response to commercial pressures, changes in the study of natural history, and corresponding changes in the communities of readers. Each title that claimed to encompass natural history as its subject offered a new space for publishers, editors, contributors, and readers to create and debate the scope and practice of natural history. In other words, the competition between natural history journals was also about competing notions of community.

The success of these new natural history publications as scientific and commercial objects depended on a number of factors: the credibility of communities of contributors, editors, proprietors, and readers that made up the audience of the periodical; the format of the periodical, which was defined by the editors and proprietors in negotiation with the audience, but which ultimately created the access point through which the content of the journal was read; and the commercial product of the periodical, which was a result of how the communities and the format performed within the periodical marketplace.[5] As we will see, establishing a lasting subject-specialist periodical required offering more than simply a space for the performance of subject expertise. Rather, success depended on a fine balance between attracting a broad community base, offering a defined structure and shape to the periodical itself, and demanding close consideration of the financial implications of publishing, printing, and mailing each issue.

The chapter will offer case studies of three periodicals that included "natural history" within their titles, to evaluate the ways in which they reflected the relationship between communities, format, and product. The *Magazine of Natural History*, the *Annals and Magazine of Natural History* (1841–1923), and the *Natural History Review* (1854–65) were edited, read, and contributed to by some of the most influential naturalists of the period. The *Magazine of Natural History* was the first specialist natural history periodical in Britain, while the *Annals and Magazine of Natural History* was one of the longest-running natural history periodicals of the nineteenth century, backed by Richard Taylor (1781–1858), the founder of the publishing house Taylor & Francis. The *Natural History Review* became, after 1861, the publishing venue for the increasingly specialized biological sciences, coedited by Thomas Henry Huxley (1825–95).[6] Focusing on these three self-defined natural history periodicals, edited and produced by the most influential naturalists of the period, allows for an understanding of three versions of a natural history periodical that were made to suit a broad, loosely defined, and changing field by the expert practitioners working within it.

While this chapter will focus on these three journals specifically, the marketplace of periodicals relating to natural history in the mid-nineteenth century was becoming rich and diverse. Periodicals such as John Lindley's *Gardeners' Chronicle* (1841–), Joseph Paxton's *Paxton's Magazine of Botany and Register of Flowering Plants* (1834–49), or William Jackson Hooker's series of periodicals variously titled *Botanical Gazette* (1830–33), *Journal of Botany* (1834–42), *London Journal of Botany* (1842–48) and *Hooker's Journal of Botany and Kew Garden Miscellany* (1849–57) were operating in the same period as the periodicals that are the subject of the case studies in this chapter. An investigation of any of these journals, with their differing "gardening," "horticulture," or "botanical" formats, would uncover multiple visions of natural history periodicals in this period, each attempting to construct its own community of readers.

From the late 1820s, differing visions for a natural history periodical were put forward as editors and publishers adapted to changing economic and social conditions as well as the development of natural history as an area of study. To this end, this chapter will follow the voices of the various editors of the three periodicals (and their previous iterations) by focusing on what they said about the content and perspectives taken within their periodicals, and how they defined a community of readers for their journals.

By using the term "natural history," the various editors of the journals dealt with in the three case studies were seeking to attract a broad audience interested in a range of scientific subjects. Similar to the increasingly specialized periodical market for subjects such as geology, physics, or astronomy, discussed elsewhere in this volume, natural history periodicals were formed around a set of practices and subjects.[7] Natural history as a subject of study in the nineteenth century encompassed a broad range of approaches—from field-based observational sciences like botany, zoology, archaeology, and geology to sciences that required observational instruments, such as microscopy, and text-based sciences like philology and antiquarianism.[8] To capture the audiences who were interested in these various subjects and practices, the natural history periodicals that emerged in the second quarter of the nineteenth century were therefore equally diffuse and varied.[9]

In order to focus on the development of natural history periodicals, the analysis will be split into three sections. The first section follows the establishment of the first specialist natural history periodical, the *Magazine of Natural History*. The second section explores the emergence of a commercially successful vision and format for a natural history periodical through the multiple establishments, failures, and mergers of titles that would ultimately emerge as the *Annals and Magazine of Natural History*. And the third section follows a single title—the *Natural History Review*—and traces the changing structure and content of this periodical between 1854 and 1865. In each section, this chapter will trace the development of one of three visions for the role of a natural history periodical—from Loudon's vision of the periodical as a communal self-improvement space to Huxley's exclusive and expert vision which aimed to separate "biology" from "natural history." In between these two poles lay the commercially successful *Annals and Magazine of Natural History*, which struck a balance between broad community participation and the performance of expert knowledge.

1828-40: ESTABLISHING A POPULAR PERIODICAL, THE *MAGAZINE OF NATURAL HISTORY*

For naturalists in the 1820s, according to David E. Allen, "there was a substantial unsatisfied demand for inexpensive works of identification."[10] This desire for accessible sites for the communication of observational information, combined with improvements in printing technology, created a spurt of short-lived natural history publications. While these new publications offered opportunities for a wider range of naturalists to publish their findings, they also ran the risk of encompassing a claim to "bogus omniscience and reckless versatility," as Allen put it, across all forms of natural history knowledge.[11] The risk of trying to skirt the boundaries between expert audiences and general appeal was that publications often ended up appealing to no particular demographic.

By the 1830s, not only natural history but also the terms of scientific discourse and the scope of participation in science were changing.[12] As Secord demonstrates in his analysis of book titles published in this same period, in the 1830s multiple visions for science were being forwarded through the publication of single-author scientific texts. The visions forwarded in these books would influence the scope of scientific practice in the nineteenth century, as well as affecting social and political reform.[13] The years just preceding the establishment of the *Magazine of Natural History*, for example, saw the beginning of the Society for the Diffusion of Useful Knowledge, which aimed to publish affordable scientific works to educate and enlighten the working classes.[14]

In a way similar to that of the authors described by Secord, Loudon was experimenting with the format of the periodical as a tool to forward his own vision of natural history. Loudon's vision was particularly influenced by the political philosophy of Jeremy Bentham, who advocated the value of education as a social good.[15] In a context where science was undergoing reform and new audiences were being developed, Loudon's periodical entered the marketplace not just as a new form of specialized periodical, but as one that attempted to capitalize on these changes by offering a space for experienced naturalists and interested members of the public to interact and develop knowledge and expertise about the natural world.

The *Magazine of Natural History* was first established in 1828 as a monthly and later bimonthly periodical priced initially at three shillings and six pence before being reduced to two shillings.[16] Addressing his readers for the first time, Loudon pointed out that "the First Volume . . . is submitted to its readers, as a fair specimen of what that Periodical is intended to be."[17] The reading community was essential for Loudon, for he was striving to establish a new form of periodical: one focused on the broad practice and subject of natural history, which would attract both expert naturalists and a broader general reading public.

Loudon's vision for the *Magazine of Natural History* posited that a natural history periodical should do a number of things. First, it should be a conduit for a "more general diffusion of a knowledge of Animals, Vegetables, and Minerals, technically and physiologically." Second, it should act as a "record of discoveries in the branches of knowledge; and of the actual state and progress of the taste for Natural History, in different parts of the British empire, and throughout the world." And third, it should encapsulate a "summary of the progress of discovery in natural science during the past year."[18] But more than anything, it was the readers who took center stage. The making of the *Natural History*

Magazine, Loudon would later state, was a community activity: "Our correspondents may regard themselves as cooperating in a Magazine of their own, for the improvement of one another, as well as for the benefit of the public."[19]

The production of a natural history periodical, at least for Loudon in the late 1820s and '30s, was about creating a space for naturalists both to exchange knowledge and expertise, and to diffuse, record, and archive natural knowledge. Moreover, this exchange was critical to—in Loudon's words, cited above—"'the improvement of one another, as well as for the benefit of the public." For Loudon, the creation of a specialized periodical did not mean excluding a large number of potential readers, but rather was an inclusive and socially beneficial endeavor. While the content of Loudon's periodical was focused on a defined set of subjects and practices, the reading community was broad and inclusive.

Loudon's goal of including both experienced and uninitiated students of natural history was embedded in the language and structure of the periodical. From 1831, he carved out a space within the *Magazine of Natural History* for "Queries and Answers," which, he argued, would allow for more content and better communication between experienced and inexperienced naturalists:

> Among other advantages which will result from this plan [establishing a Queries and Answers section] being adopted in future, will be that of pointing out to the more profound Naturalist a ready mode of ascertaining the wants of his less instructed brethren; there being many of the former, we are convinced, who would willingly answer a query, or settle a point of difficulty or criticism, who have neither leisure nor inclination to write longer articles.[20]

This section was essential for Loudon because it bridged the gap between those with knowledge and those without—offering a contained space for readers and contributors to "settle a point of difficulty or criticism." The value of the periodical for Loudon was to create a platform for knowledge to be stored and diffused, as well as a site for communication.

Loudon wanted his journal to act as a place to both express and record expert knowledge, but also as a site to develop expertise and learn the rules of scientific practice. In his farewell preface, Loudon observed, "One principal object with us has been that of exciting and promoting a *spirit of enquiry*, and a habit of observation, among those who, perhaps, did not previously possess the taste, or the means, for acquiring an insight into those delightful pursuits which are attendant upon the study of Natural History."[21] The production of his periodical was a task not just of increasing natural knowledge, but of expanding the communities of scientific practice. Reading the *Magazine of Natural History*, Loudon implied, would provide readers with the "habit of observation" that would lead to "acquiring an insight" into the workings of the natural world. Reading, here, was a very active endeavor, but also one that was encouraged by the format of the periodical itself, which invited participation. Engaging as a reader of the *Magazine of Natural History* required action—to develop an observational technique. However, reading alone was not enough to develop expertise in natural history. For Loudon, his journal was intended not just as a site for the expression of knowledge, but as an active site for communication between contributors and readers. In other words, being a reader of the *Magazine of Natural History* was intended to be a community activity.

The balance Loudon was trying to strike between creating a space for the expression of expert knowledge and catering for an inclusive community of readers and contributors was ultimately a step too far to maintain a stable reading community. According to Loudon, this vision for a periodical that encouraged interaction between experts and nonexperts improved the "general good" of readers, but ultimately resulted in a loss of the periodical's prestige within expert scientific communities. Loudon concluded his farewell preface stating, "Had our Journal been appropriated exclusively to subjects of deep research, and only open to the communications of experienced Naturalists, it might have taken a higher stand as a philosophical work, but it would not have been productive of the general good that it was our object to promote, and which has undoubtedly arisen from the course which we have followed."[22] The ultimate failure of the *Magazine of Natural History* to take "a higher stand as a philosophical work" was mitigated by the fact—in Loudon's estimation, at least—that the periodical press alone was unlikely to have the greatest influence on the development of scientific practice. For Loudon, it was the embodied community space of the scientific society that would open the doors of scientific practice to a broader community.

Alternative sites for scientific communication, discussion, and debate—particularly the scientific society—would prove particularly difficult for new natural history periodicals that were not officially connected to a scientific society. The societies acted as critical spaces for the construction of scientific communities, as the other chapters in this volume demonstrate. The naturalist society was a particularly important form of scientific community in the nineteenth century. From the 1830s, local and national natural history societies were being founded throughout the country, many of which would eventually establish their own journals to communicate the papers and proceedings read at the typically monthly meetings. Societies such as the Warwickshire Natural-ists' and Archaeologists' Field Club (1838–76), the Berwickshire Naturalists' Clubs (1834–75), and the Penzance Natural History and Antiquarian Society (1845–65) were all contemporaries with the *Magazine of Natural History* and its future iteration, the *Annals and Magazine of Natural History*, each of which printed its own journal.[23]

Toward the end of his time editing the *Magazine of Natural History*, Loudon believed that the scientific meeting—represented in the profusion of natural history societies and the establishment of the British Association for the Advancement of Science (BAAS) in 1831—had greater influence over the practice of natural history than did the journals in the field that were not affiliated with a society. In his preface in 1834, Loudon argued:

> The British Association has given a grand stimulus to natural history pursuits; and the personal intercourse, among naturalists, to which it has led, cannot fail to be highly favourable to science, and to good feeling among scientific men. By this means, also, the great object of science, viz., that of reducing it to practice, and rendering it available for the purposes of domestic and general improvement, is likely to be more immediately effected, than by the single in-fluence of the press.[24]

Writing with a tone of self-effacing encouragement, Loudon continued in the same preface to point out that many of these new natural history societies were also producing new periodicals in both London and the provinces.[25] Loudon recognized that these new societies, especially the large and successful BAAS, would have an impact on the world of natural history journals. This was es-pecially true for those periodicals, such as the *Magazine of Natural History*, that had no formal connection to these societies and would have to compete on an uneven playing field.

In 1837 Loudon officially stepped down as the editor of the *Magazine of Natural History*, and gave over control to the geologist Edward Charlesworth (1813–93). He did not leave journals altogether, however, but returned to editing his other periodical venture, the *Gardener's Magazine and Register of Rural & Domestic Improvement* (1826–44).[26] After Charlesworth took over, he quickly realized that the periodical was losing contributors and subscribers. When Loudon left, he had taken with him his own clout within the naturalist com-munity, as well as his ability to attract other naturalists to produce content for the periodical. Charlesworth's irascible personality and approach to editing—detailed in chapter 4 in this volume—also did little to help him. While Loudon had not been actively engaged in the day-to-day operation of the periodical for at least five years before officially stepping down (having handed over the primary editorial tasks to his assistant John Denson in 1831), his name still gave credibility to the publication.

Charlesworth had the credentials to enter upon expert scientific debate. He had published in the *Proceedings of the Geological Society of London* and the *Philosophical Magazine*, was a newly minted fellow of the Geological Society of London, and was employed at the British Museum. However, these credentials were recent.[27] What he did not have was established credibility within the communities of expert practitioners in natural history.

After a year under his editorial control, Charlesworth complained to his readers that he

> was unknown even by report to the subscribers: several of the more valuable Contributors had seceded to establish the 'Magazine of Zoology and Botany,' whilst another portion of them had united to establish a rival periodical, under the fallacious expectation that it would prove a source of pecuniary emolument; and no lack of solicitations and tempting proposals was wanting to win over the few who yet stood by Mr. Loudon.[28]

The "valuable Contributors" alluded to here by Charlesworth were Sir William Jardine (1800–1874), Prideaux John Selby (1788–1867), and George Johnston (1797–1855), who left to establish the rival periodical *Magazine of Zoology and Botany*. For Charlesworth, the success of a publication was ensured not only by how many subscribers one had, but also by the prestige of the individuals who contributed to the journal's content. This was a steep change from Loudon's approach to the periodical, which depended on the prestige of Loudon and his network of contributors to claim expert authority for the periodical. For Charlesworth, who had no claim to authority himself, the quality of contributors was the key to obtaining enough subscribers in the first place.

This withdrawal of the established naturalist community was a blow to Charlesworth because it not only meant a loss of prestige but also resulted in financial loss. As he bitterly pointed out in his first year, he felt "bound to acknowledge the support afforded to this Periodical, at a period when the attempt to carry it on promised to be attended with considerable difficulty, from the withdrawal in 1836 of a large number of the Contributors, and the establishment of a Journal devoted to Zoology and Botany, *by parties in no way dependent on their literary labours* [emphasis added]."[29] Jardine, Selby, and Johnston were, in Charlesworth's estimation, not just exploiting their scientific capital by establishing a competitor to the *Magazine of Natural History*, but also using their position of economic security to do so.

Charlesworth may have lost some of his expert contributors, but he still hoped to maintain a periodical that was marketable to a broad community. With the loss of prestige, however, he had to pay particular attention to the material aspects of the publication. As discussed in chapter 2 in this volume, one of the costliest aspects of publishing a journal was related to the quality and quantity of illustrations included in each volume. When Loudon was preparing to pass control of the *Magazine of Natural History* over to Charlesworth in 1836, he reflected on how the new journal would expand its visual content. Loudon told his readers:

> Mr. Edward Charlesworth . . . proposes to figure and describe some of the new and choice fossils contained in his extensive collections. With a view of enabling us to give a larger number of engravings than has yet been done, the Magazine will be reduced from three sheets and a half to three sheets.[30]

In at least Loudon's and Charlesworth's conception of a natural history periodical, the reproduction of images was becoming increasingly important—so important, in fact, that the increase in visual content required the *Magazine of Natural History* to cut down on other costs: namely, how many pages they printed in each issue. The gamble, for Charlesworth, seems to have been based on a hope that increasing the visual content of the journal might attract more long-term subscribers to the magazine.

Low subscription numbers were a source of constant anxiety for Charlesworth. In particular, he was frustrated by the increasing availability of community reading spaces where his periodical could be read without the need for individual subscription. In his first preface to readers, Charlesworth lamented:

> The circulation of the English scientific journals is so limited, that, taken in the aggregate, the sum realised by their sale falls short of the actual cost of printing and publishing; a result consequent upon their multiplicity, and perhaps still more upon the very general establishment of museums and public libraries; these institutions affording parties the means of consulting the pages of periodicals, without being obliged to have recourse to individual subscription.[31]

The production of a natural history periodical in the first half of the nineteenth century was a difficult financial prospect—especially, as Charlesworth pointed out, when actual readers might be high in number but most were not paying to read each issue. This was a consequence of increasing access to periodicals through emerging institutions such as museums, public and subscription libraries, and workingmen's clubs.[32]

1841-1923: CREATING A COMMERCIALLY SUCCESSFUL NATURAL HISTORY PERIODICAL, THE *ANNALS AND MAGAZINE OF NATURAL HISTORY*

Attempts at creating new periodical titles throughout the nineteenth century could be short-lived endeavors, with some fizzling out after one or two issues, and others lasting a couple of years before being bought up by competitors. Failures and mergers were the norm in the periodical marketplace. In the same year when Loudon was selling his interests in the *Magazine of Natural History* to Charlesworth, Richard Taylor was establishing the *Annals of Natural History*. The editors for this title would express a new vision for their natural history periodical, in which the community of natural history practitioners was more narrowly defined. Unlike the *Magazine of Natural History*, the *Annals of Natural History* focused the content and value of natural history toward expert practitioners.

The *Annals of Natural History* was itself a merger of two other titles. In 1837, Taylor purchased the rights for two competing journals—Jardine, Selby, and Johnston's *Magazine of Zoology and Botany* (1836–38) and the *Companion to the Botanical Magazine; Being a Journal Containing Such Interesting Botanical Information as Does Not Come within the Prescribed Limits of the Magazine; with Occasional Figures* (1835–36). The *Companion to the Botanical Magazine* was a periodical printed in London and published under the auspices of Samuel Curtis, whose father-in-law had established one of the most popular botanical magazines of the period, *Curtis' Botanical Magazine* (1778–). By the mid-1830s, *Curtis' Botanical Magazine* and the *Companion to the Botanical Magazine* were edited by the professor of botany, and later director of Kew Gardens, William Jackson Hooker (1785–1865). The *Magazine of Zoology and Botany*, on the other hand, was the periodical set up by those expert naturalists who had abandoned the *Magazine of Natural History*, to Charlesworth's frustration (fig. 5.1).

The credibility of these naturalist-editors, however, did not automatically mean that these two periodicals would be successful. Susan Sheets-Pyenson argues that the *Magazine of Zoology and Botany* failed for three primary reasons: the publication and circulation of the periodical was from Edinburgh rather than London, which limited its exposure; its content was focused too

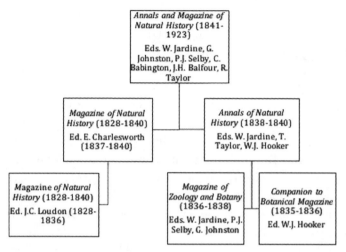

F I G . 5 . 1 . Diagrammatic tree detailing the mergers of different natural history titles—including dates, names, and editors—which ultimately became the *Annals and Magazine of Natural History*. Produced by the author.

narrowly on zoology; and the editors overextended their finances by paying for articles from well-known authors.[33] When Taylor purchased the *Magazine of Zoology and Botany* and the *Companion to the Botanical Magazine*, he consolidated the communities of readers and expert contributors into a new title, the *Annals of Natural History*. This consolidation of credibility was not just lip service. Rather, the new editors of the *Annals of Natural History* included Jardine, Selby, and Johnston as well as Hooker and Taylor.

Merging titles and the expert cast of naturalists who took over control of the *Annals of Natural History* did not ensure financial success, either. The material aspects of a periodical—how many sheets of paper each issue included, the cost of illustrations, and so on—were essential concerns for Jardine, Selby, Johnston, Hooker, and Taylor (hereafter Jardine et al.). Beyond the immediate cost of purchasing paper, paying engravers, and paying for printing, periodicals published prior to the 1840 postal reform act were expensive items to ship to readers. Writing to their readers in the preface to their third volume, Jardine et al. complained that "it is hardly possible to speak of the difficulties with which Scientific Journals have to struggle in this country in comparison with all others, without adverting to the very heavy expense of Postage, and expressing our regret and mortification that nothing has yet been done by Government to relieve Science and Literature among us from a burthen so enormously oppressive."[34] The implementation of the penny post in 1840 allowed for publishers to mail each issue published after this date for a penny, rather than being charged by weight and distance.[35]

The penny post reform act changed the marketplace—allowing for periodicals to increase their subscriber base while decreasing their operating costs. Competitors, such as the *Magazine of Natural History*, had one less thing to worry about. Taking lessons from Loudon, the *Annals of Natural History* needed to create a format that was open to a broad reading community, but which would retain its position—in Loudon's words, cited earlier—"as a philosophical work." To maintain scientific authority and focus the scope of the audience, Jardine et al. used the language and practice of binomial taxonomy as a cornerstone for the *Annals of Natural History*. This was a system, however, that limited readers to those who understood Latin and were familiar with this system of classification.

From its first issue, the language of natural history for *Annals of Natural History* was explicitly Linnean. While the model of Latin binomial taxonomy had taken root in British natural history from the start of the nineteenth century, Jardine et al. not only adopted the model but made it a centerpiece of their vision for their periodical.[36] They signaled their adherence to Linnaeus's classification model by including a quotation from the tenth edition (1785) of *Systema naturæ* (fig. 5.2). This quotation, which was printed on the verso of the title page of every volume, was taken from the end of the introduction (which Linnaeus called "Imperium naturæ") to *Systema naturæ*, and refers to the dominant theme within natural theology of the relationship between the investigation of nature and discovering the "goodness of the Creator."[37]

The inclusion of this passage at the start of each volume of *Annals of Natural History* indicated to readers that even if they did not understand Latin,

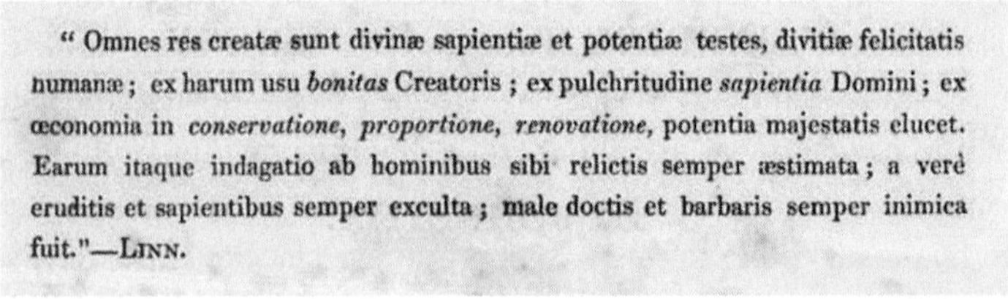

" Omnes res creatæ sunt divinæ sapientiæ et potentiæ testes, divitiæ felicitatis humanæ; ex harum usu *bonitas* Creatoris; ex pulchritudine *sapientia* Domini; ex œconomia in *conservatione, proportione, renovatione,* potentia majestatis elucet. Earum itaque indagatio ab hominibus sibi relictis semper æstimata; a verè eruditis et sapientibus semper exculta; male doctis et barbaris semper inimica fuit."—LINN.

FIG. 5.2. Quotation from the tenth edition of Carolus Linnaeus's *Systema naturæ*, printed on the verso side of the title page in every issue of the *Magazine of Zoology and Botany*, the *Annals of Natural History*, and the *Annals and Magazine of Natural History*. This example is from *Magazine of Zoology and Botany* 1 (1837): 1. Image from the Biodiversity Heritage Library, https://www.biodiversitylibrary.org; contributed by Natural History Museum Library, London.

they could expect the dominance of the Linnean system in the pages therein. The quote also reinforced—for those who did read Latin—what Jardine et al.'s vision of natural history was for the periodical. The end of the passage from Linnaeus states that the laws of nature "ha[ve] always been cultivated by men of learning and wisdom; but always inimical to the unlearned and barbarous." The community defined here was a specific one: those who "cultivated learning."

By 1840, Charlesworth's ability to carry on with the *Magazine of Natural History* had reached an end, and he sold his control of the journal to Taylor. Seeing an opportunity to combine the readerships of the *Annals* and *Magazine*, Taylor merged them into one title. The *Annals and Magazine of Natural History*, however, was a merger in name only, and for all intents and purposes it acted as a continuation of the vision established by Jardine et al., with Loudon and Charlesworth explicitly written out of this new endeavor. This erasure of the *Magazine of Natural History* was reinforced by the fact that the first volume of *Annals and Magazine of Natural History* was printed as volume 6, following from the fifth and final volume of the *Annals of Natural History*. Moreover, the new merged title would continue to emphasize the value of the Linnean system by reproducing the same Linnean quotation that had been printed at the start of each volume of the *Annals of Natural History*. While the new and final name would hint at the amalgamation of the *Annals* and *Magazine* in structure and editorial control, it was Jardine et al. who were in the driver's seat of the content and vision of the new periodical.

Editorial control of the *Annals and Magazine of Natural History* remained relatively consistent over the middle of the nineteenth century, with Jardine, Selby, Johnston, and Taylor (and later, Taylor's illegitimate son, William Francis) editing the journal until the late 1860s. In addition to these, the botanist Charles Babington[38] (1808–95) and the Scottish botanist John Hutton Balfour[39] (1808–84) were added to the editorial roster in 1842, and remained on board for around thirty years.

For the *Annals and Magazine of Natural History* to stake a claim as a site for the communication of expert natural knowledge, discussion and debate remained a critical part of Jardine et al.'s vision for their periodical. However, within the constructed spaces for debate within the journal, there was a clear hierarchy of authority that separated the editors and expert contributors from the general reader and correspondents. In other words, the audiences of the *Annals and Magazine of Natural History* were not participating in "a Magazine of their own," as Loudon's readers were being encouraged to do. Rather, the space for nonexpert participation was compartmentalized in the *Annals and*

Magazine of Natural History. The last pages of every issue contained a section titled "Miscellaneous," which included notes printed from other natural history periodicals, tables of weather observations, information such as the return or death of a traveling naturalist, and letters to the editor.

The last of these categories was the most consistent aspect of this section— every issue included at least two letters. For instance, in the first number for 1848, an entomologist named Henry Denny wrote to the editors to describe his observation of spiders and ants. He began by qualifying: "I know not whether the two accompanying scraps will be worth a line in the 'Annals of Natural History.'" He ended by asking the journal's readers whether his observation of ants attending the body of their queen when under attack could be understood as resulting from a humanlike "affection."[40] This section—where readers of the journal could pose questions and hope for a response from other naturalists—created a space where information could be actively exchanged. Not all questions were worthy, it seems, as Denny signaled by asking whether his observations were "worth a line" in the journal. Yet smaller observations, as well as questions of interpretation—which did not constitute full research articles—were becoming an essential aspect of Jardine et al.'s vision of the periodical. Loudon had encouraged interchange between experts and nonexperts as a foundational aspect of the *Magazine of Natural History*. For Jardine et al., interchange between experts and nonexperts was still important, but it was compartmentalized to one section of each issue.

As part of this desire to emphasize the role of the *Annals and Magazine of Natural History* as a site for communicating expert natural knowledge, one of the key sections of the periodical consisted of reviewing foreign journals (such as the *American Journal of Sciences and Arts* and the *Annales des sciences naturelles*). Jardine et al. made it clear that the "duty" of their natural history periodical was to bring all natural knowledge, both foreign and local, into one print space:

> The important duty of making known in this country the labours and discoveries of Foreign Naturalists, the Editors trust has hitherto been to a considerable extent fulfilled, in the great number of Translations and Abstracts from the principal Journals and Memoirs of other countries, and in Notices of Foreign Works in all branches of Natural History, which have been given with a view to enable the lovers of the science to keep pace with its progress in every stage of advancement.[41]

A natural history periodical, for Jardine and his fellow editors, was about bringing together the most learned naturalists to help readers "keep pace" with the progress of natural history. The readers, imagined here, were both international and specialized, and the periodical was a tool for establishing and communicating expert knowledge. Unlike Loudon's vision for the *Magazine of Natural History*, beginners had little place in Jardine et al.'s vision for their periodical. For the *Annals and Magazine of Natural History*, the progress of science was dependent on "making known . . . the labours and discoveries of Foreign Naturalists" to the readers, not creating a space for new naturalists to gain knowledge and expertise.

The relationship between the periodical and scientific societies also changed by the time of the merger of the *Annals and Magazine of Natural History*. Whereas Loudon and Charlesworth saw the *Magazine of Natural History* as an endeavor separate from the work of natural history societies, by the time that Jardine et al. were editing the *Annals and Magazine of Natural History*, it was becoming increasingly important to tie the communities of the journal and the societies together.

Reporting on the work of natural history societies—especially the large and eminent Linnean Society (1788–present), Zoological Society [of London] (1826–present), and Botanical Society of Edinburgh (1836–present)—was one of the primary functions of the *Annals and Magazine of Natural History*. In each issue, it dedicated fifteen to twenty pages to describing the papers given at these large and metropolitan natural history societies.

As Jardine et al. would point out at the start of their second series of the *Annals and Magazine of Natural History* in 1848, they were not just reprinting content from these three societies. Rather, "authentic reports" of those societies' meetings were being "official[ly] communicated through its pages."[42] The credibility of Jardine et al.'s journal was predicated not only on their own authority or that of their contributors, but on the societies that they were tied to. The *Annals and Magazine of Natural History*, therefore, was not just relying on the strength of its correspondents to create the value of its content, but was reaching out to multiple communities of expert practitioners—particularly, established naturalist societies—to make up its reading community.

1854–65: A DIFFERENT VISION FOR THE NATURAL HISTORY PERIODICAL, THE *NATURAL HISTORY REVIEW*

With the changes in the penny post and the establishment of successful examples, like the *Annals and Magazine of Natural History*, the number of natural history periodicals proliferated, doubling in number between 1850 and 1860.[43] This increase was in large part due to the emergence of new titles formed around the preexisting communities of natural history societies—such as the *Proceedings of the Cotteswold Naturalists' Field Club* (1853-73). The natural history society journal, moreover, was not confined to a singular format. Rather, there were also new attempts to bring groups of societies, united by geographic proximity, together into one print space.[44] The *Natural History Review* (1854-65) was a representative example of this new kind of natural history periodical. Before it became the tool of T. H. Huxley for promoting biological science, the journal had started life with a different vision, as the publication site for a consortium of Irish natural history societies.

The *Natural History Review* was a quarterly, priced initially in 1854 at eight shillings per year, then up to ten shillings and six pence in 1855 before settling on the price of fifteen shillings per year from 1856. This was comparable to the *Annals and Magazine of Natural History*, which began at twelve shillings per year before increasing to fifteen. The fundamental difference was that while the *Natural History Review* focused on the subject of natural history for the societies that made up its readership, it did not offer itself as a site for readers to contribute new findings or discuss and debate questions of theory or fact. Rather, the structure of each issue between 1854 and 1859 consisted, first, of a section reviewing books related to natural history; second, reviews of the content of other natural history periodicals; and third, publication of the transactions of the societies that published the journal. It did not include sections for the exchange of information or letters to the editor. Beginners to natural history therefore had no entry points for the *Natural History Review*. Rather, the journal was designed for the edification of the naturalist communities who were dedicated members of the specialist natural history societies around which the periodical was formed. While the reviewing of books, periodicals, and societies was an aspect of periodicals like the *Annals and Magazine of Natural History*, for the *Natural History Review* it was the only feature of its content. What it did not do was participate in the practices developed by Loudon and Jardine et al. of communicating and archiving original research, creating a site for communication, and acting as a tool for the development of observational expertise.

For its first seven years, the *Natural History Review* attempted to reflect aspects of the vision established by Loudon and Jardine et al.—the periodical was formed around a set of naturalist communities, and led by credible experts. Unlike the editors of the *Magazine of Natural History* or the *Annals and Magazine of Natural History*, however, the credible expert editors of the *Natural History Review* were invisible. In fact, until 1857, the journal did not officially have any editors listed on its front page. While the Irish botanist

Edward Percival Wright (1834–1910) is understood to be the founding editor of the *Natural History Review*, his name was not listed as such within the periodical itself.[45] Instead, the advertising cover that surrounded each issue asserted, "This periodical has been set on foot by a few well-known Naturalists, for the purpose of advancing the cause of Natural History in Great Britain and Ireland, by making the labours of their *confreres* more generally known."[46] The other "well-known Naturalists" who edited the *Natural History Review* would, in 1857, eventually give themselves more direct editorial credit. The title page for the fourth volume of the *Natural History Review* included for the first time a list of two Irish naturalists and a British naturalist who "conducted" the periodical. These were the entomologist Alexander Henry Haliday (1806–70), the botanist William Henry Harvey (1811–66), and the English clergyman and naturalist Reverend William Houghton (1828–95). All three were members of the Royal Irish Academy and Linnean Society, and taught different aspects of natural history at the University of Dublin—Haliday on zoology, Harvey on botany, and Houghton on geology.

Yet the prestige of the contributors and editors—which was so important to the *Annals and Magazine of Natural History*—was a secondary concern in the first years of the *Natural History Review*. The authority of the journal within naturalist communities was reliant predominantly on the content produced within its pages, rather than the expertise of the editors. The "well-known Naturalists" were not at first named as editors, nor were their works published in the periodical. Rather it was Wright, Haliday, Harvey, and Haughton's "confreres"—namely, British naturalists—who were given space in the periodical.

Indeed, as we can see from the index in the first volume alone, two of the editors from the *Annals and Magazine of Natural History*—George Johnston and John Hutton Balfour—had their recent books reviewed in the *Natural History Review* (fig. 5.3) While Johnston and Balfour did not contribute any new content, their names were listed as the authorities.[47] Both the editors and the reviewers for the journal remained anonymous. It was the authors of the original texts who were given special credibility as the authors of natural history knowledge—rather than the reviewers themselves or the "well-known Naturalists" who invisibly ran the journal.

While the *Natural History Review* had a specific set of audiences it catered for—namely the Irish natural history societies that the periodical represented—the chosen format did not lead to financial success. The struggle of the *Natural History Review* to attract a stable and broad readership was reflected in the ever-shifting title page of the periodical. When it was first established in

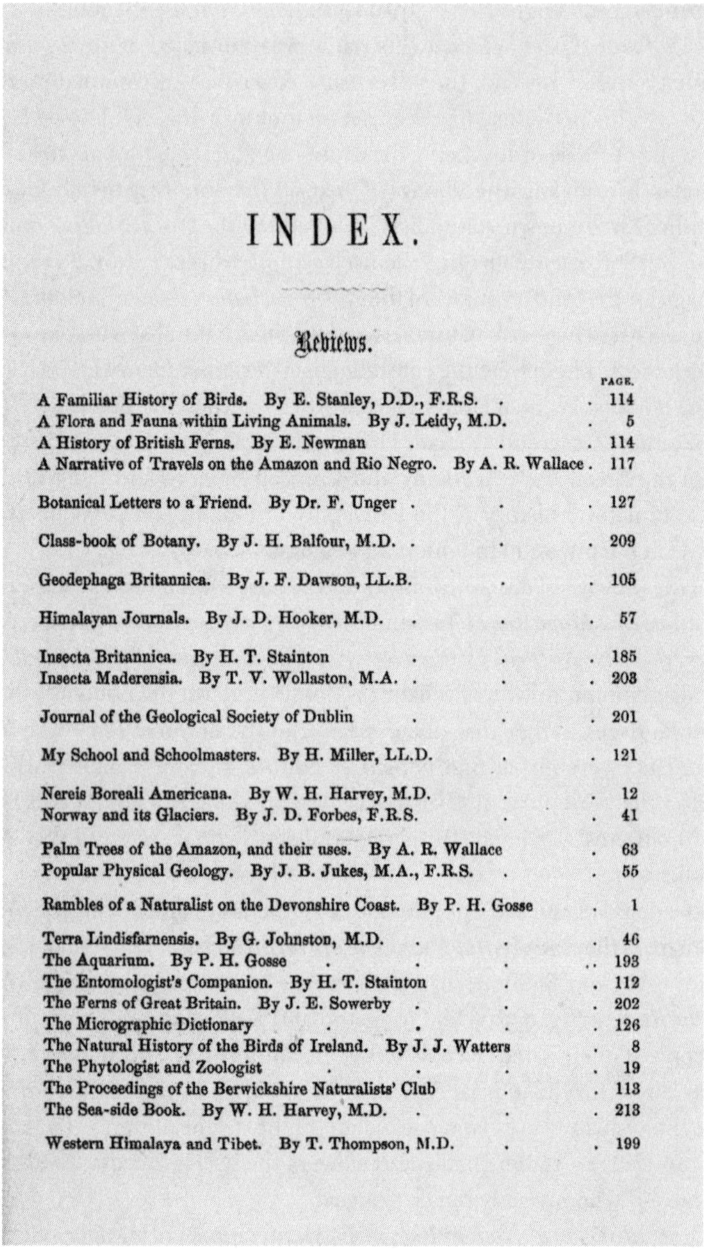
INDEX.

Reviews.

FIG. 5.3. Index page for the first volume of the *Natural History Review*, which details the books under review and the authors of those books, but not the authors of the reviews themselves. *Natural History Review* 1 (1854): iii. Image from the Biodiversity Heritage Library, https://www.biodiversitylibrary.org; contributed by Natural History Museum Library, London.

1854, the *Natural History Review* was subtitled *A Quarterly Journal, Including the Transactions of the Belfast Natural and Philosophical Society, Cork Cuiverian Society, Dublin Natural History Society, Dublin University Zoological Association, and the Literary and Scientific Institution of Kilkenny* (1854–55); then it became *Published Quarterly: Including Proceedings of the Irish Natural History Societies* (1855–56), before changing a penultimate time to a *Quarterly Journal of Zoology, Botany, Geology and Paleontology* (1857). For the first seven years of its existence, the *Natural History Review* retained one aspect of its format: as a quarterly, it was issued every three months regardless of editorial or structural changes in content. After this first seven years, however, almost everything changed, including the target audiences and the value given to natural history as a form of scientific practice.

In 1861 the biologist and advocate of Darwinian evolution T. H. Huxley took over control of the *Natural History Review*, and changed the name one final time to the more succinct and specific *Natural History Review: A Quarterly Journal of Biological Science*. In addition to this name change, the price of the periodical was lowered slightly to twelve shillings per year. In the previous year, Huxley had been approached by Wright asking if he wanted to take it over. Writing to his friend Joseph Hooker (the son of William, who had been briefly part of the editorial team for the *Annals of Natural History*), Huxley savored the opportunity to have "the effectual control" of the *Natural History Review* "pretty much in my own hands."[48] Huxley was particularly keen on taking over the journal because he saw an opportunity to fill a gap in the scientific marketplace. "Considering . . . the low condition of natural history journalisation (always excepting quarterly *Mic. Jour.*) in this country," he declared, "this seems to me to be a fine opening for a plastically minded young man, and I am decidedly inclined to close with the offer, though I shall get nothing but extra work from it."[49] Despite the "extra work" that would be required of him, Huxley saw the opportunity to make a journal that would be an organ for evolutionary theory.[50] He also saw it as an opportunity to redefine the practice of natural history more generally, by shifting away from the loosely defined set of disciplines and practices that constituted natural history, to the more disciplined and specific area of "biological science."

According to Hooker, natural history, and the periodicals that took this name, were in a critical state. "The number of badly edited and badly supported journals is quite incredible, and the present practice of cramming Zoological and Botanical researches into one periodical increases the evil many-fold."[51] When Huxley asked his friend Charles Darwin for advice on whether he should take up the *Natural History Review* to help redress this imbalance, Darwin did not think that editing a periodical was the best use of Huxley's time or energy. Nor did he agree with Huxley about the state of the periodical marketplace. Instead, Darwin argued that for natural history periodicals consisting of "original communications," there was already a "superabundance." A review, on the other hand, could have "value & utility" for the field, but Darwin worried that if Huxley took over editorial control of the *Natural History Review*, he would be "sacrificing much time which could be given to *original* research."[52]

Despite these warnings, Huxley took on the task of the *Natural History Review*, and moved it from being a society journal published in Ireland to a journal edited and produced by some of the most prominent London-based, scientific naturalists of the latter half of the nineteenth century.[53] Alongside changing the subtitle of the journal to include a focus on "biological sciences," Huxley would also add ten additional editors.[54] The new editorial team would include three original members of the *Natural History Review*—E. P. Wright; Joseph Reay Greene (1836–1903), professor of natural history at the University of Cork; and Robert McDonnell (1828–89), surgeon and lecturer at the Carmichael School of Medicine in Dublin. The Scottish naturalist Wyville Thomason (1830–82), who had taken up the chair of natural history at the University of Cork in 1853 (and was later a deep-sea explorer on HMS *Challenger*), was also added to the editorial team.[55] By including both Irish and British editors, Huxley was able to make a claim to the original Irish audiences of the *Natural History Review*, while also expanding his readership to a British and potentially international audience.

The rest of the editors were taken from Huxley's British scientific networks, who had reputations and correspondents in international scientific communities. These included the naturalist and microscopist George Busk (1807–86), the biologist William Benjamin Carpenter (1813–85), and the naturalist and politician John Lubbock (1834–1913).[56] Hooker was reticent to take part in the *Natural History Review*, but he did offer one of his botanical assistants at Kew, Daniel Oliver (1830–1916), to take his place on the editorial board.[57] The final two editors were fellows with Huxley of the Royal and Linnean Societies: the mycologist Frederick Currey (1819–81) and the zoologist Philip Lutley Sclater (1829–1913). This change in name and editorial control from the periodical's earlier Irish formation signaled a considerable shift in the type of natural history offered to its readership: while the *Natural History Review* continued to review recent books in natural history, its new focus would be on the professionalization of "biological sciences," with articles containing original content produced by a group of some of the most influential men of science of the period. Under Huxley et al. the periodical would not just act as a site for communicating expert knowledge, but would do so with the aim of creating a distinction between amateur and professional scientific expertise.

In this change, Huxley et al. were creating a new vision for a scientific periodical. This vision borrowed the name of natural history, but changed the function of the periodical to act as vehicle for the emerging biological sciences. They also changed the periodical's format. While they brought in the concept of original research—allowing the journal to be a site for the communication, recording, and archiving of knowledge—they also constructed a narrower sense of specialist scientific community to which the journal catered. This narrowing of communities, and the emphasis on the explication of evolutionary theory, would ultimately marginalize the scope of the journal's readership, and signal the demise of the *Natural History Review*.

In the first issue of the new series of the *Natural History Review*, Huxley would set out his aim for the periodical not in a preface but in an "original article" titled "On the Zoological Relations of Man with the Lower Animals."[58] In this article Huxley's goal was to press for the separation of the biological sciences from what he saw as a pernicious aspect of natural history—a belief in the uniqueness of man—that was paired with the overlooking of "facts." These two positions, Huxley argued, created opposing "schools" of natural history: "theologians and moralists, historians and poets" versus "students of the physical sciences."[59] Huxley's role in this debate, he argued, was to take sides and to highlight the facts:

> . . . there can be no doubt that the controversy [between these two schools] as to the real position of man still exists; and I have therefore thought that it would be useful to contribute my mite towards the enrichment of the armoury upon which both sides must, in the long run, be dependent for their weapons, by endeavouring to arrange and put in order the facts of the case, so far as they consist of the only matters of which the anatomist and physiologist can take cognizance.[60]

The *Natural History Review* from 1861 therefore promoted a different vision for the specialist scientific periodical, which was distinct from natural history. It brought together authoritative figures working in different fields of natural history to arrange and communicate facts, but did so in a way that pushed for a specific agenda for the practice of natural history—performed through a narrow group of practitioners.

By using the term "biological science," Huxley was attempting to create a distinct separation between the work that he—and his fellow biologists—were doing in science, and natural history. A decade after the *Natural History Review* stopped publishing, he gave a lecture called "On the Study of Biology," which outlined his position concerning biology and natural history.[61] After first tracing the invention of the term "biology" to three "great men of science" in the early nineteenth century—Marie François Xavier Bichat (1771–1802), Jean-Baptiste Lamarck (1744–1829), and Gottfried Reinhold Treviranus (1776–1837)—Huxley created clear distinctions between biology and natural history. Biology was focused on man and animals, while natural history was a catchall term for forms of observational science that did not depend on experiment or mathematics.[62] Biology was a critical discipline underpinned by fact; natural history was a "paper philosophy" in which a practitioner "contents himself with reading books on botany, zoology, and the like."[63] Natural history, according to Huxley, was practiced exclusively through the pages of the printed book and periodical—it was not a science rooted in experiment and fact.

The new *Natural History Review* was one of Huxley's main avenues for promoting this reform. Under his management the structure of the periodical changed to a format of four sections: one for "Reviews"; the second and largest for "Original Articles"; a third titled "Bibliography," which offered a list of primary and secondary sources to identify species and types; and a smaller area for "Miscellanea." The "Miscellanea" section was equivalent to the section in the *Annals and Magazine of Natural History* that offered the opportunity for the exchange of letters between experts and nonexperts. In the *Natural History Review*, however, it was the smallest part of every issue. In 1862 for instance, "Miscellanea" constituted less than 1 percent of the pages of each issue, and operated as a space for very small, mostly anonymously authored points of natural history. In this second year of Huxley's control of the *Natural History Review*, only one letter appeared in this section. The letter, written by Philip Norman to John Lubbock, was a description of young vipers collected and examined in situ.[64] It was not directed toward the *Natural History Review* itself, but rather was used as a piece of evidence by one of the periodical's editors, and set within a small article on the subject of young vipers. Correspondence, for Huxley and the other editors of the journal, was a useful form of evidence, but was not a site for establishing debate within the periodical itself.

By minimizing the space for the circulation of readers' letters in the *Natural History Review*, Huxley et al. were undermining a long-standing practice in the production of natural history knowledge. The exchange of information between naturalists through their correspondence had been a key element of natural history from at least the sixteenth century.[65] For naturalists in the nineteenth century, the exchange of letters remained a critical practice for exchanging knowledge and specimens on a national and international scale. The scientific periodical, moreover, offered an emerging parallel site for naturalists to exchange the content of their correspondence on a larger scale.[66] Periodicals like the *Magazine of Natural History*, for instance, had been critical sources for Darwin in gaining information about plant and animal species variation for supporting his theory of evolution. Moreover, Darwin was a regular subscriber and correspondent to periodicals like the *Gardener's Chronicle*, which offered a large community of fellow naturalists who could supply him with relevant species information he did not have access to himself.[67] Huxley et al. were therefore not just balking at a practice within natural history by minimizing the space for correspondence in the *Natural History Review*, but were undermining a practice that had been critical to the work in developing evolutionary theory itself. For Huxley, the need for a site for scientists to communicate was fulfilled by other journals, particularly those that wanted to carry on the antiquated practices of natural history. By changing the structure of the content in this way, he removed the critical aspect of community participation, which had been an essential aspect of the vision developed through the *Annals and Magazine of Natural History*.

As an example of this narrowed restriction of readership, within the first two years of Huxley et al.'s editorial control the *Natural History Review* published forty-two "original articles." Of these, sixteen were written by the editors of the journal, and the other twenty-six were written by naturalists, professors, and medical doctors.[68] The first and only article in this period written by Huxley was the one mentioned above. The other editors contributed varying numbers of articles, with Lubbock writing the most—four articles on the ancient history of man and the earth.[69] Carpenter, Oliver, and Currey each wrote two articles over the period, all on subjects of botany.[70] Busk, Slater, Thomson, and McDonnell all contributed one article each on ancient man, the kiwi bird, human skulls, and the electrical organs in fish.[71] Each of these articles promoted or debated the theory of evolution. In this early period, just two years after the publication of *On the Origin of Species*, the *Natural History Review* was not just promoting the biological sciences but was doing so with a pro-evolutionary stance. Considering that Huxley was at the helm of this publishing endeavor, this is not surprising; but it does signal that the communities who read and contributed to the *Natural History Review* were narrowly defined.

This change in direction for the *Natural History Review*—from natural history to biology, orchestrated by Huxley—did not bring financial success. As Dawson points out, after being taken over by Huxley, the *Natural History Review* "became both increasingly specialist and narrowly partisan."[72] Reflecting upon the the periodical's demise in 1865, the *Berkshire Chronicle* noted dryly, "The *Natural History Review* has come to an end. It was the able advocate of the peculiar views of Mr. Darwin."[73] The journal would ultimately fail not because of a lack of prestige, but because of its ideological stance. Its new focus on content for exclusive audiences proved ineffective for producing a large subscription base in the 1860s.

CONCLUSION

This chapter has followed three broadly differing visions for the natural history periodical between 1828 and 1865, through three emerging specialist periodicals: the *Magazine of Natural History*, the *Annals and Magazine of Natural History*, and the *Natural History Review*. For Loudon, a natural history periodical was a communal space where the work was "supported by the voluntary contributions of their readers." For Huxley et al., natural history was an antiquated practice in need of reform. In between these two visions lay the *Annals and Magazine of Natural History*, which struck a balance between these two poles. Jardine et al., through their network of expert contributors and their emphasis on Linnean taxonomy, were able to maintain the scientific credibility of their periodical while also creating a space for discussion and debate.

In many ways, Loudon's early vision for a natural history periodical that included both beginners and experts, which was reformed by Jardine et al., would be taken up in popular natural history periodicals that emerged in the mid-1860s, when distinctions between amateur and professional practitioners of science were starting to develop.[74] The monthly, richly illustrated *Hardwicke's Science Gossip: An Illustrated Medium of Interchange and Gossip for Students and Lovers of Nature* (1865–1910) is a representative example. It was sold for four pence an issue and was edited by the mycologist Mordecai Cubitt Cooke (1825–1914), and later the geologist John Ellor Taylor (1837–95). It would offer a space for a wide community of naturalists to print their new research, read about and establish scientific societies, and communicate with each other.[75] With language similar to that of Loudon and Jardine et al., Cooke and Taylor asserted that their periodical was aimed at "students and readers of every class [. . .] we hope with some success." They also noted "large progress which Science is making, and [. . .] the greater number of students its fascinating discoveries are attracting."[76]

Jardine et al.'s vision, on the other hand, to establish a space for expert communities to publish and communicate, continued throughout the 1860s when Huxley and the astronomer and editor Norman Lockyer (1836–1920) first established the *Reader* (1863–66), and then later the ultimately very successful *Nature* (1869–present). While these two periodicals would borrow heavily from visions of the natural history periodical articulated by Loudon, Jardine et al., and Huxley et al.—influential contributors, reports of meetings and other periodicals, and the intercommunication of readers—this, as Melinda Baldwin demonstrates, had little to do with Huxley.[77] Moreover, as Lightman points out, when *Nature* was first established, it tried to reach both popular and specialist audiences, and only after its first decade (over which time it was losing rather than making money) did it position itself as a place for professionalizing, specialist scientific knowledge.[78]

Huxley's vision for a specialized periodical to advocate for biological science would be adopted by two specialist periodicals after the close of the *Natural History Review*. In 1867, Wright would join four anatomists and zoologists to establish the *Journal of Anatomy and Physiology* (1867–1916). The team of editors included three Cambridge University professionals—the professor of anatomy George Murray Humphry, the professor of zoology and comparative anatomy Alfred Newton, and the superintendent of the Museum of Zoology and Comparative Anatomy John Willis Clark—and a professor of anatomy at Edinburgh University, William Turner. A decade later, a Cambridge physiologist and acolyte of Huxley's, Michael Foster, established the *Journal of Physiology* (1878–1988). Foster had worked as a demonstrator on Huxley's courses on biology in South Kensington, and it was thanks to Huxley's support that he was offered a position at Trinity College, Cambridge, as lecturer in physiology in 1870. In all his work, including editing the *Journal of Physiology*, Foster echoed Huxley's desire for the need to professionalize communities of scientific practice, especially within the biological sciences.[79] Through these two titles, Huxley's desire for a long-standing specialist periodical for pro-evolution anatomists and biologists would come to fruition. Predominantly supported by professional biologists working in dominant universities, the *Journal of Anatomy and Physiology* and the *Journal of Physiology* were distinguished from the practices of natural history that Huxley had been keen to separate from the work of biology.

The work of natural history in the nineteenth century, this chapter has shown, became increasingly tied to the publication of specialist periodicals. To produce natural knowledge continued to require observation, evaluation, and communication between other knowledgeable naturalists. However, while these practices were carried out previously through personal interaction in the field, club, or society, or by written communication through correspondence, by the middle of the nineteenth century participation in natural history also meant reading and contributing to periodical publications. As this chapter has shown, the range of participants and the role of expert knowledge changed under every new title or new editor. But the success of a particular periodical title did not depend primarily on the fame or credibility of those who ran the journal. Rather, establishing a successful natural history periodical required a balance between the vision established by the editor(s); the price, structure, and availability of the journal; and the scope of participation offered to the community of subscribers and readers. In the middle nineteenth century, in other words, natural history became increasingly tied to the construction of periodical communities. And, even after natural history was no longer considered an accepted form of professional scientific practice, the legacy of the periodical visions established in the *Magazine of Natural History*, the *Annals and Magazine of Natural History*, and the *Natural History Review* would remain. A mixture of these visions would become foundational aspects of the specialist scientific periodical culture of the late nineteenth and early twentieth centuries.

NOTES

1. *Magazine of Natural History* 3 (1830): iii.

2. For comparison, see chapter 6 in this volume, which examines the role of correspondence columns as a form of scientific authorship.

3. James A. Secord, *Visions of Science: Books and Readers at the Dawn of the Victorian Age* (Oxford, UK: Oxford University Press, 2014), 3.

4. David E. Allen, *The Naturalist in Britain. A Social History* (Princeton, NJ: Princeton University Press, 1976), 86.

5. Stern notes, "*Format* denotes a whole range of decisions that affect the look, feel, experience and workings of a medium." Jonathan Stern. *Mp3: The Meaning of a Format* (Durham, NC: Duke University Press, 2012), 7.

6. For a description of the establishment of *Annals and Magazine of Natural History*, see William Brock, "The Development of Commercial Science Journals in Victorian Britain," in *Development of Science Publishing in Europe*, ed. A. J. Meadows (Amsterdam: Elsevier Science Publishers, 1980), 105–11; William Brock and A. J. Meadows, *The Lamp of Learning: Taylor & Francis and the Development of Science Publishing* (London: Taylor & Francis, 1984). For

a discussion of the *Natural History Review*, see Gowan Dawson, "*Natural History Review* (1854–1965)," in *Dictionary of Nineteenth-Century Journalism in Great Britain and Ireland*, ed. Laurel Brake, Marysa Demoor, and Margaret Beetham (Ghent: Academia Press; London: British Library, 2009), 441; and Susan Sheets-Pyenson, "From the North to Red Lion Court: The Creation and Early Years of the *Annals of Natural History*," *Archives of Natural History* 10 (1982): 221–49.

7. The literature pertaining specifically to natural history publication is fairly limited. See Jonathan R. Topham, "Technicians of Print and the Making of Natural Knowledge," *Studies in History and Philosophy of Science* 35 (2004): 391–400, for the most thorough historiographic examination of the area. The most comprehensive work is by Susan Sheets-Pyenson. See Susan Sheets-Pyenson, "A Measure of Success: The Publication of Natural History Journals in Early Victorian Britain," *Publishing History* 9 (1981); 21–36; Susan Sheets-Pyenson, "Darwin's Data: His Reading of Natural History Journals, 1837–1942," *Journal of the History of Biology* 14 (1981): 231–48; Susan Sheets-Pyenson, "War and Peace in Natural History Publishing: *The Naturalist's Library*," *Isis* 72 (1981): 50–72; Susan Sheets-Pyenson, "From the North to Red Lion Court: The Creation and Early Years of the *Annals of Natural History*," *Archives of Natural History* 10 (1982): 221–49; and Susan Sheets-Pyenson, "The Effect of a Change in Printing Technology on the Development of Natural History Sciences During the Nineteenth Century," in *The Advent of Printing: Historians of Science Respond to Elizabeth Eisenstein's "The Printing Press as an Agent of Change*," ed. P. F. McNally (Montreal: Graduate School of Library and Information Studies, McGill University, 1987), 21–26. David Allen's work in this field is also foundational. See David E. Allen, "The Struggle for Specialist Journals: Natural History in the British Periodicals Market in the First Half of the Nineteenth Century," *Archives of Natural History* 23 (1996): 107–23. For other work relating to natural history periodicals, see James Mussell, "Bug-Hunting Editors: Competing Interpretations of Nature in Late Nineteenth-Century Natural History Periodicals," in *(Re)Creating Science in Nineteenth-Century Britain: An Interdisciplinary Approach*, ed. Amanda Mordavsky Caleb (Newcastle, UK: Cambridge Scholars Publishing, 2007), 81–96. For a critical history of natural history practices within nineteenth-century science, see Jim Endersby, *Imperial Nature: Joseph Hooker and the Practices of Victorian Science* (Chicago: University of Chicago Press, 2010).

8. For the most comprehensive work on natural history in Britain, see Allen, *The Naturalist in Britain*, and Nick Jardine, James A. Secord, and E. C Spary, eds., *Cultures of Natural History* (Cambridge: Cambridge University Press, 2008).

9. For a comprehensive list of natural history periodicals produced in the nineteenth century, see *The Natural History Museum: Serial Titles Held in the Department of Library and Information Services*, 5 vols. (London: Natural History Museum, 1996). For a more general list of scientific periodicals in the period, see S. H. Scudder, *Catalogue of Scientific Serials of All Countries Including the Transactions of Learned Societies in Natural, Physical and Mathematical Sciences, 1633–1876* (Cambridge, MA: Library of Harvard University, 1879).

10. Allen, *The Naturalist in Britain*, 84.

11. Ibid., 85.

12. See, for instance, Alborn's discussion of Charles Babbage and William Whewell's efforts to wrest science (and the terminology used to define it) away from the influence of the aristoc-

racy within the contexts of the participation in and running of the Royal Society. Timothy L. Alborn, "The Business of Induction: Industry and Genius in the Language of British Scientific Reform, 1820–1840," *History of Science* 34 (1996): 91–121. Also see Secord's discussion of working-class participation in natural history in this period. Anne Secord, "Botany on a Plate: Pleasure and the Power of Pictures in Promoting Early Nineteenth-Century Scientific Knowledge," *Isis* 34 (2002).

13. Secord, *Visions of Science.*

14. James A. Secord, *Victorian Sensation: The Extraordinary Publication, Reception, and Secret Authorship of the Vestiges of the Natural History of Creation* (Chicago: University of Chicago Press, 2000), 48.

15. Dewis, *The Loudons and the Gardening Press,* 11.

16. For a decription of Loudon's work with periodicals, see Ray Desmond, "Loudon and Nineteenth-Century Horticultural Journalism," in *John Claudius Loudon and the Early Nineteenth Century in Great Britain,* ed. Elisabeth B. MacDougall (Washington: Dumbarton Oaks, 1980); Sarah Dewis, *The Loudons and the Gardening Press* (Farnham, UK: Ashgate, 2014).

17. *Magazine of Natural History* 1 (1829): iii.

18. Ibid.

19. *Magazine of Natural History* 2 (1829): iii.

20. *Magazine of Natural History* 4 (1831): iii.

21. *Magazine of Natural History* 9 (1836): iii.

22. Ibid.

23. *Proceedings of the Warwickshire Naturalists' and Archaeological Field Club* (1838–76), *History of the Berwickshire Naturalists' Club* (1834–75), and *Transactions of the Penzance Natural History and Antiquarian Society* (1845–65).

24. *Magazine of Natural History* 7 (1834): iii.

25. Ibid.

26. For a discussion of Loudon as an editor, see chapter 2 of Dewis, *The Loudons and the Gardening Press,* 33–79.

27. See Sheets-Pyenson, "A Measure of Success," 22.

28. *Magazine of Natural History,* New Series, 2 (1838): iii.

29. *Magazine of Natural History,* New Series, 1 (1837): iii.

30. *Magazine of Natural History* 9 (1836): iv.

31. *Magazine of Natural History,* New Series, 1 (1837): iii.

32. For a discussion of the changes in cultures through an analysis of subscription reading libraries, see Rebecca Bowd, "Useful Knowledge or Polite Learning? A Reappraisal of Approaches to Subscription Library History," *Library and Information History* 29 (2013): 182–95.

33. Sheets-Pyenson, "A Measure of Success," 23.

34. *Annals of Natural History* 2 (1839): iv.

35. For a discussion of the cost of printing and the penny reform, see Aileen Fyfe, *Steam-Powered Knowledge: William Chambers and the Business of Publishing, 1820–1860* (Chicago: University of Chicago Press, 2012). Also see Secord, *Victorian Sensation.*

36. For a discussion of the adoption of the Linnean system in Britain in the nineteenth

century, see Paul Lawrence Farber, *Finding Order in Nature: The Naturalist Tradition from Linnaeus to E. O. Wilson* (Baltimore: Johns Hopkins University Press, 2000).

37. The passage is translated in John Lewis Bradley and Ian Ousby, eds., *The Correspondence of John Ruskin and Charles Eliot Norton* (Cambridge: Cambridge University Press, 2011), 172n3: "All created things bear witness to divine wisdom and power, and to the wealth of human felicity; in their utility the goodness of the Creator is displayed; in their beauty the wisdom of the Lord; in their economy in Conservation, Proportion, Renovation, the power of His Majesty. And so the investigation of them . . . has always been cultivated by men of learning and wisdom; but always inimical to the unlearned and barbarous."

38. David E. Allen, "Babington, (Charles) Cardale (1808–1895)," *Oxford Dictionary of National Biography* (Oxford, UK: Oxford University Press, 2004), http://www.oxforddnb.com /view/article/970, accessed 27 February 2017.

39. David E. Allen, "Balfour, John Hutton (1808–1884)," *Oxford Dictionary of National Biography* (Oxford, UK: Oxford University Press, 2004), http://www.oxforddnb.com/view /article/1192, accessed 27 February 2017.

40. *Annals and Magazine of Natural History*, Second Series, 1 (1848): 75–76.

41. *Annals and Magazine of Natural History*, Second Series, 1 (1848): iii.

42. *Annals and Magazine of Natural History*, Second Series, 1 (1848): iv.

43. See Susan Sheets-Pyenson, "Popular Science Periodicals in Paris and London: The Emergence of a Low Scientific Culture, 1820–1875," *Annals of Science* 42 (1985): 549–72 (551); and Ruth Barton, "Just before *Nature*: The Purposes of Science and the Purposes of Popularization in Some English Popular Science Journals of the 1860s," *Annals of Science* 55, 1 (1998): 1–33 (2–3).

44. The *Midland Naturalist* (1878–94) was a later example of this endeavor. It proposed to publish the transactions of sixteen natural history societies operating in British cities and towns in the triangle between Birmingham, Nottingham, and Northampton.

45. See Wright's obituary in H. H, Dixon, *Proceedings of the Linnean Society of London* 122 (1910): 102–4.

46. *Natural History Review* 4, no. 9 (1856): advertisement cover.

47. The two books reviewed here are George Johnston, *Terra Lindisfarnensis: The Natural History of the Eastern Borders* (London: John van Voorst, 1853); and John Hutton Balfour, *Class-Book of Botany, Being an Introduction to the Study of the Vegetable Kingdom* (Edinburgh: Adam and Charles Black, 1854).

48. Leonard Huxley, *Life and Letters of Thomas Henry Huxley*. 2 vols. (London: Macmillan, 1913): 302.

49. Ibid., 303. The "Mic Jour" referred to in this letter was likely the *Transactions of the Microscopical Society of London* (1844–66).

50. Huxley told Hooker that he hoped to make the tone of the periodical "mildly episcopophagous" or "Bishop eating." This was no mild turn of phrase for Huxley. Consider that a month earlier, Huxley and Hooker had participated in a heated debate with Bishop Samuel Wilberforce over evolutionary theory. Huxley, *Life and Letters of Thomas Henry Huxley*, 303. See Nanna Katrine Lüders Kaalund, "Oxford Serialized: Revisiting the Huxley-Wilberforce Debate through the Periodical Press," *History of Science* 52, no. 4 (2014): 429–53, for a discussion of the centrality of the periodical press to this debate. For further context, see also

Edward Caudill, "The Bishop-Eaters: The Publicity Campaign for Darwin and *On the Origin of Species*," *Journal of the History of Ideas* 55 (1994): 441–60; and Adrian Desmond, *Huxley: The Devil's Disciple* (London: M. Joseph, 1994), 284.

51. Leonard Huxley, *Life and Letters of Sir Joseph Dalton Hooker*, 2 vols. (London, John Murray, 1918): 409.

52. "Letter no. 2873," Darwin Correspondence Project, accessed on 7 August 2017, http://www.darwinproject.ac.uk/DCP-LETT-2873.

53. For a discussion of the term "scientific naturalism," see Bernard Lightman and Gowan Dawson, eds., *Victorian Scientific Naturalism: Community, Identity, Continuity* (Chicago: University of Chicago Press, 2014).

54. For a historiographical discussion of the implications of the creation of "biology' as a concrete form of science, see Joseph A Caron, "'Biology' in the Life Sciences: A Historiographical Contribution," *History of Science* 26 (1988): 223–68. For Caron's particular description of Huxley's adoption and promotion of the "biological sciences" in the late 1850s and '60s, see pp. 247–54.

55. "Sir C. Wyville Thomson," Encyclopaedia Britannica, accessed 7 December 2017 at https://www.britannica.com/biography/C-Wyville-Thomson.

56. For a discussion of the X Club, see Ruth Barton, "'An Influential Set of Chaps': The X Club and Royal Society Politics 1864–85," *British Journal for the History of Science* 23 (1990): 53–81; Ruth Barton, "Scientific Authority and Scientific Controversy in Nature: North Britain against the X Club," in *Culture and Science in Nineteenth-Century Media*, ed. Louise Henson, Geoffrey Cantor, Gowan Dawson, Richard Noakes, Sally Shuttleworth, and Jonathan R. Topham (Aldershot, UK: Aldgate, 2004), 223–35.

57. Huxley, *Life and Letters of Thomas Henry Huxley*, 303.

58. Adrian Desmond called this article Huxley's "vote-catching masterpiece." Adrian Desmond, "Huxley, Thomas Henry (1825–1895), biologist and science educationist," *Oxford Dictionary of National Biography*, accessed 7 December 2017, http://www.oxforddnb.com/view/10.1093/ref:odnb/9780198614128.001.0001/odnb-9780198614128-e-14320.

59. Thomas Henry Huxley, "On the Zoological Relations of Man with the Lower Animals," *Natural History Review*, New Series 1 (1861): 67–84 (67–68).

60. Ibid., 68.

61. Thomas Henry Huxley, "On the Study of Biology [1876] (A Lecture in Connection with the Loan Collection of Scientific Apparatus, South Kensington Museum)," in *Science and Education: Essays* (London: Macmillan, 1925).

62. Ibid., 265.

63. Ibid., 282.

64. *Natural History Review* 2, (1862): 120.

65. Elizabeth Yale, *Sociable Knowledge: Natural History and the Nation in Early Modern Britain* (Philadelphia: University of Pennsylvania Press, 2016).

66. For recent research on this, see Matthew Wale, "'The Sympathy of the Crowd': Periodicals and the Practices of Natural History in Nineteenth-Century Britain" (PhD thesis, University of Leicester, 2018), especially chapter 2.

67. Sheets-Pyenson, "Darwin's Data," 244; Janet Browne, "Corresponding Naturalists," in

The Age of Scientific Naturalism: Tyndall and His Contemporaries, ed. Bernard Lightman and Michael S. Reidy (London: Pickering and Chatto, 2014), 157–69.

68. The contributing authors (who were not editors) to these two volumes fell into three categories: Darwinian naturalists (George Bentham, George Henry Lewis, George Rolleston, J. D. Hooker, Albany Hancock, Robert Caspary, Edward Lartet, William Houghton, and John Scouler), surgeons and anatomists (John Cleland, John Marshall, William Selby Church, John Denis Macdonald, William Turner, and William Bedford Kesteven), and foreign naturalists whose articles were originally published elsewhere (Josef Hyrtl, Hermann Schlegel, Jacobus Ludovicus, Conradus Schroeder van der Kolk, and Willem Vrolik). In this whole community of contributors there were only two outliers: an article listing the objects collected by the chief trader for the Hudson's Bay Company, Bernard Rogan Ross, and an article by the antievolutionist Joseph Bernard Davis titled "Crania of Ancient Britons."

69. See John Lubbock, "On Sphaerularia Bonbi," *Natural History Review*, New Series 1 (1861): 44–56; John Lubbock, "On the Kjokkenmoddings: Recent Geological-Archaeological Researches in Denmark," *Natural History Review*, New Series 1 (1861): 489–503; John Lubbock, "On Sphaerularia Bonbi," *Natural History Review*, New Series 1 (1861): 44–56; John Lubbock, "On the Ancient Lake Habitations of Switzerland," *Natural History Review*, New Series 2 (1862): 26–52; and John Lubbock, "On the Evidence of the Antiquity of Man, Afforded by the Physical Structure of the Somme Valley," *Natural History Review*, New Series 2 (862): 244–68.

70. See W. B. Carpenter, "General Results of the Study of Typical Forms of Frominifera," *Natural History Review*, New Series 1 (1861): 185–200; W. B. Carpenter, "On the Systemic Arrangement of Rhizopoda," *Natural History Review*, New Series 1 (1861): 456–72; D. Oliver, "The Atlantis Hypothesis in Its Botanical Aspect," *Natural History Review*, New Series 2 (1862): 149–69; D. Oliver, "The Structure of the Stem in Dicotyledons," *Natural History Review*, New Series 2 (1862): 298–328; F. Currey, "Report on Vegetable Partenogenesis," *Natural History Review*, New Series 1 (1861): 447–55; and F. Currey, "On the Germination of Reticularia Umbrina," *Natural History Review*, New Series 2 (1862): 406–8.

71. See G. Busk, "On the Crania of the Most Ancient Races of Man," *Natural History Review*, New Series 1 (1861): 155–75; P. L. Sclater, "Report on the Present State of Our Knowledge of the Species of Apteryx Living in New Zealand," *Natural History Review*, New Series 1 (1861): 504–6; W. Thomson, "On Distorted Human Skulls," *Natural History Review*, New Series 2 (1862): 397–405; and R. McDonnell, "On an Organ in the Skate Which Appears to Be the Homologue of the Electrical Organ of the Torpedo," *Natural History Review*, New Series 1 (1861): 57–59.

72. Dawson, "Natural History Review (1854–1965)," 16.

73. *Berkshire Chronicle*, 10 February 1866, 7.

74. For a discussion of the complicated relationship between the terms "amateur" and "professional" in natural history, see Ruth Barton, "'Men of Science' Language, Identity and Professionalization in the Mid-Victorian Scientific Community," *History of Science* 41, no. 1 (2003): 73–117; and Endersby, *Imperial Nature*.

75. The Quekett Microscopical Club (1865–present) established itself first within the pages of *Hardwicke's Science Gossip*. For a discussion of this, see Geoffrey Belknap, "Illustrating Nat-

ural History: Images, Periodicals, and the Making of Nineteenth-Century Scientific Communities," *British Journal for the History of Science* 51 (2018): 395–422.

76. *Hardwicke's Science Gossip* (1876): iii.

77. Melinda Baldwin, *Making Nature: The History of a Scientific Journal* (Chicago: University of Chicago Press, 2015).

78. Bernard Lightman, *Victorian Popularizers of Science: Designing Nature for New Audiences* (Chicago: University of Chicago Press, 2007), 325–27.

79. Gerald L. Geison, *Michael Foster and the Cambridge School of Physiology: The Scientific Enterprise in Late Victorian Britain* (Princeton, NJ: Princeton University Press, 1978); T. Romano, "Foster, Sir Michael (1836–1907), physiologist and politician," *Oxford Dictionary of National Biography,* accessed 15 January 2018 at http://www.oxforddnb.com/view/10.1093 /ref:odnb/9780198614128.001.0001/odnb-9780198614128-e-33218.

CHAPTER 6

"The Sympathy of a Crowd": Imagining Scientific Communities in Mid-Nineteenth-Century Entomology Periodicals

Matthew Wale

If you had gone down to the woods near Maltby, a town in the south of York-shire, on 4 June 1858 at around nine in the morning, two men could have been observed vigorously beating some large elm and oak trees. This seemingly eccentric behavior had a perfectly reasonable explanation, as these individuals were entomologists, catching the larvae that fell from among the leaves. One was James Batty, a razor grinder from nearby Sheffield, who earned his living making the cutlery for which his hometown was famous, but who also enjoyed collecting moths and butterflies during his leisure hours. He wrote a brief account of this insect-hunting trip and sent it to his favorite periodical, the *Entomologist's Weekly Intelligencer* (1856–61), in which it was published.[1] The *Intelligencer* was among the earliest of the many natural history periodicals that proliferated from the late 1850s onward, following the final repeal of the infamous "taxes on knowledge." As this chapter will show, such periodicals played a significant role in shaping and managing communities of practice in natural history, not only changing the ways in which practitioners exchanged information, opinions, and specimens, but also changing the character and social composition of the communities themselves.

The *Entomologist's Weekly Intelligencer* was established and edited by Henry Tibbats Stainton (1822–92; fig. 6.1), one of the foremost entomologists of the nineteenth century. Possessed of a personal fortune that permitted him to devote all his energy to the pursuit of science, Stainton was a leading figure in the Entomological Society of London, and was elected a fellow of the Royal Society in 1867, a rare honor for those who studied insects. He specialized in

F I G . 6 . 1 . Henry Tibbats Stainton. Carte de visite, date unknown. © Trustees of the Natural History Museum, London.

microlepidoptera, a grouping of moth families characterized by their small, sometimes microscopic size, though his knowledge in all branches of entomology was highly regarded. Alongside his own research, Stainton dedicated much of his life and money to producing entomological periodicals. In addition to the *Intelligencer*, he was proprietor and editor of the *Entomologist's Annual* (1855–74), and was among the original founders of the *Entomologist's Monthly Magazine* (1864–present). An obituary records that even a few days before his death, Stainton was busily engaged in correcting proofs for this last publication.[2] He remains an important figure for current entomology, and both his specimens and his correspondence are now held at the Natural History Museum in London. The fourteen thousand letters contained in this archive represent one of the largest collections of material relating to the running of scientific periodicals outside of the Royal Society, and they have not hitherto been subject to sustained analysis by historians. This chapter draws on my larger study of Stainton's editorial practice to explore how he used his periodicals to develop and manage communities of entomological practice.[3]

Stainton and James Batty represent two extremes of those who practiced natural history in nineteenth-century Britain. The former was a gentleman of independent wealth, and a member of the metropolitan scientific elite. The latter was thoroughly working-class, spending his entire life engaged in an arduous industrial occupation. However, through the *Intelligencer* and the practice of correspondence, they formed part of a wider community of practitioners. In a now classic study, Susan Sheets-Pyenson identified an "ideology of amateur participation" in British popular natural history periodicals of the mid-nineteenth century, describing a "low scientific culture" in which socially diverse individuals were actively encouraged to take part in the production of knowledge. No expertise was required, "but simply an eagerness to participate and communicate."[4] More recent scholarship has perpetuated Sheets-Pyenson's formulation, and the term "low science" (sometimes used interchangeably with other terms such as "ethno-science" or "vernacular science") is still used to denote "an expectation of being involved in the creation of new knowledge."[5] However, to describe the contents of such periodicals as "low" science sets up an unsustainable contrast with a supposedly distinct "high" science. As shown in this chapter, natural history periodicals like the *Intelligencer*, in which the practice of correspondence brought together diverse communities of naturalists, complicate any attempt on the part of historians to achieve a clear demarcation between "low" and "high" scientific cultures in this period.

Within the pages of a periodical like the *Entomologist's Weekly Intelligencer*, the poorest of working men, such as Batty, and the most respected of genteel naturalists, such as Stainton, could engage in detailed interaction concerning many aspects of the practice of entomology, especially in relation to the location, identification, capture, breeding, and exchange of specimens. Such interactions were not new, but had long taken place through private correspondence, with artisans and gentlemen satisfying what Anne Secord has called "corresponding interests" in their complex exchanges.[6] As this chapter will

show, however, the interaction that took place through natural history periodicals was significantly different. The deference that had characterized letters
written by artisan naturalists to their gentlemanly counterparts was no longer
such a prevalent and socially necessary feature of correspondence intended for
the wider readership of a periodical (as opposed to a single, known recipient).
Not only did the mechanism of the journal make the communicative network
much larger and in some ways more efficient, but in doing so it also offered the
possibility of a more expanded conception of the entomological community
and of the individual practitioners' place within it.

Through a focused study of both the *Entomologist's Weekly Intelligencer*
and Stainton's correspondence, this chapter will demonstrate how the periodical created a scientific community that encompassed a wide range of
individuals with divergent motives, skills, and interests, who nevertheless engaged with each other in ways that were scientifically productive. It thus re-
sponds to Jonathan Topham's call to rethink our accepted notions of "popular
science" and the hierarchies of scientific participation.[7] The chapter will begin
by setting the *Intelligencer* within the context of the mid-nineteenth-century
natural history periodical market, emphasizing what was new and distinctive
about Stainton's publication. In particular, it will show that the highly seasonal practices of natural history, and especially entomology, made the rapid
communication of information a necessity. The next section will demonstrate
how the periodical functioned, arguing that the *Intelligencer* applied printing technologies to the existing mode of scientific correspondence, leading
to new patterns in the circulation of knowledge. As a result, the periodical
enabled the imaginative process through which a community of entomologists
could be constructed, creating what Stainton described as the "sympathy of
a crowd." The third section will explore the motivations of periodical editors
such as Stainton, who actively pursued the creation of such a diverse community of practitioners through these publications, effecting a division of labor
in the creation of knowledge and the advancement of science. Stainton was
keen to encourage the pursuit of entomology among the working classes, as
he believed these collectors and observers were a vital component of an effective scientific community. The final section will explore the participation of
working-class naturalists from their own perspective, using the razor grinder
James Batty as a key example. The periodical provided individuals such as
Batty with a greater degree of agency within the scientific community than had
hitherto been possible. Through its focus on the *Intelligencer*, the chapter will
thus demonstrate that new types of natural history periodicals at mid-century

served to support the construction of diverse communities of practitioners that cannot be captured by a distinction between "high" and "low" science.

PUBLISHING AN ENTOMOLOGICAL PERIODICAL

The first number of *Intelligencer* arrived in the hands of entomologists on Saturday 5 April 1856, costing a single penny, a price that never varied throughout its existence. "It was by some considered an extremely wild-goose speculation to attempt to bring out a penny weekly journal in any degree scientific," observed Stainton in 1857, a year later. He admitted, "A few years ago this would not have been practicable," but that "thanks to Mr. Milner Gibson and his colleagues, their endeavours to remove the taxes on knowledge, and their success in obtaining the repeal of the newspaper stamp and advertisement duty," it had been made possible.[8] The publication of natural history periodicals, particularly those dedicated to a special branch of study such as entomology, was generally characterized by failure in this period.[9] High production costs and the difficulty in securing a sufficient readership were a constant struggle. The two preceding attempts to establish a commercial journal in this particular field, the *Entomological Magazine* (1832–38) and the *Entomologist* (1840–42), were both monthlies that were discontinued due to lack of support. The *Intelligencer* was therefore a bold departure from these previous efforts, as it was not only specialized but also cheap and weekly. As Stainton acknowledged, the abolition of advertising and stamp duties in 1853 and 1855 respectively, the result of a campaign led by the politician Thomas Milner Gibson (1806–84), had been a critical turning point.

The *Intelligencer* was printed and published by Edward Newman (1801–76), another renowned entomologist, periodical editor, and author of numerous books and articles.[10] He was a friend of Stainton, and they would have seen each other frequently at meetings of the Entomological Society of London. Newman also visited Stainton's home at least once, accidentally leaving behind a pair of "entomological spectacles" that he requested be returned should Stainton "have to come to town."[11] The son of a Quaker manufacturer of morocco leather, Newman lacked the inherited wealth that sustained Stainton. After a brief spell in his father's rope-manufacturing business, Newman became a partner in the printing company of George Luxford (1807–54) during the late 1830s, and by 1841 had taken sole ownership. In addition to printing, Newman edited the monthly *Zoologist* (1843–1916), a "popular miscellany of natural history" that enjoyed a remarkably long life, surviving him by forty years. Significantly, Newman had conducted both the *Entomological Magazine* and the *Entomologist*, and the latter was subsumed by the more broadly construed *Zoologist* in 1843.[12] Furthermore, Newman established and managed (but did not edit) another monthly periodical at his own expense, the *Phytologist* (1841–54), a "popular botanical miscellany." Another key player in this network was the publisher John Van Voorst (1804–98), whose company specialized in works on natural history, including those of such luminaries as Richard Owen. Van Voorst published the *Zoologist* (up to 1886), the *Entomologist*, the *Phytologist*, both Stainton and Newman's own books, and the *Entomologist's Annual*. He also acted as a retailer for the *Intelligencer* from his premises on Paternoster Row.[13]

Newman was at first a little skeptical that another periodical was needed, and worried about a rival publication harming the sales of his own *Zoologist*. Despite these reservations, however, he wrote to Stainton promising to

> stitch the advertisement [for the *Intelligencer*] in the *Zoologist* although I am well aware it is what is technically called "cutting my own throat" for I am thoroughly aware how the *Intelligencer* will interfere with the sale of the *Zoologist*.

In return for this selfless act, Newman requested that Stainton "stitch up" an advertisement in one of the latter's forthcoming books.[14] The *Zoologist* was clearly equal to the challenge of the *Intelligencer*, and indeed any of the considerable number of rival natural history periodicals that sprang up over the next half century and beyond. While on the subject of financial considerations, it is worth noting that Stainton and the *Intelligencer* were largely above such concerns. Although "the circulation was not sufficiently extensive [at least in 1857] to make the sale at a penny remunerative," fully bound volumes were made available at the cost of nine shillings each to ensure that "the loss [to Stainton] would be *nil*, or nearly so."[15] Unlike Newman, whose livelihood rested upon his business, Stainton was not concerned about turning a profit. His personal fortune allowed him to afford a degree of loss, which he presumably considered a price worth paying if entomology was significantly advanced by such an enterprise. This chimes with Stainton's noted philanthropic disposition, and sets the *Intelligencer* apart from many natural history periodicals whose utility to science was contingent on their commercial viability. Newman's *Phytologist*, for instance, in contrast to his *Zoologist*, "never was successful" as a "speculation," and consequently folded.[16]

It seems that others shared Edward Newman's doubts regarding the need for a periodical such as the *Intelligencer*. The *Zoologist* already fulfilled the function of announcing captures and sharing observations among naturalists, though its contents were not limited to entomology alone. Stainton was adamant that an alternative was required. The opening article of the first issue was titled "Why Do Entomologists Want a Weekly Newspaper?"—a question Stainton answered as follows:

> Those who discover a fact in the economy of insect-life don't like to keep their discovery to themselves till the end of the season, yet to write to each of their intimate correspondents [. . .] requires more time than they are disposed to spare; now each discoverer has but to write one full notice of his discovery and forward it to us, and in ten days, at the very outside, it is in print and in the hands of nearly every Entomologist in the kingdom.

Stainton hoped that, as a result of this swift exchange of information, "each entomologist will find that he can live quicker and do more in a season, from the instantaneous intercommunication of ideas, than he could formerly do in two or three seasons."[17] In contrast to the individualized letter with a single recipient, the *Intelligencer* reproduced each communication hundreds of times and thereby permitted rapid circulation of knowledge. A full-page advert for the periodical was included in a newly published book, *Practical Hints Respecting Moths and Butterflies* (1856), by Richard Shield, and this notice explained the *Intelligencer*'s utility in relation to the need for rapid and widespread communication:

> No existing publication supplies this want; at present a rarity, caught on the 29th June, cannot be published till the 1st of August, when the information comes too late to be of use to others.[18]

All letters received by Wednesday were considered for inclusion in the issue of Saturday that same week, which was an unprecedented speed for communication on such a scale.[19]

Natural history was highly seasonal, as climatic variations from week to week could drastically alter the populations of specific organisms. Insects in particular were sensitive to such changes, with the larvae or imago phases of certain species only emerging for a short time when conditions were ideal. A weekly periodical was therefore far more effective at communicating these fluctuations than a monthly such as the *Zoologist*. The impact of weather upon the fieldwork of individuals is illustrated by the example of the Reverend Hugh A. Stowell, of Faversham, whose offer of insects published in the *Intelligencer* brought "a perfect flood of correspondence, which is quite beyond all my previous calculations." The clergyman lamented:

> That offer was dated May 22nd, but unfortunately did not appear till June 6th. When I wrote *Argiolus* was in full beauty; now I can find none but wind and rain-worn specimens. [. . .] With the other three Lep[idoptera]s. named, I hope to be able to supply most of those who want them; but this boisterous weather has come most importunately.[20]

This seasonal variation was a consideration particular to natural history. It is worth noting that the *Intelligencer* predated other significant weekly science periodicals, including the *Chemical News* (1859–1932), the *English Mechanic* (1865–1926), and *Nature* (1869–). Melinda Baldwin observes that speed of publication allowed for a sense of immediacy among correspondents not afforded by a monthly or a quarterly, which was much more akin to the meeting of a scientific society, but with the added benefit that far more people could read the periodical than attend such gatherings. The sense of community engendered by a weekly publication was consequently often much stronger. Additionally, establishing precedence of discovery was increasingly becoming a concern among a new generation of practitioners, and a weekly was the best way to stake a claim.[21] Nevertheless, it was the immediate practical application of information regarding the capture of insects that was of greatest importance to Stainton and the *Intelligencer*'s readers.

The degree to which the *Intelligencer* was tied to seasonal fieldwork practices is further illustrated by Stainton's decision to temporarily discontinue the periodical in September 1856, having begun in April that year. Stainton did not believe that such a publication was required during the winter, as most entomologists ceased to collect at that time. There would no longer be any information to share regarding the capture of insects, with the majority of species only active during the warmer months. The *Intelligencer* therefore hibernated until the following April, when the collecting season recommenced. However, such was the popularity that Stainton's periodical had gained in this short time that a plan was put in place to fill the void with another publication, titled the *Substitute*. It was edited by John William Douglas (1814–1905), a close friend and collaborator of Stainton's, and printed by Edward Newman. The *Substitute*'s title had two meanings: it was a replacement for the *Intelligencer*, but it also pointed to the periodical's purpose as an "entomological exchange facilitator."[22] Collectors did not spend the winter in complete inactivity, but rather used this time to sort through the insects they had captured in the summer and arrange them suitably in display cabinets. It was often during this process that duplicates in their collection would became apparent, and the periodical enabled them to exchange (or "substitute") these surplus insects. In subsequent years it was not considered necessary for the *Intelligencer* to repeat this hiatus, as its growing readership thereby kept it supplied with sufficient material.

Published every Saturday, a typical issue of the *Intelligencer* followed a reasonably predictable format, consisting of eight pages printed in two columns. There was no cover, but simply the title emblazoned at the head of the first page, which also featured a short leading article written by Stainton (fig. 6.2). This is where his editorial voice was most evident, as he used this space to address his readers on a wide range of subjects, making his opinions known and attempting to shape those of others. The tone ranged from gently humorous and whimsical to more serious and occasionally even angry. For example, the second issue asked, "Why did Mr. Westwood get the Royal Medal?" The entomologist John Obadiah Westwood had received this prestigious award from the Royal Society in 1855, and while Stainton did not dispute that the prize was deserved, he questioned why the field of entomology had not previously been honored thus.[23] These editorials played a key role in shaping the community of the *Intelligencer*, as he addressed his readership as a whole, emphasizing cohesion and familiarity. The remainder of each issue was devoted almost entirely to correspondence—mostly "communications," letters written by insect collectors from across Britain and selected by Stainton for publication. These were sometimes no more than a sentence or two in length, and generally no more than a paragraph, each headed with a short title to indicate its subject. The *Intelligencer* thus replicated the mode of personal correspondence in some respects, while operating in a significantly different way in others. As the following section will show, it was this development above all that was key to the formation of a new kind of scientific community through the journal's pages.

THE INDUSTRIALIZATION OF CORRESPONDENCE

Letter writing and the practices of natural history share a long and close association. Elizabeth Yale has shed light on the use of correspondence by early modern naturalists in conducting and communicating their researches, and how this manuscript culture related to the burgeoning print media of the period.[24] Furthermore, one of the most famous and influential works of nature writing, Gilbert White's *Natural History and Antiquities of Selbourne* (1789), took the form of letters addressed by the parson-naturalist to his peers.[25] The

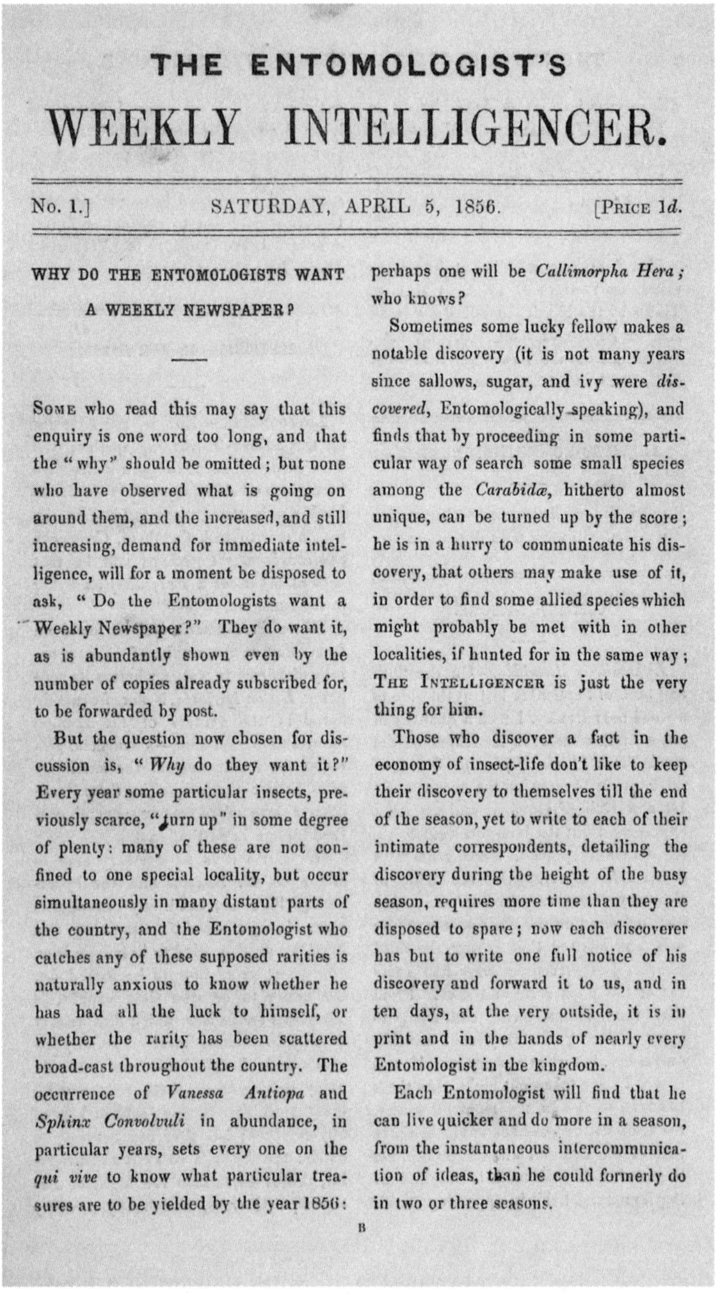

THE ENTOMOLOGIST'S

WEEKLY INTELLIGENCER.

No. 1.]   SATURDAY, APRIL 5, 1856.   [PRICE 1*d*.

WHY DO THE ENTOMOLOGISTS WANT A WEEKLY NEWSPAPER?

SOME who read this may say that this enquiry is one word too long, and that the "why" should be omitted; but none who have observed what is going on around them, and the increased, and still increasing, demand for immediate intelligence, will for a moment be disposed to ask, "Do the Entomologists want a Weekly Newspaper?" They do want it, as is abundantly shown even by the number of copies already subscribed for, to be forwarded by post.

But the question now chosen for discussion is, "*Why* do they want it?" Every year some particular insects, previously scarce, "turn up" in some degree of plenty: many of these are not confined to one special locality, but occur simultaneously in many distant parts of the country, and the Entomologist who catches any of these supposed rarities is naturally anxious to know whether he has had all the luck to himself, or whether the rarity has been scattered broad-cast throughout the country. The occurrence of *Vanessa Antiopa* and *Sphinx Convolvuli* in abundance, in particular years, sets every one on the *qui vive* to know what particular treasures are to be yielded by the year 1856:

perhaps one will be *Callimorpha Hera*; who knows?

Sometimes some lucky fellow makes a notable discovery (it is not many years since sallows, sugar, and ivy were *discovered*, Entomologically speaking), and finds that by proceeding in some particular way of search some small species among the *Carabidæ*, hitherto almost unique, can be turned up by the score; he is in a hurry to communicate his discovery, that others may make use of it, in order to find some allied species which might probably be met with in other localities, if hunted for in the same way; THE INTELLIGENCER is just the very thing for him.

Those who discover a fact in the economy of insect-life don't like to keep their discovery to themselves till the end of the season, yet to write to each of their intimate correspondents, detailing the discovery during the height of the busy season, requires more time than they are disposed to spare; now each discoverer has but to write one full notice of his discovery and forward it to us, and in ten days, at the very outside, it is in print and in the hands of nearly every Entomologist in the kingdom.

Each Entomologist will find that he can live quicker and do more in a season, from the instantaneous intercommunication of ideas, than he could formerly do in two or three seasons.

B

FIG. 6.2. *Entomologist's Weekly Intelligencer* 1 (1856): 1. The layout of the front page varied little throughout its existence. Courtesy of the Trustees of the Natural History Museum, London.

great growth of natural history in nineteenth-century Britain was accompanied by a further expansion in the use of correspondence. Indeed, Janet Browne has written that "for some Victorians, a large-scale correspondence network constituted a scientific method."[26] Stainton pitched his new periodical squarely in relation to such practices.

The very title of the *Intelligencer* referenced a long-standing tradition in scientific correspondence. In the seventeenth century, an "intelligencer" was an individual possessed of an extensive international correspondence network, such as Henry Oldenburg, secretary of the Royal Society. Acting as "information brokers," intelligencers served as intermediaries in the transmission of letters and broadcast information they deemed of interest to a wider community of practitioners. With the early modern national postage system far from efficient, and international mail delivery unreliable at best, this was a vital function in the transit of knowledge. Furthermore, naturalists only required a single correspondent's address in order to participate.[27] The parallels with Stainton a few centuries later are clear, as his own voluminous correspondence archive demonstrates. His establishment of the *Intelligencer* strongly suggests that Stainton was self-conscious regarding his position, and also shows the degree to which the periodical was considered a logical extension of this role. The *Intelligencer* performed this function far more efficiently than any single person could hope to achieve, effectively industrializing the process of scientific correspondence among entomologists through the application of printing technology and the exploitation of the periodical form.

Entomologists with such extensive correspondence networks as Stainton were relatively rare, and those located outside major metropolitan areas were particularly unlikely to cultivate a wide range of entomological acquaintances. Yet, without a sense of a wider scientific community, there could be no consciousness of how their individual activities contributed to a greater project. Anne Secord has argued that private correspondence between two individuals should be considered as evidence of participation in a community, but the industrialized form of correspondence facilitated by the periodical considerably expanded the opportunity to engage in such community participation.[28] Whether they contributed letters or not, the readers of correspondence published within the *Intelligencer* could imagine themselves as part of a larger entomological collective. Relevant to this argument is the work of Benedict Anderson regarding the role of newspapers in the imaginative construction of a nation-state, which can be fruitfully applied to the analysis of scientific communities. The "mass ceremony" of immediate and simultaneous consumption of a newspaper by its readers is key to the imaginative process of community, as each individual is aware that many others are enacting the same ritual (repeated at regular, set intervals), though the vast majority of their fellow devotees remain personally unknown to them. Furthermore, a fundamental characteristic of the newspaper, Anderson argues, lies in its presentation of unconnected stories as somehow linked. The concurrence of the events related by a newspaper—embodied by the date printed at the head of its front page—is the only connection, which is fundamentally arbitrary and fictive.[29]

The consumption of the *Intelligencer* by its diffuse readership permitted the imaginative leap necessary for a community to cohere, as the periodical presented otherwise unrelated narratives as connected, if only because they had occurred at the same time. A single issue circulated information regarding the insects captured in Britain during the week preceding publication. Most of its contributors and readers were distant strangers, geographically and socially dispersed, much like the citizens of a nation-state. Included on the same page as James Batty's report of his trip to Maltby were similar accounts of insect collecting at around the same date from correspondents based in London, York, and Glasgow (fig. 6.3). The disparate activities of naturalists from across the country were thereby unified in their minds. In a comparable case, Michael Brown points to the role of the *Lancet* in constructing a distinct medical community in early nineteenth-century Britain. He describes periodicals as "technologies of the imagination," which enabled readers to conceive of themselves as part of a collective that was temporally and spatially coextensive.[30]

In addition to entomologists' newfound awareness of each other's existence, Stainton emphasized the importance of shared work in bringing about an entomological community:

> [The] publication of each other's movements reacts favourably upon Entomologists [. . .]: it produces the sympathy of a crowd; each finding himself no longer isolated, and working only for his own amusement, finds himself placed in a higher and more unselfish position; he works now for the amusement and instruction of others as well as his own.[31]

The terms used by Stainton to express this sense of community suggests a strong affective dimension to the use of periodicals. The term "sympathy" suggests an imaginative act by which individuals placed themselves in a state of shared feeling and experience. This "unselfish position" of exchange and cooperation, which Stainton believed was necessary for the advancement of entomological knowledge, is something that we shall return to in the following section.

Captures at West Wickham.—I made the following captures at West Wickham on the 6th inst.:—

Sesia Fuciformis (3),
Trochilium Culiciforme (1),
Macroglossa Stellatarum (1),
Nemeophila Plantaginis (2),
Pyrausta purpuralis (2).

—CHARLES HEALY, 4, *Bath Place, Haggerstone, N.E.; June* 8.

Captures near Sheffield.—On the 4th inst. I and a friend, Mr. Moore, being provided with a two-yards-square sheet, took the route for Maltby Woods, where we arrived at 9 A.M. We put the sheet together, and began to beat some large elms, oaks, &c.: we beat about two hours, and took 100 larvæ of *Thecla W-album*, two of *T. Quercus* and three more, which no doubt will be *X. Gilvago* (*C. trapezina* was of course at every corner of the sheet), one *Geometra Papilionaria*, one *P. Syringaria*, three *N. Hispidaria*, and a few other strange Geometræ. We then wrapped up the sheet, and took to our nets and worked hard till three in the afternoon: we took *N. Lucina, T. Alveolus, T. Tages, P. Sylvanus, P. Statices, C. Jacobææ, P. Purpuralis*, one *B. Pandalis* (this insect seems rare this year) and *E. Decoloraria.* We returned home quite satisfied with our journey.—JAMES BATTY, 133, *Bath Street, Park, Sheffield; June* 7.

Captures in Argyleshire.—Last week, on new ground, four miles north of Kilmun, I met, for the first time, with *M. Stellatarum, T. Laricaria* and *Botys Decrepitalis* (see Ent. An., 1856, p. 32). *Laricaria* sits closely pressed on the bark of larches, and is very unwilling to move, requiring a pull to take them off. The females are equally unwilling to lay their eggs in confinement, having obtained only nine eggs from five individuals. The eggs are oval, dull and of a delicate green.—THOMAS CHAPMAN, *Glasgow; June* 7.

Duplicates.—I have been very successful in taking *A. Euphrosyne* and *M. Artemis*, and shall be glad to exchange for almost any local or South-country butterflies or moths: those who are in want of them had better write first, and state what they have to spare.—J. ROBINSON, *Jackson Street, Groves, York; June* 8.

Entomological Excursion to Roche Abbey.—On the 2nd inst., in company with Mr. W. Green, I visited the above charming spot, which for beauty of wooded scenery has few equals, and where a great many of the British butterflies love to disport themselves on the margins of the beautiful wooded hills and vales. The following is a summary of our captures:—*N. Lucina* (60), *A. Euphrosyne, T. Tages, T. Alveolus, E. Glyphica, P. Purpuralis, A. Cardamines, E. Angularia, S. Ribesaria, E. Decoloraria*, and 95 larvæ of *T. W-Album*, just emerging from the chrysalis, and several larvæ unknown to us. Mr. Green has at present about 400 of the larvæ of *H. Dispar*, which he will be glad to exchange with any entomologist for other local species; he has also 112 *S. Carpini*, male, which he took by the attraction of one female.—W. H. SMITH, *Eccleshall New Road, Sheffield; June* 7.

Eggs for Distribution. — My notice of last autumn respecting *Ptilodontis palpina* and *Notodonta dictæa* brought many requests for eggs or larva of one or both species, if I obtained any. Such of these as I entertained I answered, promising to place the names on the list. I beg now to say that I have obtained impregnated eggs of both insects, and before this notice appears each person on the list will have received a small batch of such as they respectively wished for. The matter has been some time on hand, and the applications dropping in throughout the winter, possibly some names may have been omitted: any one therefore who holds a promise from me, and who has not already received a supply will

---

FIG. 6.3. *Entomologist's Weekly Intelligencer* 4 (1858): 85. A typical page of the periodical, featuring various accounts of insect collecting from correspondents around the country, including James Batty's trip to Maltby Woods. Courtesy of the Trustees of the Natural History Museum, London.

The *Intelligencer* fulfilled its projected purpose, at least in the opinion of the editor, who later asserted: "As an instantaneous *medium* of communication between Entomologists in all parts of the country it has proved most serviceable."[32] Claims of instantaneity were an exaggeration, of course, but they serve to emphasize the novelty of such rapidity. As railway networks spread and became more efficient during this period, it was possible to ensure that periodicals arrived at booksellers throughout the country, as evinced by the growing list of vendors included in many issues of the *Intelligencer*.[33] Following Stainton's death in 1892, the *British Naturalist* noted:

> Just at the right moment, when extra postal facilities, and the extension of the railway system gave greater opportunities for the inter-communication among Entomologists, he [Stainton] brought out his *Entomologist's Annual* (1855), his *Manual of British Butterflies and Moths* (1856), and the *Entomologist Weekly Intelligence* [sic]. These gave the impetus wanting, and made Entomology what it is to-day.[34]

Chief among the "extra postal facilities" was the introduction of the Uniform Penny Post in 1840, which was a key development in the industrialization of correspondence. As the title suggests, this reform set the cost of sending a single letter at one penny, paid in advance and irrespective of distance, thus rendering it much more affordable. Previously, the responsibility of payment had usually been placed upon those receiving the letter, with prepayment by the sender "often considered an indirect social slur."[35] In 1839, the year before the reform was passed, an estimated 76 million letters were delivered in the United Kingdom. By 1856, the year the *Intelligencer* began publication, this figure had risen to around 478 million.[36] Furthermore, it was not only letters that were being sent through the post, as attested by John Steven Henslow, the distinguished professor of botany who had mentored Charles Darwin at the University of Cambridge.

> To the importance of the penny postage to those who cultivate science, I can bear most unequivocal testimony, as I am continually receiving and transmitting a variety of specimens by post. Among them, you will laugh to hear that I have received three living carnivorous slugs, which arrived safely in a pill-box![37]

The advent of the penny post must be considered a key factor in facilitating periodicals that relied on correspondence for their content. The sheer volume of letters and specimens circulated in response to these publications would have been considerably more limited with a more expensive and less efficient postal service.

The creation of a community based upon the rapid circulation of knowledge was considered necessary to the advancement of science. As Stainton remarked, "The knowledge attained by an *individual*, unless rendered available to others, may be no gain to science: at his death all his thoughts perish, and all his knowledge is lost for ever." Stainton went on to implore his readers: "We look upon it as the bounden duty of all who acquire information at once to render it available to others."[38] This statement is a recognition that the production of scientific knowledge necessarily entailed its communication. Exactly what warranted publication and what did not was very much down to Stainton's personal judgement. Most commonly, a correspondent would announce the capture of a particular insect. If this were a rare species, or had been found to occur in a location or habitat previously unrecorded, then the interest in such news was much greater. Unlike other natural history journals of the period, the focus of the *Intelligencer* was almost exclusively upon the fieldwork of its correspondents. Rather than presenting extended descriptions of new species, it more simply gave practical accounts of where and how such insects had been captured. One correspondent, A. Wallace of Clerkenwell (not Alfred Russel Wallace, who was in the Malay archipelago at the time), wrote to the *Intelligencer* expressing his opinion on why such information was useful:

The whole question of collecting lies in a nut-shell: it is the old game of "How, when and where?" Answer these three questions with reference to any one insect, and then the right man in the right place, at the right time, is sure to realize, —*viz.* let Mr. Samuel Stephens go down to West Wickham the first fortnight in May. The result is self-evident: *Carmelita* is taken, eggs; larvae obtained; our cabinets supplied.[39]

To this end, more than the insect itself must be collected, as the information relating to its capture was of great importance. The entomologist should note "where he went, the name of the capture, and whether by sugar, light, flight or their capture as larva or pupa, [. . .] the condition of the wind, weather, whether cold or warm, dry or wet." The date was of particular importance, as "many insects appear true to time, from year to year, even on the same day."[40] The reference to Mr. Stephens is most likely a misspelling of Samuel Stevens (1817–99), the noted natural history agent and keen entomological collector. His business on Bloomsbury Street in London sold ready-prepared microscopic slides, in addition to exotic specimens collected by men such as Alfred

Russel Wallace and Henry Walter Bates.[41] It is not clear from the above note whether Stevens would have sold or exchanged the specimens he collected himself, but nevertheless it demonstrates why the sharing of such information was useful to a wider community of collectors.

The *Intelligencer* increasingly became a forum for the exchange of specimens themselves, in addition to information. Collectors with multiple specimens of a certain species would publish lists of their "duplicates" and desiderata in the periodical, allowing other readers to apply directly to the correspondent and thereby arrange a mutually beneficial swap. Although many collectors did sell specimens, which no doubt provided a vital source of income for some, such transactions were not carried out through the *Intelligencer*. Exchanges were intended to be liberal and disinterested rather than hard-driven bargains, with many experienced collectors happily distributing specimens to beginners with no expectation of receiving anything in return.[42] Given the value of insect specimens, particularly those of rare species, there were many who did not conform to this gentlemanly aspiration. In botany, by contrast, dried plant specimens lacked the market value of insects, and therefore such monetary considerations were less of an issue. A more centralized approach was adopted by a number of botanical societies, such as the Thirsk Exchange Club in 1857, in which plant specimens were sent to these organizations and then redistributed in an evenhanded fashion among subscribers.[43] Perhaps due to the persistence of specimen dealing, a similar approach was not adopted by entomologists at this time.

One key difference between exchanging letters one-on-one and corresponding through a periodical is illustrated by the case of Robert Burns of Edmund Street, Birmingham. Burns was a newsagent, and it was from his business that the *Intelligencer* could be obtained by residents of this city. He also collected insects, and placed a notice in the periodical advertising his willingness to distribute specimens of the striking green-and-pink elephant hawk-moth. As a result, "such a flood of correspondence quite alarmed my little home, eight or ten letters arriving each day,—untiring, unceasing,—more in one week than in all my life before!"[44] It seems this startling "flood of correspondence" was quite a common occurrence among those who advertised in the periodical, as it was also experienced (and described in the same terms) by the Reverend Stowell, as quoted above. A wry note in one issue of the *Intelligencer* advised:

> Our correspondents should bear in mind that an offer of duplicate Lepidoptera which includes any of the less common species is pretty sure to produce

from 80 to 100 applications; offers of Coleoptera from 40 to 60. We are never surprised at entomologists being overwhelmed with applications, but each new correspondent appears thunderstruck at the result of his announcement.[45]

The *Intelligencer* served exactly as its early modern namesakes, forming a key link in a network that permitted individuals to engage in correspondence with a larger community of practitioners. The numbers of applications received points toward the industrial scale of this correspondence network. Previously, few if any collectors of insects could have broadcast news of their duplicates to so many. Moreover, the novelty of this is apparent from the shock experienced by those who received such floods of letters.

The circulation of the *Intelligencer* reached a peak of six hundred, with readers distributed across the country. This figure may seem relatively small, but the largest entomological community that had previously existed, the Entomological Society of London, had fewer than two hundred members at this time, and these were primarily concentrated in the capital.[46] The imagined community that cohered through the pages of the *Intelligencer* drew upon a more diverse range of individuals. With the making of entomological knowledge contingent upon the participation of collectors from around Britain, it was the expressed aim of Stainton and others to create the "sympathy of a crowd" among a socially varied range of naturalists. It is to this ambition that we now turn.

DEVELOPING THE ENTOMOLOGICAL COMMUNITY

In an address to the Entomological Society of London titled "How May the Onward Progress of the Study of Entomology Be Best Furthered?" read on 4 February 1856, Stainton outlined his communal vision of science using an appropriately entomological analogy:

> The bee rifles the flower of its honey not for its own immediate pleasure and enjoyment, but in order that it may be carried home and added to the common store for the use of the community: [. . .] so must it be with the scientific student.[47]

This statement is an earlier formulation of Stainton's "sympathy of a crowd," in which the individual works for the good of the community rather than their own selfish gain. The comparison of human society with that of bees (and of other social insects such as wasps and ants) is one with a rich tradition, stretching back to antiquity. It was a prevalent trope in nineteenth-century political discourse, employed by individuals of widely divergent allegiances, as a beehive could be seen to exemplify either a well-ordered and industrious constitutional monarchy or a perfect socialist society.[48] Stainton conceived of entomologists as members of a community among whom the work of scientific progress was divided, although his own utopian vision is likely to have been rooted in ideals of gentlemanly disinterest rather than more radical notions of cooperative labor. In the same address to the Entomological Society, he observed:

> On looking through Mr. [Frederick] Smith's *Monograph of the British Bees*, we find that it condenses not merely his own observations during twenty years, but also a mass of extraneous observations made by others, themselves unaware of their value, but which, being communicated to Mr. Smith, were at once recognised by him as supplying some important link in the chain of information he was collecting.[49]

It is significant that Stainton expressed these views only a few months before he commenced the *Intelligencer*. Although he did not expressly mention periodicals in his address to the Entomological Society, it is highly suggestive that he shortly went on to establish a publication that in many ways embodied the cooperative endeavor he espoused. The *Intelligencer* served to accumulate "extraneous observations" through the printing of short notices, whereas Frederick Smith and others would have previously relied on personal correspondence.

The *Intelligencer* soon found an admirer in Germany, and in the sincerest form of flattery, Dr. Gottlieb August Wilhelm Herrich-Schäffer (1799–1874) began his own imitation entitled *Correspondenzblatt für Sammler von Insecten, Insbesondere von Schmetterlingen* ("Journal for Insect Collectors, Especially for Collectors of Lepidoptera").[50] Herrich-Schäffer was a physician by profession, but also an entomologist of considerable European repute, having written a highly influential work on the taxonomic classification of Lepidoptera.[51] Stainton remarked upon the *Intelligencer*'s new continental counterpart in an editorial titled "A Good Move," quoting from Herrich-Schäffer's comments concerning his rationale in starting the journal:

> The demand for periodical entomological literature would appear to be supplied already by the *Stettin Entomologische Zeitung*, the *Berlin Entomologische Zeitschrift*, and the *Vienna Monatschrift*, but the two former only appear quar-

terly, and that though the last-named is a monthly publication, yet all the three are more restricted to works of a purely scientific character.

Herrich-Schäffer continued:

> The appearance in London of the *Weekly Intelligencer* first suggested to me [. . .] the idea of establishing a similar journal for Germany, which, like its London prototype, without pretending to learned investigations, should serve as a medium of intercommunication for the amateurs and collectors of insects [. . .]. We possess in the three above-named periodicals, and in the *Linnaea* and some other works, more than sufficient for scientific and longer treatises, but for some short notices on single species and genera, especially on points of difference between allied species, for observations on their local or periodical occurrence, or on their habits, and especially for notices of which the usefulness consists in their immediate circulation, we have at present no suitable channel.[52]

Although we should be aware of cultural differences between the British and German contexts, Herrich-Schäffer's use of the word "amateur" here is instructive, as it is juxtaposed with more rarefied "learned investigations."

At this period, "learned" or "scientific" entomology generally meant work of classification, of distinguishing one species from another and naming new discoveries. This process was largely based upon the examination of specimens—that is, dead and pinned insects—rather than a study of living creatures, just as botanists employed herbarium sheets to achieve the same end. How exactly to determine a species was a vexed question, with many practitioners employing their own idiosyncratic systems. Insects are among the most numerous organisms on the planet, and they exist in a multitude of varieties that can often only be differentiated through the study of microscopic differences. At the time of the *Intelligencer*'s publication, new species were still being discovered in Britain relatively frequently, and more intrepid collectors were supplying a constant stream of specimens from around the globe. Entomologists, for the most part, were less concerned with biological questions of physiology and evolutionary descent than with classification. Charles Darwin's *Origin of Species* (1859), published at the midpoint of the *Intelligencer*'s lifespan, received scant and far from favorable notice in the periodical.[53] For these readers and correspondents, such lofty questions were considered beyond the purview of the working naturalist, whose role was to steadily accumulate facts without indulging in theoretical speculation. Stainton himself expressed such a view, remarking that the advancement of entomology had been retarded by the time wasted in "elaborating *theories*, whilst a collection of *facts* on which alone theories ought to have been founded was disregarded."[54] The necessary groundwork of natural history was carried out by localized collectors, who obtained the specimens and observations that were essential material upon which the methodical endeavor of taxonomy was based.

The more informal character of the *Intelligencer*, with its focus on exchanging news and short notices rather than on publishing more thoroughly researched "scientific" articles, was clearly considered to be less forbidding to novices in the field and those unfortunate enough to lack extensive (or expensive) educations. Furthermore, a significant distinction was drawn between "scientific" entomologists and those who primarily worked at amassing specimens. Herrich-Schäffer used the terms "amateur" and "collector" to denote those who acquired insects and other specimens for purposes that were not "'purely scientific" (at least according to self-described "learned" entomologists such as himself). A great number of people considered collecting and displaying insects, particularly aesthetically pleasing specimens of Lepidoptera, as an end in itself—but we must be wary of any suggestion that such individuals were not participants in the broader project of natural history. The term "amateur," in this sense, does not necessarily hold any pejorative meaning. The very suggestion that "amateurs and collectors" required a periodical of their own is an admission of their value to natural history. The *Intelligencer*, and the publications that imitated it, sought to cultivate such "practical" workers, as Stainton and others who considered themselves to be of a more "scientific" bent were fully aware of the importance of such men and women to their field. Even if the collectors themselves did not attach any scientific importance to the information or specimens they shared through the periodical, the very act of circulating such information regarding the "local or periodical occurrence" of insects among a wider community was one that produced scientific knowledge.

Stainton was particularly committed to encouraging entomology among those less socially fortunate than himself, boldly stating that "entomologists are not drawn from the wealthy, but rather from the working classes."[55] Elsewhere, he claimed that "the Spitalfield weavers, the Sheffield cutlers and the Manchester cotton-spinners are amongst the most successful collectors of insects, as well as great amateurs of birds and flowers."[56] Stainton may not have envisioned science as entirely classless, but he acknowledged the worth of such individuals to the project of natural history:

An Entomologist is none the less one because he wears fustian, and "labours, working with his hands"; and in very many of this class the innate love of these beautiful objects of creation, the Butterflies and Moths, supplies them with one of their purest pleasures. Should not such tastes and such pursuits be encouraged? An observation, if new, is as important by whomsoever made; and a Spitalfields weaver may supply some important gap in our knowledge, which Oxford and Cambridge put together would fail to elucidate.[57]

Thus, while Stainton was always eager to recruit beginners into entomology, his work should not be considered simply as popularization. He was fully aware that various kinds of practical natural history were being undertaken among the working classes, and his efforts through the *Intelligencer* were therefore an attempt to acknowledge this fact, and to recruit these workers toward the useful purpose of advancing entomological knowledge. Although he points to the moral benefits of such pursuits among working men, this does not appear to have been his primary motivation in cultivating natural history as a form of rational recreation. Earlier in the century, in the face of Chartist agitation and the threat of radical print, natural history (when presented or interpreted correctly) was considered a "safe" form of knowledge calculated to defuse potential revolution by occupying the minds of discontented laborers.[58] This anxiety had waned somewhat by the 1850s, as demonstrated by the repeal of the taxes on knowledge. Stainton was a staunch Liberal, who apparently only differed from the party line on the vexed subject of Ireland. He was very likely, therefore, to have been in favor of the franchise reform that was debated in the 1850s and 1860s, which acknowledged that the working classes were worthy of participating in democracy as well as science.

Stainton was by no means alone in his attempts to cultivate scientific workers by encouraging active participation through periodicals. Edward Newman stated in the introductory address of the *Zoologist*: "Every one who subscribes a single fact is welcome—nay, more than that—has a direct claim to be admitted as a contributor."[59] Similarly, the *Phytologist* asked for "FACTS, OBSERVATIONS and OPINIONS" that would be considered "trifling" by those of "high scientific pretensions." Practical skills were therefore favored over learnedness, as Newman wished to harness the efforts of "field-botanists—these observers—these labourers in the delightful fields of botanical enquiry."[60] At the heart of these publications lay the centrality of fieldwork to the accumulation of information in natural history. However, if we are to understand how these periodicals functioned in practice, we must consider the perspective of the working-class naturalists themselves, rather than merely the intentions of editors.

MAKING ENTOMOLOGICAL WORKERS

The role of the *Intelligencer* in facilitating the production of scientific knowledge among a wide range of individuals is exemplified by the case of James Batty, the razor grinder from Sheffield whom we encountered at the beginning of this chapter. Batty earned his living within the thriving cutlery industry of nineteenth-century Sheffield, an occupation that Frederick Engels described in the *Condition of the Working Class in England* (1845) as being particularly injurious to the workers' health, with many grinders lucky to reach the age of forty.[61] This does not seem to have been the case for Batty, who was born c. 1831 and lived into his sixties despite remaining in the same profession all his life.[62]

Notwithstanding the disadvantages of his situation, James Batty acquired sufficient education to read and write. Furthermore, in his leisure time he took to the study of Lepidoptera. Batty was an active member of the Sheffield Entomological Society, a group that has left little record aside from a few scattered references in the *Intelligencer*, including a brief account of the society's annual "feast," held on 20 April 1858, at which Batty exhibited specimens. Hosted by the Hen & Chickens Inn, a Sheffield public house, this event is reminiscent of the working-class scientific meetings described by Anne Secord.[63] In addition to these associational activities, Batty struck up a lively correspondence through the pages of the *Intelligencer*. His letters demonstrate the range of his entomological skill and knowledge, and reveal much about the practices in which he engaged. The original notes Batty submitted to the journal survive, and are written in a large but careful hand on small, cheap pieces of paper that contrast with much of the personalized stationary employed by Stainton's more affluent correspondents (fig. 6.4). Batty's first published letter was in the issue for Saturday 20 June 1857, announcing to the world his pleasure upon attaining a "fine female specimen of *Ceropacha fluctuosa*" from his "breeding-cage."[64] The breeding cage was a small boxlike structure with the sides covered in a fine gauze that prevented the insects within from escaping, but allowed them to breathe. It was usually kept in a darkened space in order to simulate the underground conditions in which many Lepidoptera species pupate, transforming from caterpillars into butterflies or moths. This was a mode of acquiring specimens very unlike pursuing them in the wild, and it required a different kind of expertise and dexterity, as not all species responded well

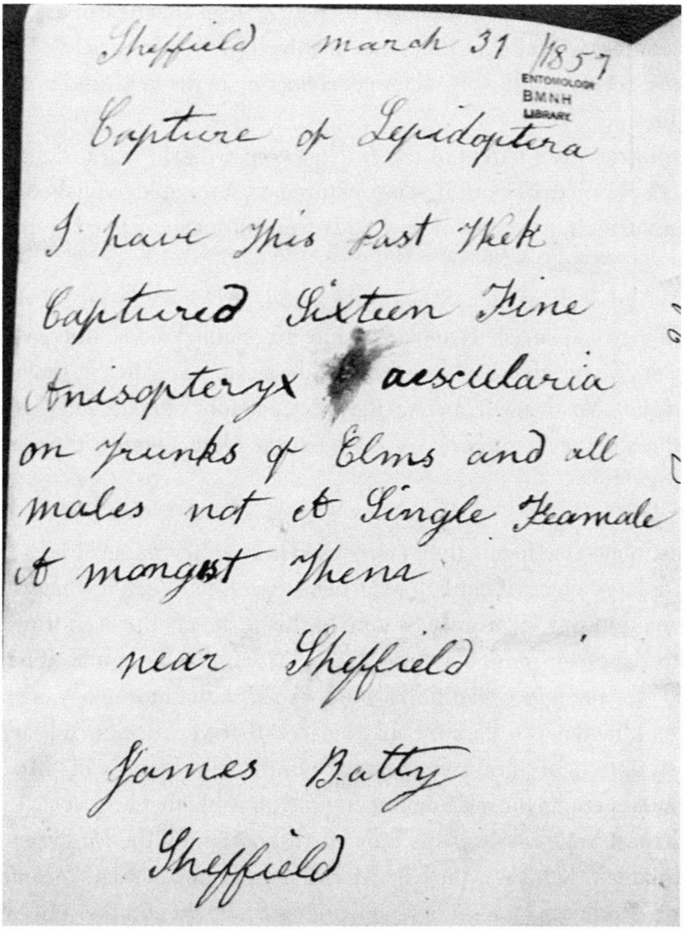

Sheffield march 31 1857

ENTOMOLOGY BMNH LIBRARY

Capture of Lepidoptera

I have This past Week
Captured Sixteen Fine
Anisopteryx aescularia
on trunks of Elms and all
males not a Single Feamale
amongst Them

near Sheffield

James Batty
Sheffield

FIG. 6.4. James Batty to H. T. Stainton, 31 March 1857, London, Natural History Museum, H. T. Stainton correspondence from British Entomologists, MSS STA E 118:118, STAINT 4:118. This is a note announcing the "capture of Lepidoptera," presumably intended for inclusion in the *Intelligencer* but not published. It reads: "I have this past week captured sixteen fine *Anisopteryx aescularia* on trunks of elms and all males not a single female is amongst them." By permission of the Trustees of the Natural History Museum, London.

to this treatment. Being the first to successfully breed a particularly difficult species ensured renown among the entomological community. Batty's second notice was a report from the field, noting that he had come across the species *Margaritia augustalis* "in a meadow near Maltby Woods."[65] Around a month later, in early August, Batty entered into the process of specimen exchange, advertising his willingness to part with a few "fine specimens" of the aforementioned *Ceropacha fluctuosa* in return for species that he listed.[66] Through the periodical we see that Batty was involved in a wide variety of scientific practices: corresponding, associating, collecting in the field, and breeding insects at home.

Through Batty's letters to the *Intelligencer*, we gain real insight into the fieldwork practices of working-class naturalists. A notable published narrative gives an intriguing account of how Batty went about acquiring his specimens:

> On the 4th inst. [of June 1858] I and a friend, Mr. Moore, being provided with a two-yards-square sheet, took the route for Maltby Woods, where we arrived at 9am. We put the sheet together, and began to beat some large elms, oaks, &c. [. . .]. We then wrapped up the sheet, and took our nets and worked hard till three in the afternoon [. . .]. We returned home quite satisfied with our journey.[67]

Two distinct collecting methods were used here. The first involved a "beating sheet," a large piece of cloth spread beneath a tree to catch whatever fell as the vegetation was vigorously beaten. In this instance it proved fruitful, with "100 larvae" of one particular butterfly species among the results. The second method was one more traditionally associated with entomology, namely the use of handheld nets to capture imago insects that were presumably active and relatively abundant during the summer months. The identity of "Mr. Moore" must remain enigmatic, although it is quite possible that he was B. J. Moore, a resident of York who was the only correspondent of the *Intelligencer* with that surname.[68] If this was the "Mr. Moore" in question, it raises the interesting question of how the two men, one living in Sheffield and the other in York, became acquainted. It is quite possible that the *Intelligencer* facilitated their friendship, but this must remain largely a matter of speculation.

Maltby Woods are located more than ten miles to the northeast of Sheffield, near the town of Maltby. Due to its unique geology and the species occurring there, it was an area much frequented by individual naturalists and society excursions during the nineteenth century. The woods remain to the present, with the nearby Maltby Low Common being a nature reserve that continues to attract enthusiasts. The area was a hunting ground particularly favored by Batty, who might have made the whole journey from Sheffield on foot, as no railways served the town of Maltby at that time. Batty's account therefore gives some indication of the range a naturalist could cover, but also demonstrates the kind of localized information the *Intelligencer* was effective in transmitting, particularly during the height of the collecting season, when others could profit from up-to-date news of which species were occurring in very specific parts of the country.

The periodical gave no indication of Batty's social status, with his name and address the only information provided as to his identity. However, his skill and expertise as an entomologist was clearly displayed through the *Intelligencer*, as illustrated by his ability to capture, breed, and identify numerous species of Lepidoptera. Batty's messages become more detailed and self-confident, giving meticulous descriptions of the appearance and habits of larvae, using more specialized terminology: "larva rigid rugose [. . .]; head slightly bifid [. . .]."[69] Writing a letter for publication in a periodical was a public, performative act, quite different from the more private practice of corresponding with an individual. This distinction became apparent when Batty's skill and knowledge was questioned in the *Intelligencer* by another Sheffield collector, William Thomas, who cast doubt over Batty's account of breeding *Acidalia inornata* (a small brown moth now known as *Idaea straminata*, or the "plain wave"). Batty had found this species to be particularly unreceptive to his attempts at hand rearing, as the larvae refused to eat any greenery other than a single type of shrub. He enquired of the other readers as to the moth's occurrence, hoping to settle the question of its preferred foodstuff.[70] In answer, Thomas, a fellow working man listed in the 1861 census as a "furnace builder," asserted: "I and several of my correspondents have bred the above-named species this year [. . .] it can be successfully reared on almost any low plant." Acknowledging that this moth was frequently confused with the very similar *Acidalia aversata*, or "ribband wave," Thomas magnanimously noted that it was Batty who had "first pointed out the differences of the two species to myself," revealing that the two men knew each other (it is likely that they had met at meetings of the Sheffield Entomological Society, though it has not been possible to verify this).[71] Thomas's reference to his "correspondents" gives an indication that such men already operated within networks of their own outside the periodical, albeit on a more limited scale.

A disagreement over the feeding habits of specific moth larvae may seem a minor issue, even among devoted lepidopterists, but Batty clearly felt that his reputation within the community was at stake. His response to Thomas, sent to Stainton but never published in the *Intelligencer*, claimed that "Mr. T. wants to make his self appear very large." Batty went on to remark, "I should think if Mr. T. had bred a [sic] *Inornata* he would have been in a great hurry to publish it to have the first claim but I think he's not yet on the throne."[72] That this dispute was carried out in the public forum of a nationally distributed periodical, rather than privately between two individuals who were already acquainted and living in close proximity, suggests that the publication had come to play an important role in the self-identity of the two men. Choosing to publicly dispute a claim was a provocative act calculated to call into question another naturalist's aptitude, and Batty's ire is palpable in his reply. It may have been a matter of space that influenced Stainton's decision not to print the rebuttal, or quite possibly a distaste for such disputes, which could be perceived as having much more to do with ego than with the interests of science. While Thomas's letter may have disputed Batty's assertion, it nevertheless contained pertinent information regarding the breeding of the *Inornata*, while Batty's response was one of wounded pride. Stainton seemed to care more about knowledge itself than about those who were first to discover it, and this perhaps points to divergent conceptions of science and skills between working-class and gentleman naturalists, as identified by Anne Secord.[73]

Batty died in 1893, less than a year after Stainton. The razor grinder had clearly gained enough repute within the entomological community that his obituary was published in at least four periodicals, including the *Entomologist's Monthly Magazine*. The short notice, replicated verbatim in each publication, described Batty as "a useful worker," and "an excellent type of the working-man lepidopterist":

> Batty had an excellent knowledge of larvae, and was the discoverer of the larvae of *Tapinostola elymi* and *Celæna haworthii*. He was a regular correspondent of the late Mr. Wm. Buckler and the late Rev. Joseph Hellins, and used to keep them well supplied with material for description.[74]

Here we have Batty cast in the mold of a self-made working-class scientific hero, a trope enshrined in the biographical works of Samuel Smiles.[75] The razor grinder was "useful" in providing specimens to others, but it is notable that, though he "discovered" the larvae of *Tapinostola elymi* (a moth known as a lyme grass), he was not the one who described it. He sent the specimen on to his "regular correspondent," the noted entomological illustrator William Buckler, who wrote a full scientific notice of it in the *Entomologist's Monthly Magazine*. Buckler thanked Batty in print, remarking that the latter "took a long journey during inclement weather, that he might search for the larva of this species."[76] Once again, Batty was portrayed in heroic terms, battling the elements for the greater good of science. It was not mentioned exactly where Batty had traveled to, but the larvae of this particular moth feed exclusively upon *Leymus arenarius*, a type of grass that only occurs on the eastern coast of Britain, at least seventy miles from Sheffield. Presumably his specimens of

Celæna haworthii (Haworth's minor) were collected a little closer to home, as this moth occurs most commonly in wild moorland of the kind abundant close to Sheffield. These larvae were also sent to Buckler, who undertook their description.[77]

Through the periodical, Batty participated in a correspondence network far larger than any he could have cultivated on an individual basis. Like some of the correspondents cited earlier in this chapter, he also experienced an unexpected deluge of letters in response to an advertisement of specimens:

> I have received so many applications for *L[iparis]. dispar*, &c., in consequence of the notice of my duplicates in the *Intelligencer*, that I cannot possibly answer all; those who do not hear from me must therefore conclude that my stock of duplicates is exhausted.[78]

However, Batty should not be dismissed as simply a provider of specimens to others, but should be seen instead as an entomologist in his own right, taking an active part in the making of scientific knowledge. Although he was perhaps one of the more exceptional working-class men of this kind, he was by no means alone in his pursuit, as demonstrated by his bricklaying acquaintance William Thomas. Individuals such as these played an important function within a loosely defined community of scientific practitioners, and an approach grounded in communicative practices reveals their role in greater detail.

Those who practiced natural history in the nineteenth century were diverse in terms of geography, social class, educational attainment, and scientific standing. The *Intelligencer* and other periodicals became important media through which these individuals cohered into a community, and the study of these publications offers us an opportunity to recover the ways in which naturalists such as James Batty participated in the circulation of scientific knowledge, and how this participation was negotiated. However, we should be wary of suggesting that this made natural history a classless endeavor, as many disadvantages remained to be faced by those who lacked social status or wealth. A number of factors limited the inclusivity of a periodical such as the *Intelligencer*. There was the obvious preclusion of all those unable to read and write from direct participation in the correspondence carried out within its pages. Furthermore, it should be noted that the correspondents to the *Intelligencer* were almost exclusively male. In the few places where a woman's name can be found, she is almost invariably married and referred to in a letter written by her husband, even if she captured the insect herself. In June 1858, Mr. J. P. Duncan, from Troon on the west coast of Scotland, reported that

Mrs. Duncan captured, yesterday afternoon, a fine female specimen of what I conceive to be *Micra ostrina*. In examining a clump of thistle on the sand hills it started up, and she gave it pursuit; twice it alighted, and having nothing but a small pill-box to take it with, it was at last secured.[79]

This notice suggests a number of things. It seems Mrs. Duncan was involved in fieldwork, paying greater attention to thistles than if she had simply been out walking. Furthermore, she carried a pill box, if nothing else—a standard piece of entomological equipment for retaining captured insects. However, it was Mr. Duncan who identified the specimen, and he who wrote the letter to the *Intelligencer*. Mrs. Duncan's individuality was elided, identified only by her husband's surname. Such women were not afforded agency by the periodical, as it was to their husbands that any further correspondence was addressed.

Stainton himself remarked upon the paucity of women corresponding directly with his periodical:

Until a female will make herself known her communication must remain unnoticed. Among our many valuable correspondents may be reckoned several eminent lady entomologists (who have furnished us with several useful hints), but it is not necessary that we should advertise their names and addresses, unless they wish it.[80]

This implies that Stainton had received letters from women wishing to be published in the *Intelligencer*, but who wished to do so anonymously, and therefore withheld the details of their identity. As a result, they could not be contacted by other readers of the periodical, and were effectively shut out of participation in any form of direct correspondence and to that extent excluded from its community. This anxiety is perhaps understandable, given the potential impropriety of a single or even a married woman exchanging letters with an unknown man.[81] However, Stainton was clearly unwilling to make an exception to his requirement that at least a name and address be provided, if not necessarily published, by any correspondent as some guarantee of veracity. As demonstrated by Mrs. Duncan, women were very much active practitioners in this period, even if the extent of their contribution is not evident in periodicals such as the *Intelligencer*. Later in the nineteenth century, the visibility of women in entomology did begin to increase, with Eleanor Ormerod (1828–1901) being a particularly notable example. However, the *Entomologist's Monthly Magazine* and other leading publications in the field continued to be dominated by men.

CONCLUSION

A correspondent to the *Intelligencer*, identified only as "K. T. L.," recounted how he was a frequent traveler along a particular railway route and often shared a carriage with another gentleman, though they were "ignorant of each other's names and tastes." By coincidence, both were summoned to jury duty, and it was there that they discovered a mutual interest:

> Before business was commenced, reading the paper was the fashion of the hour, and I pulled out the new number of the *Intelligencer*. My companion, looking at my paper, exclaimed, "Are you an entomologist?" "Yes," said I, "a little in that way." "Did you come to town by train yesterday morning at 8.30? For I had a box of insects in my pocket just received from Mr. A.," &c., &c. Of course we shall now be able to converse whilst travelling in the train together, instead of looking glumly at each other, in the orthodox English fashion.[82]

This anecdote indicates a shared sense of community, or the "sympathy of a crowd," between these two men. The reference to a third party—"Mr. A."— shows how the *Intelligencer* allowed its readers to conceive of themselves as part of a group. It is unclear whether Mr. A. was personally known to these gentlemen, but reading the periodical nevertheless provided the equivalent of mutual acquaintances, thereby allowing the imaginative leap required to form a community.

The *Entomologist's Weekly Intelligencer* reached ten volumes by 1861, at which point Stainton chose to discontinue the periodical, citing the proliferation of such publications as his primary reason for doing so.[83] His decision likely had as much to do with the strain of singlehandedly producing an issue per week, which must have demanded considerable time and energy from a man who did not enjoy the most robust health. The outpouring of dismay from many correspondents at Stainton's announcement demonstrates the central place the periodical had come to hold in the practices of entomologists. William Gates, a London-based collector and "entomological apparatus maker" whose business was advertised in the *Intelligencer*, wrote to Stainton expressing his disappointment. Gates stated his opinion that the periodical was "as much a desiderata as any books upon the subject [of entomology]," but more significantly, he characterized it as "our weekly newspaper which we cannot do without."[84] Once again, this nicely expresses the collective sensibility inspired by the *Intelligencer*, encapsulated by Gates's insistence that it is "*our* weekly newspaper," and his fear that the entomological community simply could not exist if the periodical ceased publication.

Through the rapid and widespread intercommunication made possible by the *Intelligencer*, a scientific community emerged on a scale that had not hitherto been possible. While older networks of correspondence of the kind described by Anne Secord and Janet Browne had permitted interaction bound by constraints of the handwritten letter, the application of print to this method in the second half of the nineteenth century had major implications. In addition to the more tangible benefits of efficiency, the more significant result took place in the minds of the periodical's readers. It enabled the imaginative process by which communities are constructed, permitting naturalists to conceive of their activities in relation to those of others. This imagined community consisted of individuals from across the social spectrum, and afforded working-class naturalists the opportunity to engage with a far greater number of correspondents. Stainton and other editors such as Edward Newman wished to recruit these practitioners in the process of collating observations that incrementally added to the store of knowledge, and periodicals such as the *Intelligencer* consequently represent a significant reshaping of scientific communities, challenging us to reconsider our accepted notions of "popular" science in this period. We must therefore move beyond simplistic notions of "high" and "low" cultures of science, instead looking to the complex and multivocal ways in which the social topography of natural knowledge was negotiated and defined.

NOTES

1. *Entomologist's Weekly Intelligencer* 4 (1858): 85 (hereinafter *Intelligencer*).

2. *Entomologist's Monthly Magazine* 29 (1893): 3.

3. Matthew Wale, "The Sympathy of a Crowd: Periodicals and the Practices of Natural History in Nineteenth-Century Britain," PhD diss., University of Leicester, 2018.

4. Susan Sheets-Pyenson, "Popular Science Periodicals in Paris and London: The Emergence of a Low Scientific Culture, 1820–1875," *Annals of Science* 42 (1985): 549–72 (554).

5. For example, Aileen Fyfe and Bernard Lightman, "Science in the Marketplace: An Introduction," in *Science in the Marketplace: Nineteenth Century Sites and Experiences*, ed. Aileen Fyfe and Bernard Lightman (Chicago: University of Chicago Press, 2007), 1–19 (4).

6. Anne Secord, "Corresponding Interests: Artisans and Gentlemen in Nineteenth-Century Natural History," *British Journal for the History of Science* 27 (1994): 383–408.

7. Jonathan Topham, "Rethinking the History of Science Popularization / Popular Science," in *Popularizing Science and Technology in the European Periphery, 1800–2000*, ed. Faidra Papanelopoulou, Agusti Nieto-Galan, and Enrique Perdiguero (Aldershot, UK: Ashgate, 2009), 1–20 (20).

8. *Entomologist's Annual* (1857): 174.

9. David Allen, "The Struggle for Specialist Journals: Natural History in the British Periodicals Market in the First Half of the Nineteenth Century," *Archives of Natural History* 23 (1996): 107–23.

10. T. P. Newman, *Memoir of the Life and Works of Edward Newman* (London: John Van Voorst, 1876). See also Sheets-Pyenson, "Popular Science Periodicals," 561.

11. Edward Newman to H. T. Stainton, 26 November 1857, H. T. Stainton Correspondence from British Entomologists (MSS STA E 118:118), Natural History Museum, London. STAINT 78:118.

12. *Entomologist* 1 (1840–42). Newman would revive this periodical in 1864.

13. *Intelligencer* 1 (1856): 42.

14. Newman to Stainton, 19 October 1855, STAINT 78:118. The enclosed advertisement referred to is not extant, but was probably for the *Zoologist*, or possibly one of Newman's numerous books.

15. *Entomologist's Annual* (1857): 173.

16. *Phytologist* 5 (1854): vi.

17. *Intelligencer* 1 (1856): 1.

18. Richard Shield, *Practical Hints Regarding Moths and Butterflies, with Notices of Their Localities* (London: John Van Voorst, 1856), inside rear cover.

19. *Intelligencer* 1 (1856): 18.

20. *Intelligencer* 2 (1857): 93.

21. Melinda Baldwin, *Making Nature: The History of a Scientific Journal* (Chicago: University of Chicago Press, 2015), 64–65.

22. *Substitute; or Entomological Exchange Facilitator, and Entomologist's Fire-Side Companion* (1856–57).

23. *Intelligencer* 1 (1856): 9–10.

24. Elizabeth Yale, *Sociable Knowledge: Natural History and the Nation in Early Modern Britain* (Philadelphia: University of Pennsylvania Press, 2016).

25. Gilbert White, *The Natural History and Antiquities of Selbourne* (Oxford, UK: Oxford University Press, 2016).

26. Janet Browne, "Corresponding Naturalists," in *The Age of Scientific Naturalism: Tyndall and His Contemporaries*, ed. by Bernard Lightman and Michael S. Reidy (London: Pickering & Chatto, 2014), 157–69 (158).

27. Brian Campbell Vickery, *Scientific Communication in History* (Lanham, MD: Scarecrow Press, 2000), 69–72; Yale, *Sociable Knowledge*, 65–66.

28. Secord, "Corresponding Interests," 400.

29. Benedict Anderson, *Imagined Communities: Reflections on the Origins and Spread of Nationalism* (London: Verso, 1983; rev. edn, 2016), 22–36.

30. Michael Brown, *Performing Medicine: Medical Culture and Identity in Provincial England, c. 1760–1850* (Manchester, UK: Manchester University Press, 2011), 159–60.

31. *Entomologist's Annual* (1857): 7.

32. *Entomologist's Annual* (1857): 173.

33. For a detailed overview of the changes in periodical distribution, see Graham Law, "Distribution," in *The Routledge Handbook to Nineteenth-Century British Periodical and*

Newspapers, ed. Alexis Easley, Andrew King, and John Morton (London: Taylor & Francis, 2016), 42–59.

34. *British Naturalist* 3 (1893–94): 18.

35. Catherine Golden, *Posting It: The Victorian Revolution in Letter Writing* (Gainesville: University Press of Florida, 2009), 43.

36. *Third Report of the Postmaster General on the Post Office* (London: 1857), 36.

37. William Lewins, *Her Majesty's Mails: An Historical and Descriptive Account of the British Post-Office* (London: Sampson Low, Son, and Marston, 1864), 135–36.

38. *Intelligencer* 1 (1856): 121–22.

39. *Intelligencer* 5 (1858–59): 166.

40. Ibid.

41. Stevens's obituaries: *Zoologist* 3 (4th series, 1899): 479; *Entomologist* 32 (1899): 264.

42. For a more in-depth discussion of specimen exchange in the *Intelligencer*, see Wale, "The Sympathy of a Crowd," 75–118.

43. *Phytologist* 2 (1857–58): 294–96.

44. *Intelligencer* 5 (1858): 38. Burns is listed as a newsagent in the *General and Commercial Directory of the Borough of Birmingham* (Sheffield, UK: W. H. Dix, 1858), 82. Furthermore, his name and address was included in the list of newsagents and booksellers from whom the *Intelligencer* could be purchased, which was printed in almost every issue of the periodical. It appears that Burns was the sole retailer of the *Intelligencer* in Birmingham.

45. *Intelligencer* 3 (1857): 7.

46. See list of members in *Transactions of the Entomological Society of London* 4 (second series, 1856–58), xvii–xxiv.

47. *Transactions of the Entomological Society of London* 4 (series 2, 1856–58): 39.

48. J. F. M. Clarke, *Bugs and the Victorians* (New Haven: Yale University Press, 2009), 54–104.

49. *Transactions of the Entomological Society of London* 4 (second series, 1856–58): 40.

50. *Correspondenzblatt für Sammler von Insecten, Insbesondere von Schmetterlingen* (1860– 61).

51. *Entomologist's Monthly Magazine* 11 (1874): 20.

52. *Intelligencer* 7 (1859): 153–54.

53. For example, the *Intelligencer* (or rather, Stainton) gave a favorable review of a book by Charles Robert Bree titled *Species Not Transmutable, nor the Result of Secondary Causes* (London: Groombridge and Sons, 1860), which argued against Darwin's theories. See *Intelligencer* 10 (1861): 78–79. For an account of how entomologists' views on Darwin's theories changed after this period, see J. F. M. Clarke, *Bugs and the Victorians*, 105–31.

54. *Transactions of the Entomological Society of London* 4 (second series, 1856–58), 41.

55. *Entomologist's Annual* (1855): 3.

56. *Intelligencer* 5 (1858–59): 183.

57. *Entomologist's Annual* (1855): 3.

58. James Secord, *Victorian Sensation: The Extraordinary Publication, Reception, and Secret Authorship of Vestiges of the Natural History of Creation* (Chicago: University of Chicago Press, 2001), 40–76.

59. *Zoologist* 1 (1843): vi.

60. *Phytologist* 1 (1844): v.

61. Frederick Engels, *The Condition of the Working Class in England* (Oxford, UK: Oxford University Press, 1999), 211–12.

62. *Census Returns of England and Wales, 1861, 3492,* folio 39, 8; *Census Returns of England and Wales, 1891,* 3820, folio 43, 35.

63. *Intelligencer* 4 (1858): 44; Anne Secord, "Science in the Pub: Artisan Botanists in Early Nineteenth-Century Lancashire," *History of Science* 32 (1994): 269–315. It is difficult to verify the social makeup of the Sheffield Entomological Society with any degree of certainty, as the addresses provided to the *Intelligencer* by known members other than James Batty do not match census records.

64. *Intelligencer* 2 (1857): 91.

65. Ibid., 100.

66. Ibid., 151.

67. *Intelligencer* 4 (1858): 85.

68. Moore's single contribution in 1858: *Intelligencer* 4 (1858): 139.

69. *Intelligencer* 8 (1860): 172.

70. *Intelligencer* 10 (1861): 84.

71. Ibid., 92; *Census Returns of England and Wales, 1861,* 3494, folio 105, 18.

72. James Batty to H. T. Stainton, 25 June 1861, STAINT 4:118.

73. Secord, "Science in the Pub."

74. *Entomologist's Monthly Magazine* 29 (1893): 287–88. See also *British Naturalist* 3 (1893–94): 248; *Naturalist* 18 (1893): 355; *Entomologist* 26 (1893): 368.

75. Anne Secord, "'Be What You Would Seem to Be': Samuel Smiles, Thomas Edward and the Making of a Working-Class Scientific Hero," *Science in Context* 16 (2003): 147–73.

76. *Entomologist's Monthly Magazine* 8 (1871): 68–69.

77. *Entomologist's Monthly Magazine* 10 (1873): 195–96.

78. *Intelligencer* 3 (1857): 46–47.

79. *Intelligencer* 4 (1858): 99.

80. *Intelligencer* 1 (1856): 58–59.

81. For a (predominantly literary) study of the penny post's potential for facilitating impropriety, see Kate Thomas, *Postal Pleasures: Sex, Scandal, and Victorian Letters* (Oxford: Oxford University Press, 2012).

82. *Intelligencer* 3 (1857–58): 43.

83. *Intelligencer* 10 (1861): 161–62.

84. William Gates to Stainton, undated, STAINT 36:118. The advert for Gates's business is placed in the *Intelligencer* 10 (1861): 206.

CHAPTER 7

Periodical Physics in Britain: Institutional and Industrial Contexts, 1870–1900

Graeme Gooday

Industrial developments in the 1870s precipitated new journals that evolved symbiotically with new professional audiences for physics. The rise of the international telegraph industry in the decade after the successful Atlantic cable-laying of 1866, followed by the advent of telephony and incandescent lighting, led to a huge growth of interest in the technological uses of electricity among both scholars and technicians. These communities, dispersed across Britain and the empire, needed frequent up-to-date information in an affordable format about useful developments in electrical theory and practice as well as related financial and safety issues. Just as medical practitioners might conceivably have saved lives or ameliorated suffering from perusing reports and innovations in the weekly *Lancet* (launched 1823),[1] so might engineers and technicians hope to improve the prospects of new telecommunications enterprises or other electrotechnical ventures by receiving regular weekly or fortnightly information from their specialist publications. What they thus needed were weekly journals to deliver the most up-to-date news in electrical science and its applications. Commercial journals such as the *Electrician* were set up to accomplish on a seven-day cycle what sedate society journals could not do with their annual or biannual volumes, paid for by membership fees, and refereeing practices that could bring considerable inertia to publication processes.

Concurrent with the rising interest in technical electricity was a new educational demand for practical physics. Since the UK's 1870 Education Act required national school provision in all main subjects, the ensuing decade saw a growth both in the number of schoolteachers and in associated teacher training activities at burgeoning universities and colleges. The new demand for practical physics teaching generated a growth of practitioners who viewed their specialist task as that of developing new forms of accessible experimental demonstration techniques. To meet those needs, the Physical Society of London was founded in 1873, with its first meeting in 1874 and its associated *Proceedings* launched in 1875, to disseminate a culture of educational and empirical physics beyond the remit of older Royal Society journals and the newer technical journals on electricity. While dissatisfaction with existing periodicals that published on physics was a key factor in the rise of these new institutions and periodicals, we shall see that the Georgian *Philosophical Magazine* (1798–) continued nevertheless to operate alongside both commercial and academic journals in ways that reveal overlaps in both the audiences for and content of those journals.

The first part of this chapter examines the nonspecialist cultures of physics publishing in the mid-nineteenth century, to show the considerable diversity not only among the journals' constituencies but also in their economic operations and their refereeing and editing practices. Successive parts thereafter look at the publication history of the main periodicals outlined above. In the context of Fyfe's recent study of the enormous economic challenges facing the Royal Society's *Philosophical Transactions*,[2] it becomes clear why that particular journal's refereeing practices had to exclude a significant body of papers read to the society's membership—especially of a speculative or otherwise unconventional kind. If it had not done so, the publishing costs involved could eventually have bankrupted the Royal Society. I extend Alex Csiszar's analysis of the troubled advent of refereeing in the nineteenth century as a costly and time-consuming practice that was by no means self-evidently necessary or progressive in its effects. Whereas refereeing could delay publication of articles for months or even years, new technical journals such as the *Electrician* instead commissioned pieces from house journalists or trusted freelance authors in order to meet the weekly or fortnightly publication schedules expected by those working on commercial physics. And while the Physical Society of London nurtured its new educationally oriented membership in its *Proceedings*, this community-building aim was clearly not accomplished by rigorously scrutinizing and then turning away willing contributors to the new specialist group.[3]

To understand the emerging differences between the physics communities involved, this chapter looks in more detail at the contrasting editorial styles and contexts of the Royal Society's *Philosophical Transactions* and the *Electrician*. First, it examines George Gabriel Stokes's politically and intellectually conservative cosecretaryship of the Royal Society (1854–85), with its associated editorial role. I explore how he used external refereeing practices in a "gatekeeping" capacity, to allow at least some contentious innovative researches to be published in the elite *Transactions* while others appeared only as abstracts in the Royal Society's *Proceedings*. This dialectical refereeing culture is then contrasted with the unilateral interventive role of the editor in the nonrefereeing *Electrician*—a periodical of positively Franklinian breadth that openly solicited newer, more elegant treatments of electrical circuitry for high-speed publication.[4] Priced at sixpence an issue, this journal had a readership constituency and circulation that grew to be much greater than those of the "philosophical" journals. We will see how its first editor, Charles Biggs, could secure innovative articles from less orthodox writers such as Oliver Heaviside to develop more practically useful formulations of James Clerk Maxwell's theory of electromagnetism,[5] which were only later published in the *Philosophical Transactions* once Heaviside's innovative mathematical techniques were acknowledged in his Fellowship of the Royal Society.

In the final part of this chapter I will show how Stokes's editorial strategy, combined with unsympathetic referees, led to the rejection of novel researches by the former chemist Frederick Guthrie from the Royal Society's *Philosophical Transactions* in 1873. This rebuff to what Guthrie considered as highly original work was one of the main catalysts for setting up the *Proceedings of the Physical Society* to publicize "incomplete" experimental and pedagogical researches. To round off the chapter, we will see that the creative ties in practical physics, exemplified in the relationship between the Physical Society and the Institution of Electrical Engineers, generated international "science abstracts" of foreign journals from 1895 involving editorial staff and experts associated with the *Electrician*. This, I suggest, was recognition that British journals alone were insufficient to meet the communal demands for knowledge of new physics generated worldwide by the early twentieth century.

While this chapter does not map detailed readership profiles for the journals discussed, it will show how initially marginal figures such as Guthrie and Heaviside could secure recognition for their endeavors through the innovations allowed within the clustering of newer physics journals. In both cases, the Royal Society's *Philosophical Transactions* looms large: Guthrie parted company with it to pursue his more inclusive form of experimentally oriented physics, while Heaviside's eccentric, autodidactic brilliance was eventually embraced by it, albeit with diplomatic and practical challenges that highlight the complexities of publishing mathematical physics. Both cases reveal the tensions between the intersecting communities—academic, educational, and industrial—that read journals for physics.

OVERVIEW OF PHYSICS IN PRE-
1870S VICTORIAN PERIODICALS

The specialist, quasiprofessional nature of the new periodicals I discuss marks a clear break with the characteristics of those launched in preceding decades, and especially with those of older "philosophical" periodicals from earlier centuries. Principal among these was the Royal Society of London's elite *Philosophical Transactions*, which from 1776 served as that society's official journal. Whether it concerned "natural philosophy" or any other branch of natural science, a paper read by a fellow at one of the society's meetings thereby became eligible for publication in the *Philosophical Transactions* if approved as such by the society's publishing committee. In 1854 the society launched its less formal *Proceedings*, previously *Abstracts of the Papers Communicated to the Royal Society of London*, as an independent journal in which fellows had an effectively automatic right of publication. As Csiszar notes in chapter 3 of this book, publication of shorter papers in the *Proceedings* was advantageous for a scholar who wanted to bring research promptly to the attention of relevant audiences. Yet, as I show below, for those whose work was rejected by *Transactions* referees—on grounds of insufficient rigor, gentility, or orthodoxy—publication in the *Proceedings* was nevertheless unmistakably less prestigious.[6]

Somewhat more specialist in its dedication to the physical sciences was the *Philosophical Magazine*, the origins of which date back to 1798. This was run by publishers Taylor and Francis as a commercial periodical produced on a monthly schedule. As Topham has recently shown, the *Philosophical Magazine*'s aim was to secure a financially self-sustaining operation via a wide purchasing readership—very different from the Royal Society's project of subsidized publishing of elite research. Formally, the *Philosophical Magazine* was titled the *London, Edinburgh and Dublin Philosophical Magazine and Journal of Science*, a nomenclature that reflected the complicated history of amalgamations with other serial publications since its foundation. A clue to the *Philosophical Magazine*'s working practice is to be found in the Latin epigraph on the title page, translated as "The spider's web is no whit the better because it spins it from its own entrails; and my text no whit the worse because, as does the bee, I gather its components from other authors' flowers."[7] This was a wry allusion to the *Philosophical Magazine*'s pragmatic practice of gathering its material from many quarters in order to meet its monthly production schedule.

Such material typically included papers submitted directly to the journal or via scientific societies, notably the Royal Society. In fact, from 1874 the *Philo-*

sophical Magazine was the preferred journal for publishing select papers from the meetings of the new Physical Society of London, the council of which only formalized and prioritized its own *Proceedings* somewhat later (see below). Of course, to maximize its readership, there was much more to the *Philosophical Magazine* than just records of papers read before learned societies. It also covered "intelligence and miscellaneous articles" on physical science calculated to interest its dispersed readership, as well as letters to the editor, reports from meetings of scientific societies, selected research published in foreign journals, local reports in observational sciences such as meteorology and astronomy, and schedules of public lectures in a range of scientific subjects including medicine. While the *Philosophical Magazine*'s "conductors" in the late nineteenth century were, like the secretaries of the Royal Society, typically university academics, their editorial role in the *Philosophical Magazine* was traditionally less interventive than that of the editors of the *Philosophical Transactions*. As Fyfe and Moxham note, by the 1820s a short submission to the *Philosophical Magazine* could be published within a month—a pace that the sedate machinery of the Royal Society could not match.[8] Although we know that Taylor and Francis sought out practitioners such as John Tyndall to act as "readers" for papers submitted, Clarke and Mussell suggest that it was not until these publishers appointed Oliver Lodge, George Carey Foster, and J. J. Thomson as expert editors in 1911 that the *Philosophical Magazine* adopted a formal refereeing process comparable to that of the Royal Society.[9]

While the *Philosophical Magazine* survived the entire nineteenth century, specializing mostly in the physical sciences, before the transition period of the 1870s no other journal in this subject domain had comparable longevity. Specialization in the physical sciences was not, it seems, a financially viable option without an extensive readership grounded in successful industrial applications. For example, William Sturgeon's journal *Annals of Electricity, Magnetism, and Chemistry; and Guardian of Experimental Science* was read between 1836 and 1843 by a mixed audience of experimenters and entrepreneurial projectors. However, the small-scale early networks of electroplating and telegraph specialists were clearly insufficient as a stable readership; thus, the *Annals* collapsed after seven years.[10] Once the first short-lived transatlantic telegraph cable began communications in summer 1858 between Britain and North America, a community of international telegraphic specialists emerged to exploit its possibilities. Notwithstanding the cable's permanent breakdown after just three weeks' operation, there was enough interest to prompt the first run of the short-lived periodical known as the *Electrician*. This first appeared in November 1861, edited by Desmond Fitzgerald, and was dedicated to both telegraphy in particular and electricity in general. But with no great growth in the international industry, this journal was discontinued three and a half years later—just before the great boom that attended the first fully operational transatlantic telegraph cable laid in 1865–66.[11]

Overall it would seem that nonspecialist publications covering broader territory that included physics appealed more generally to periodical audiences than those focusing on physics alone. Two in particular that were launched in the second half of the 1860s also covered physics but intermixed it with much else, and therein perhaps lay their success. The *English Mechanic* (1865–1926) was a popular weekly magazine initially subtitled "A Record of Mechanical Inventions, Scientific and Industrial Progress, Building, Engineering, Manufactures, and Arts"; latterly the subtitle changed to "Mirror of Science and Art," covering in addition electricity, photography, and chemistry. This magazine was directed at broad but not necessarily vocational audiences keen to share adventures in applied science with fellow readers.[12] More scholarly in tone, and focused more on natural sciences than on technology, Macmillan's generalist weekly *Nature* was launched in 1869. This was long subsidized by Macmillan so that it could offer the attractive prospect of publishing up-to-date material from across the physical, biological, mathematical, and human sciences to a smaller readership than that of the *English Mechanic*.[13] As a specialist himself in the sciences of the physical observatory, *Nature*'s first editor, Norman Lockyer, ensured that much physics filled the pages of this broad-ranging general science periodical.[14]

By this time, moves were afoot in a rapidly expanding international telegraph industry to secure specialist representation in both institutions and journals. This fertile territory was first marked in 1871 by the creation of the Society of Telegraph Engineers, founded as a breakaway from the Institution of Civil Engineers (founded in 1818) which treated this new cable-laying enterprise as being only of minority interest.[15] The *Journal of the Society of Telegraph Engineers* (*JSTE*) was published from 1872, and included advanced electrical measurement techniques derived from transoceanic operations, especially for locating remote faults of seabed cable to be repaired by the most advanced techniques of contemporary electrical science. In ensuing decades the specialist methods and apparatus developed in such electrical measurement were widely adopted in physics teaching and research laboratories. As Simon Schaffer notes, some of these were very swiftly included by the *JSTE* subscriber James Clerk Maxwell in his *Treatise on Electricity and Magnetism* in 1873.[16] The gesture was returned by a struggling freelance writer and erstwhile telegraph clerk, Oliver Heaviside, who elaborated in the *JSTE* upon

Maxwell's theories in more readily usable forms for telegraphers and physics teachers alike.[17] As we shall see, however, Heaviside's most significant contributions to electrical technoscience had been published in a newly revived form of the *Electrician* a decade later.

A revealing precursor to the story I tell below about physics journals can be found in mid-nineteenth-century developments in chemistry periodicals. The *Chemist and Druggist* was launched in 1859 and the *Chemical News* in 1862; both catered to an expanding market of commercial and some academic chemical practitioners with similarly specialist interests. Following the earlier founding of the Pharmaceutical Society in 1841 and its organ the *Pharmaceutical Journal* in 1842, these developments collectively served as precedents for the technical organizations and journals dedicated to electrical knowledge that emerged in the 1870s. In a complementary sense, the Chemical Society of London, which first published its periodical *Memoirs* in 1841, was the obvious model for the Physical Society of London, founded one-quarter of a century later, and indeed with some common membership.[18]

It should be emphasized that the journals and institutions of physics discussed below were not neatly demarcated enterprises. Given the rise of technical-industrial journals presenting new forms of physics, especially the electrical variety, it would be anachronistic to treat this period as one in which physics was an autonomous discipline of unworldly abstractions; that would be much more the story of the mid-twentieth century. By contrast, in the late nineteenth century physics was still a mutable assemblage of subspecialties, still sharing with chemistry an interest in heat and latterly radioactivity; with electrical engineering an interest in electromagnetism and precision electrical measurement; with mechanical engineering a concern with mechanics and thermodynamics; and with astronomy an interest in both remote heavenly spaces and mathematical dynamics. Significantly, then, none of the newer journals involving "physics" discussed below actually included that specific term in their titles. Equally significantly, physicists read and published in electrical engineering journals, and chemists were leading members of the Physical Society of London. Indeed, the boundaries of physics remained permeable until the First World War brought an urgent imperative for specialized and compartmentalized technical expertise.[19]

THE CHANGING LANDSCAPE OF "PHYSICS" JOURNALS

Another anachronism to be avoided in the period covered by this chapter is the tendency to differentiate physics sharply from the traditional study of "natural philosophy." After all, the older terminology has long persisted both in the curricula of the Scottish universities and in the titles of the longest-established periodicals that published papers in this domain: the *Philosophical Transactions* and the *Philosophical Magazine*. In the period covered by this chapter, the broad coverage of these periodicals still matched the "philosophical" breadth with which they had been founded, and thus to some extent "physics" and "natural philosophy" were largely synonymous at least until the early twentieth century. During the later nineteenth and early twentieth centuries, however, changes in these older periodicals reflected a changing publishing landscape in which newer groups of physics practitioners demanded new kinds of specialist journals.

By contrast, in the early nineteenth century the Royal Society's *Philosophical Transactions* was by no means a venue for preponderantly innovative research. As is well known, Charles Babbage's complaints about the "decline of science in England" eventually prompted a limiting of aristocratic influences in its annual operations.[20] Thus as the Royal Society's historian Henry Lyons points out, after the mid-nineteenth century degentrification of the society, the content of the *Philosophical Transactions* increased threefold.[21] During the period of George Stokes's tenure as cosecretary of the Royal Society (1854–86), the *Philosophical Transactions* was published in a unitary volume covering the entire spectrum of both physical and life sciences. While Stokes was appointed as "physical secretary" and a succession of co-appointments as secretaries covered the life sciences, notably Thomas Henry Huxley (1870–81) and Michael Foster (1881–1903), their division of labor for the Royal Society was not discipline-specific. While all candidate papers to be published in the *Transactions* had first to be read out to a society meeting, if the author was not a fellow this task typically fell to the available secretary. Thus, Stokes at times strained to read out biological papers, as is illustrated in this letter written to his wife Mary, at home in Cambridge, in January 1857 after one such arduous performance:

> . . . Last night was a R. S. night. I had a paper to read out on the minute anatomy of the earth-worm and another on a kindred subject. I told General Sabine [P. R. S.] that in reading such papers I felt as if I were a pair of bellows to be blown for the benefit of the Society. I felt I was a sort of reading machine, the papers were so completely out of my line.[22]

This non-discipline-specific labor extended also to the editorial roles for the society's periodicals. While Stokes took on the "internal scientific work, both physical and biological," involving correspondence with all prospective or actual authors for society journals, Huxley and then Foster took on the "external scientific work" of dealing with government inquiries, communications with other national or overseas bodies, and scientific expeditions.[23] Presumably it was because of the impossibility of replicating Stokes's omniscience and extraordinary dedication in editorial work that, after he relinquished the role of physical secretary in 1885 to take on the presidency following Huxley's resignation, the refereeing process was managed by sectional committees appointed from among Royal Society fellows.

Up to this point, a characteristic of the "philosophical" breadth of both Royal Society publications was the inclusion of physics papers among all other kinds of natural science papers. But this ultimately proved unsustainable as a result of the gradually increasing size of successive journal issues, combined with the expense of circulating a full journal issue to all members and all communicating societies. Given the diverging specialist interests of its readerships into subcommunities in the physical and biological sciences, the Royal Society's committees decided to split the hitherto unitary journals to reflect this change. From 1887, two separate specialist series, physical sciences (A) and biological sciences (B), were introduced to the *Philosophical Transactions*, with the *Proceedings* following the same bifurcation in 1905.[24] Economic pragmatism was indeed as much a force as reader specialization in the emerging distinctive publication patterns of "physics."[25] As a background to that transition, we will see that in the 1870s new specialist publications, especially for electricity, helped to shape the new landscape for periodical physics.

By then, one of the most important British locations for the publishing of new research in electrical technoscience was the *Electrician*, relaunched in May 1878 with the subtitle "A Weekly Journal of Theoretical and Applied Electricity and Chemical Physics." Like the previous version of this journal that ran briefly in the early 1860s, this was published under the joint proprietorship of Sir James Anderson, former captain of the cable-laying "Great Eastern," and the Atlantic telegraph magnate John Pender. It was initially edited (1878–87) by Charles Henry Walker Biggs, with a broad-ranging ambition to publish technical and business articles on applied electricity, but also translations of overseas research publications, technical correspondence, students' pages, book reviews, reports from scientific society meetings, and patent specifications.[26] As an example of how closely the *Electrician* reported innovations in physics, we can note that in 1897 it was the only mainstream science periodical to report J. J. Thomson's identification of a unique small mass-to-charge ratio for "corpuscles" in cathode ray tubes—these later being redesignated as "electrons."[27] Eight years later the *Electrician* was also the first periodical in Britain to comment on Einstein's new theory of relativity in its weekly column "Contemporary Electrical Science," by its regular paid international columnist E. E. Fournier D'Albe.[28]

As a sixpenny weekly, the *Electrician* had a circulation in the thousands by the 1880s, and reached out to an anticipated audience of telegraphists, electrical engineers, entrepreneurs, teachers, and technological researchers. And with plentiful commercial advertising of new equipment to underwrite its expenses, the journal could afford to do more than just report recent news in electrical science and technology. As editor, Biggs was in a position to invest in commissioning more exploratory and theoretical articles on topics of potential future value for the development of electrical communications, power, and lighting. Commencing in 1882, his commissions were for various series of thematically related articles on interrelated aspects of physics and engineering. These were sufficiently extensive in scope that they could last up to a year or more, accompanied by correspondence from readers engaging critically with the commissioned articles. In Strange's analysis of the *Electrician*'s "learned journal" function, he notes that among the most significant among these were numerous series of articles commissioned from Oliver Heaviside between 1882 and 1906, which I will discuss further below.[29] So successful was this format that by 1885 the *Electrician*'s more industrially focused competitor, the *Telegraphic Journal and Electrical Review* (founded in 1872), began to emulate Biggs's successful strategy of commissioning series of articles.[30]

The *Philosophical Transactions* and the *Electrician* thus provide important examples of the contingent power of the editorial role to adopt either refereeing or commissioning processes to shape the content of periodicals and their associated communities. In what follows, the consequence of that divergence of editorial practice will be spelled out in terms of how Royal Society referees worked with Stokes to maintain an exclusive elite culture in the *Philosophical Transactions*, but also how commissioned authors worked with Biggs and his successors to nurture a more diverse community of practitioners engaging with the *Electrician*.

REFEREEING PHYSICS IN THE *PHILOSOPHICAL TRANSACTIONS*

To understand the particular expectation of physicists aiming to publish in the *Philosophical Transactions* and the fate of their submissions in its exclusive refereeing processes, we need to understand the overall pattern of developments in Royal Society periodical publishing during the nineteenth century. Moxham and Fyfe have mapped the evolving patterns of refereeing across the history of the *Philosophical Transactions*, and they document the broader changes that prompted the society's formalized recourse in 1832 to expert refereeing by fellows of the Royal Society. Since the papers published there were substantial in length and often lavishly illustrated, the heavy financial commitment entailed a significant selectivity: only half of the papers communicated to the society were sent to a pair of referees, and just 30 percent of paper submissions were actually published. Even for papers that survived the critical scrutiny to check that their contributions were substantial and original, referees could request significant changes before publication.[31] Aileen Fyfe notes that the refereeing process entailed that many months could elapse before a paper was published in the *Philosophical Transactions*; this was thus not a journal organized to address the needs of communities seeking quick and easy access to cutting-edge research.[32]

Although refereeing was commonplace before his appointment, a rigorous system of conferring with two referees (rather than deferring to a society committee) was largely introduced by the Royal Society's cosecretary George Gabriel Stokes, Lucasian Professor of Mathematics at Cambridge. This system dated from his appointment in 1854 as de facto editor of the *Transactions* and *Proceedings*. This system of refereeing was not only designed to filter out all submissions judged to be below sufficient quality, but also to ensure that the production costs of the *Philosophical Transactions* did not become astronomically large. Policing the distribution of articles between these house journals was a major preoccupation for Stokes and T. H. Huxley during their time as joint secretaries of the Royal Society, as they sought the judgment of referees to make the key decisions.

Stokes's implementation of this framework for the publication of physics articles depended considerably upon the particularities of his editorial predilections and his personal trusted subcommunity of elite referees. The scope of Stokes's commitment to editing is evident from the sheer number of letters in his correspondence: about seventeen thousand extant items. Much of Stokes's correspondence—first as secretary, then as president from 1885 to 1890—related to four decades of the Royal Society's incoming and outgoing business.[33] Stokes would write so many letters a day that by 1878, even he could not always read his own handwriting. Indeed, until he got a typewriting machine in 1879, his correspondents might be forgiven for finding it difficult to tell whether a paper submitted by them had been accepted or rejected.[34] One of the difficulties, as his daughter Isabella (Mrs. Laurence Humphry) later reported, was that Stokes never allowed anyone to assist him in his labors, since he worked at his Cambridge home. Another was that he retained everything he received by post, even advertisements: "It may be imagined that, keeping everything, he could find nothing."[35] Often the extra work and delay was of Stokes's own making. For example, he attempted in May 1882 to help a prospective contributor to the *Philosophical Transactions* improve the stylistic aspects of a paper in response to referees' criticisms, but then forgot about it so that the author had to prompt him to follow it through to publication.[36]

Generally, however, the referees to whom Stokes appealed to adjudicate papers for the *Philosophical Transactions* replied within several weeks. And while standards for acceptance were high in terms of rigorous exegesis and detailed explanations, there was some latitude given to matters of controversy. As noted by Fyfe and Moxham, one correspondent who served as a frequent and liberal-minded referee for the *Philosophical Transactions* was William Thomson (later Baronet/Lord Kelvin). It was to Thomson, closely trusted as a friend, that Stokes often turned on difficult questions and in matters concerning newer directions of natural philosophy. Thomson proved to be a ready respondent, generous in his recommendations even when he disagreed with authors' conclusions. Consider, for example, submissions to the Royal Society by John Tyndall, a fellow of the Royal Society (FRS) from 1852 and professor of natural philosophy at the Royal Institution from 1853.[37] In his second phase of research on the new and still somewhat controversial topic of diamagnetism, Tyndall submitted his paper "On the Nature of the Force by Which Bodies Are Repelled from the Pole of a Magnet" to the Royal Society on 31 October 1854. This was eventually read to a society meeting on 25 January 1855 and, following society committee recommendations, delivered under the rubric of the society's prestigious Bakerian Lecture.

As Roland Jackson notes, in seeking referees for potential publication of the lecture in the *Transactions*, Stokes turned to the mineralogist William H. Miller and to William Thomson. While Miller unhesitatingly recommended publication, Thomson swiftly returned his more critical report to Stokes on 12 February.[38] Following several years of debate in which Tyndall had allied himself with his mentor Michael Faraday to criticize Thomson's work, Thomson as a referee was in turn critical of Tyndall's claims about the nature of magnetic influence, wishing "much modification to be made in the controversial parts of the communication." Yet, as a latitudinarian in this controversy, Thomson added: "Should Mr. Tyndall be disposed to make no change, I should advise publication as it stands." And thus after a robust response by Tyndall to Thomson's critique, the Bakerian lecture was published, only lightly revised, in the *Philosophical Transactions* in May 1855. For the benefit of broader audiences, it was reprinted in two parts in the September and October issues of the *Philosophical Magazine*.[39] By contrast, in May of the following year, Thomson agonized rather longer over a paper by Reuben Philips on aurora, fireball lightning, and shooting stars, which Thomson found to be ill-judged rather than controversial. Feeling "very much regret" at having "kept the letter so long," Thomson explained to Stokes that there was "a good deal objectionable" in the paper. Rare was it for the Royal Society's editorial authorities to go against Thomson's judgements; hence, this particular article ended up only abstracted in the Royal Society's *Proceedings* for 1856–57.[40]

John Tyndall in turn took it upon himself to support the ambitions that less established (non-FRS) members of the physics and natural philosophy community had for publication in the Royal Society's prestigious *Transactions*. One such was the chemist-turned-physicist Frederick Guthrie, who returned briefly to London in the late 1860s from isolation teaching at a college in Mauritius. Guthrie sought assistance from Tyndall as an FRS to secure a reading and then a publication for his researches on the thermal resistance of liquids. Having already secured a publication on some preliminary notes in the *Philosophical Magazine*,[41] Guthrie submitted a fuller version of his paper to Stokes, informing him somewhat presumptuously that "Professor Tyndall has read and agreed to communicate it to the Royal Society, and should he be referee would advise publication in the Philosophical Transactions."[42] Tyndall subsequently communicated the paper to the society in October 1868, where it was read on 21 January 1869 and thence published in the *Proceedings*.[43] Nonetheless, it is clear that the referees secured by Stokes did not immediately concur with Tyndall's purported recommendation. This we can surmise from Guthrie's letter to Stokes of July 1869 asking him either to publish his paper promptly in the *Transactions* or to return it to him immediately, commenting tersely: "I am bound to have my observations made useful to the public with as little further delay as possible."[44]

One of the Royal Society's sceptical referees for this paper was very probably James Clerk Maxwell, as we can discern from the well informed letter he wrote in November 1869 when William Thomson asked for his wisdom on the conductivity of liquids for heat. As Maxwell recalled to Thomson:

> The last thing on the subject is "on the thermal resistance of liquids." Guthrie states in his paper (I do not know if it is to be printed . . .) [*sic*] previous results. His experimental methods seem very good. His chief defect is that he never seems to know what he is going to measure. He works at the Royal Institution

and has been so Tyndallized that he describes the specific resistance of a liquid to be "the ratio" of the quantity of heat arrested by the liquid to that arrested by an equal thickness of water. . . .[45]

Notwithstanding Maxwell's negative views, a version of Guthrie's paper was indeed published as Tyndall had recommended in the *Philosophical Transactions* later in 1869.[46] At around that time, also on Tyndall's recommendation, Guthrie took over Tyndall's lectureship at the Government School of Mines.

In his next paper submitted to the Royal Society, "Approach Caused by Vibration," matters turned out somewhat differently. For this qualitative study, Guthrie secured Thomson's sympathy. He read Guthrie's abstract with "great interest" once it had appeared in the *Proceedings* for autumn 1870,[47] and wrote to thank Guthrie for supplying him with a physical mechanism to explain the mutual interactions of atoms, conceived by Thomson as fluid vortices. Such was the tenor of Thomson's enthusiasm that he wrote three further letters to Guthrie on the subject during the following ten days. At Guthrie's request, each of these letters from Thomson was published by Stokes in the *Proceedings* for November 1870.[48] Thomson's high-level interest in Guthrie's work soon secured him a new status too: on 9 June 1871, Stokes wrote to Guthrie informing him of his election as a fellow of the Royal Society.[49] Yet, even with this new status and Thomson's support, no fuller version of Guthrie's largely empirical paper appeared in the more prestigious *Transactions*. Since Royal Society referees expected claims for originality to be supported by substantial theoretical, quantitative, or mathematical analysis, this was not an unexpected result. And, as we will see later in this chapter, the critical response of Royal Society referees to Guthrie's next eclectic empirical offering prompted him to launch the Physical Society of London instead.

Managing such decisions to decline publication of innovative or heterodox articles in the Royal Society's *Transactions* was a customary matter for Stokes, acting as the Royal Society's editor in all but name.[50] This tendency to publish only innovative work that met the society's most traditional standards resulted in it rejecting from the *Transactions* original work that was only later considered very important. For now, let us note that Stokes himself, described aptly by David B. Wilson as "cautious," epitomized the most conservative elements of this tradition in his own refereeing work for the Royal Society. When reviewing Osborne Reynolds's research on the theory of turbulent fluid flow in 1883, Stokes encountered a work written in the Cambridge University idiom of hydrodynamics with which Stokes had been familiar since undergraduate days. Accordingly, Stokes directly recommended publication in the

Philosophical Transactions with only minor revisions needed. By contrast, for Reynolds's more radical paper on this topic eleven years later, which involved significant terminological and methodological innovations, Stokes expressed great hesitancy. In declining to offer a recommendation, Stokes implied his disapproval of Reynolds's paper. Four years after he retired from the Royal Society's presidency, Stokes's scholarly reputation as a conservative in both science and politics was thus sealed.[51] In fact, Osborne's radical paper was indeed published in the *Philosophical Transactions*, since by then the Royal Society's publication of papers framed in new mathematical styles had a precedent in Oliver Heaviside's vector calculus published several years earlier. Overall, then, we can see that the Royal Society's refereeing processes, especially when Stokes was involved, tended to favor contributions from authors with more conventional approaches and elite education backgrounds. Unsurprisingly, we will see that practitioners of newer kinds of experimental or technical physics found it easier to get speedier and sympathetic publication in other newer periodicals.

CHARLES BIGGS: EDITORIAL DISCRETION AT THE *ELECTRICIAN*

The contrast between Stokes's conservative approach to editing within the institutional demands of the Royal Society and the entrepreneurial strategy of Charles Henry Walker Biggs in seeking new writers to contribute to commercial publications on electrical science could not have been greater. Biggs was the *Electrician*'s first editor at its relaunch in 1878, explicitly picking up the title from the short-lived precursor in the early 1860s, discussed above. Like Guthrie and Heaviside, Biggs was something of an outsider in the genteel world of metropolitan science publishing. Before joining the *Electrician* in 1878, he had been a schoolteacher and freelance tutor with no previous experience in journal editing or formal qualifications in science, though he had submitted papers on educational and electrical matters to the British Association for the Advancement of Science.[52] Untrammelled by most orthodox educational attainments, Biggs was not especially inclined to privilege writers with formal academic credentials, as was evidently the case with Royal Society publications. Instead, in producing weekly news publications he sought out sources on both practical and theoretical developments in electricity, especially for the new fields of telephony, electric power, and incandescent lighting. It was, after all, for the community of professional specialists in these areas that Biggs edited the *Electrician*, as well as for the growing communities of academic teachers and city entrepreneurs whose interests he also sought to meet, so as to maximize the journal's circulation figures.

For example, early issues of the *Electrician* in 1878 covered reports of anticipations of, and then experiments in, new forms of electrical illumination: the dazzling Jablochkoff arc lights and the domestic Edison and Swan incandescent lamps.[53] Developments in microphone technology developed for the telephone industry were also early discussed by Biggs, leading swiftly to the recognition by *Nature* of the *Electrician* as a key up-to-date "scientific journal" that could discuss the science of microphones most authoritatively.[54] Biggs's editorials every January thereafter documented the developments in technologies over the preceding year, as well as the broader financial and political climate in which those innovations emerged, drawing in both academic commentators and practically experienced workers to discuss the developments in short pieces or via correspondence pages. At other times Biggs was approached by writers from the community of telegraph engineering, seeking advice on publishing their hard-earned wisdom.

One such writer was the Irish telegraphist John Joseph Fahie, an overseas employee of the Indo-European Government Telegraph Department since 1867. Fahie had used his spare time to compile an early history of the telegraph, to show how much had been accomplished prior to William Cooke and Charles Wheatstone's famous and allegedly foundational patent of 1837. Finding no other publisher willing to take on such a project, Fahie approached the *Electrician* in 1883. As editor, Biggs swiftly agreed to publish those parts of Fahie's manuscript that dealt strictly with electrical matters, while declining sections on the mechanical aspects of telegraphy. Fahie thus reedited the volume into a series of papers that Biggs published in the *Electrician* throughout 1883. The following year, Fahie was able to capitalize upon the recognition thus achieved to secure a contract with the London and New York publisher E. & F. N. Spon to reproduce these articles as *A History of Electric Telegraphy, to the Year 1837*.[55]

As well as being speedily reactive to journalistic copy, Biggs was well placed to pay for the commissioning of articles, given the advertising income that the *Electrician* received from commercial manufacturers and suppliers of electrical goods from its very first volume. Specifically, he had the capacity to commission various series of cutting-edge articles from authors whom he considered to be well placed to advise the *Electrician*'s wider readership of electrical specialists in academic life and industry on future developments.[56] In this he recognized both academic writers and innovative thinkers situated outside academic circles, especially those who offered fresh approaches to the novel challenges of making electric telephony, power, and lighting reliable and affordable enough to be both a commercial and a technical success. Biggs evidently saw the importance for telecommunications of the work by Oliver Heaviside, the freelance mathematical writer on Maxwellian physics.

After retiring from his uncongenial work as a telegraph clerk in 1873, Oliver Heaviside had scraped together a living from freelance writing. Much of his subsequent reclusive career was spent writing on James Clerk Maxwell's *Treatise on Electricity and Magnetism* (1873), to clarify its more obscure passages and to develop practical applications from it for telegraphy and telephony. Up to 1882, Heaviside's densely analytical writings had appeared in the *Philosophical Magazine* and the *Journal of the Society of Telegraph Engineers*. After seeing a few carefully crafted pieces that Heaviside submitted to the *Electrician* in 1882, Biggs quickly commissioned a series of paid articles, reassuring Heaviside: "So long as you remain good to write, so long shall I be pleased to receive and insert your MS. . . . I hope this period will be a period of infinite duration."[57] What would turn out to be only five years of such commissions began with a fortnightly series titled "The Relations between Magnetic Force and Current," which ran from November 1882 to March 1883. Even without giving evidence of communal enthusiasm for Heaviside's articles, Biggs commissioned further series from him: "Some Electrostatic and Magnetic Relations" (1883), "The Energy of the Electric Current" (1884), "The Induction of Currents in Cores" (1884–85), and "Electromagnetic Propagation" (1885–87).[58]

As Heaviside later dryly noted, the *Electrician* paid him around forty pounds a year for these commissioned series: "less than a hodman."[59] While the fellow Maxwellians Oliver Lodge and George Fitzgerald regretted his decision to publish this material in a commercial journal, this opportunity at least gave Heaviside both a living and an opportunity to publish untrammeled by referees. It was thus that he could introduce such neologisms as "reluctance" and "impedance" without fear of objection.[60] He evidently relished not only the way that Biggs gave him free rein to write in his own wryly idiosyncratic style, but also the access he gave him to very accurate typesetting for his elaborate mathematical expression. Later he recalled that the *Electrician*'s compositors were "very intelligent, read mathematics like winking and carry out all instructions by [the] author." Moreover, Heaviside's submitted copy appeared in the *Electrician* within just two weeks, far quicker than in the Royal Society's journals or even the *Philosophical Magazine*.[61] Thus, as Hunt and Nahin have shown, it was in the *Electrician* that Heaviside chose to present his Maxwellian perspective, that optimizing self-induction (electromagnetic inertia) was the critical factor for telegraph and telephone engineers to secure effective long-distance communications.[62]

Yet Heaviside's freedom of expression in the *Electrician*, unfettered by referees or editorial interventions, threatened the continuation of his long-term commissions. His Maxwellian exegeses sometimes involved undiplomatic assaults on senior figures in the telegraphic community, such as William Preece, the Post Office's "chief electrician," who refused to accept that anything other than electrical resistance mattered in accomplishing long-distance telephony. The barely disguised insults that Heaviside flung at Preece on this point precipitated considerable embarrassment to John Pender and others as proprietors of the *Electrician*. Hence Biggs wrote to Heaviside on 20 September 1887 indicating that, to be able to continue the speedy publishing of his articles, "I must beg of you to leave the hot controversial discussions for a future opportunity." Even after several such enjoinders to Heaviside, Biggs seemed unable to resolve the matter further. So he resigned the *Electrician*'s editorship in early October, promptly taking up the editorship of the newly founded *Electrical Engineer*.[63]

Biggs's successor in editing the *Electrician*, William Henry Snell, initially advised all invited contributors on 14 October 1887 to continue submissions as before. Yet Heaviside was an exception: on 30 November Snell wrote to him canceling his series on electromagnetism, claiming that nobody, not even students, actually read them. Until Snell's conservative editorship ended upon his death in March 1890, the *Electrician* carried nothing more by Heaviside and made little effort to court new expertise that looked beyond the immediate needs of the electrical subcommunities that made up the readership of this journal.[64]

Shunned by the *Electrician*, Heaviside instead began to develop a new "vector calculus" using mathematical "operators." This articulated a more succinct formulation of Maxwell's sprawling algebraic formulas than was possible with the conventional technique of "quaternions" developed originally by William Rowan Hamilton in Dublin.[65] While Heaviside's approach was better suited to capturing the interrelated behavior of multiple changing electromagnetic variables, it also required new methodology and typography of a sort alien to readers of the *Philosophical Transactions* (see further below). In the meantime, however, Sir William Thomson's public endorsement, as president of the Institution of Electrical Engineers, of Heaviside's Maxwellian analysis of long-distance telephony brought the shy, reclusive autodidact back into the mainstream in 1889.[66] The editor of the *Electrician* from 1890, Cambridge Natural Sciences graduate Alexander Pelham Trotter, was more sympathetic too; Trotter was "extremely constructive in thought with a dislike for methods which he considered prohibitive and likely to delay progress."[67]

Heaviside thus received new commissions for the *Electrician* just as he was elected a fellow of the Royal Society in 1891 with support from Thomson and Oliver Lodge.[68]

The following year saw Heaviside's first-ever publication in the *Philosophical Transactions*. This was his cue to introduce to the Royal Society's readership his own distinctive recasting of Maxwell, and his new notation of vector calculus.[69] As A. P. Trotter noted, not a few FRSs grumbled at the introduction to the *Philosophical Transactions* of an alien form of mathematics that lacked the "rigor" of the Cambridge Mathematics Tripos.[70] Without the freedoms in which he had been indulged by the *Electrician*'s editors, Heaviside was obliged to be more formal in his writing for *Philosophical Transactions*, but he still took the opportunity to attack the journal's traditional conventions on mathematical notation. Concurrently, the Royal Society's printers found it rather more challenging to handle his mathematical notation than had those at the *Electrician* and *Philosophical Magazine*. When Heaviside's paper "On the Forces, Stresses, and Fluxes of Energy in the Electromagnetic Field" was finally published on 1 January 1892, he complained of four months of delays for the paper, caused by "typographical troubles." To satisfy him, the society's printers had had to cut a whole new series of type.[71] Worse still, in 1894 he found that his attempt to use the FRS privilege of publishing papers directly (without refereeing) in the Royal Society's *Proceedings* was thwarted by its publishing committee's concerns about his lack of rigor in his incomplete formulation of results.[72]

It was around this time that his fellow FRS, the eminent aristocratic engineer James Swinburne, wrote to Heaviside about the new sourness emerging in his relationship with Royal Society publishing. Of his own accord, Swinburne recommended the Physical Society, not the Royal Society's allegedly little-read journals, as the institution most appropriate for disseminating news of practical applications:

> Why do you secrete important papers at the Royal Society? You have hinted that you had difficulty in finding a good outlet; the Royal Society is not an outlet, it is an inlet or sink. If you send your papers to the Physical Society there will be a [divulgence] of information.[73]

Despite Swinburne's advice, Heaviside appears to have had no subsequent dealings with the Physical Society, since his work did not match its empirical agenda. He continued instead to publish in the *Electrician* and *Philosophical Magazine*. Yet, as we shall see in the next section, the Physical Society of

London was indeed able to publish certain kinds of innovation more readily than was the Royal Society. Swinburne's recommendation was based upon personal knowledge that the newer society would be more welcoming than its royal counterpart to important but incomplete physical research.

BETWEEN THE *PHILOSOPHICAL TRANSACTIONS* AND THE *PHILOSOPHICAL MAGAZINE*: THE PHYSICAL SOCIETY'S *PROCEEDINGS*

To understand the advent of the Physical Society of London and the subsequent launch of its *Proceedings* over the period from 1873 to 1875, we need to consider two factors that shaped both its characteristically inclusive membership and its focus on demonstrations of useful practical physics. Both of these factors were mediated through the role of the relatively little-known chemist-turned-physicist Frederick Guthrie, mentioned above. First, following the Education Act of 1870, there was a general expansion in school education in Britain across all subjects, and this indirectly increased the demand for physics teachers across the country. One major center for training science teachers created soon afterwards was the Science Schools in South Kensington, where Guthrie was professor of physics alongside (inter alia) Thomas Henry Huxley as professor of biology. Not only were these science school professors all national examiners in their subject for the Department of Science and Art: in their respective South Kensington laboratories, each professor taught experimental methods to trainee science teachers. In managing and shaping the demand for a new generation of physics teachers, Guthrie came to know well that constituency and its demands.[74]

Looking back fifty years after its first meeting at the Physical Society's jubilee in 1924, Guthrie's erstwhile laboratory assistant (and later professor of physics at the Royal College of Science) William Fletcher Barrett vividly recalled this stimulus:

> It was, no doubt, contact with these science teachers from all parts of the country which showed the need of a society for diffusing knowledge of the progress of physical research at home and abroad, and for the publication of original work, and this led to the conception of a Physical Society along the lines of the Chemical Society.[75]

Barrett vividly remembered his first presentation to a Physical Society meeting in Guthrie's laboratory. This was a demonstration of a trombone adapted to show the interference effects of sound waves, Barrett's brief notes on this being published in the first volume of the society's *Proceedings*, covering the years 1874 to 1875. This kind of teaching demonstration apparatus was, as the society's president noted, still widely in use in schools and colleges, and thus was implicitly a vindication of the society's original rationale for sharing educational techniques among teachers of physics. And indeed Barrett's recollection effectively epitomized the society's view of itself in its jubilee year as a natural outgrowth of educational developments.[76] While this pedagogical context is important for understanding the main constituency of the Physical Society's membership, that alone does not explain the timing of why Guthrie solicited support for the new society in the summer of 1873, nor why others at the jubilee hinted at less comfortable dealings with the Royal Society and its fellows at the new society's founding.

Guthrie's own motivations in setting up the Physical Society relate to its second major role: to demonstrate incomplete or novel researches that were not yet fully amenable to orthodox explanations as required by the Royal Society's referees for the *Philosophical Transactions*. By 1873, Guthrie had enjoyed substantial use of the Science Schools laboratories to research a hitherto unreported asymmetrical relationship between heat and static electricity. He ascribed the peculiar differences he observed in the behavior of negative and positive charges to a hot metal to a new kind of "electro-coercitive" force. The abstract, "On a New Relation between Heat and Electricity," submitted to the Royal Society on 10 January, was published in its *Proceedings* of 13 February, in accordance with his prerogative as a fellow.[77] Guthrie also considered this finding to be of sufficient originality and importance to warrant publication of a full version in the *Philosophical Transactions*.

The Royal Society's referees disagreed, doubting that Guthrie had differentiated his purported "new relation" from other well-known electrothermal phenomena. Fleeming Jenkin, a telegraphic expert and professor of engineering at Edinburgh University, wrote briskly to Stokes in May 1873 that Guthrie's piece was "unsuitable," since all the allegedly new results were "clearly explicable on well-established principles," such as the discharge of a conductor by a flame or point held opposite it. Jenkin advised Stokes that "a judicious friend" should give Guthrie a "hint to withdraw the paper" and thus spare him from any further embarrassment.[78] James Clerk Maxwell was more sympathetic, but argued that Guthrie's theoretical position was "nowhere very clearly stated," so that it was not clear he had demonstrated any "new relation" between heat and electricity. Although Maxwell could not recommend publication of the paper in the *Philosophical Transactions*, he charitably concluded that a more carefully formulated project, "if successfully carried out," might furnish matter for a "very valuable communication."[79] Perhaps chastened by these referees' views and by rejection from the *Philosophical Transactions*, Guthrie published a purely experimental account of his researches in the *Philosophical Magazine* of October 1873, eschewing any claims to have discovered a new kind of force or relation.[80]

One of Guthrie's student associates at this time was John Ambrose Fleming. As an employee of the Marconi Company thirty years later, Fleming drew upon Guthrie's thermoelectric research in his invention of the thermionic valve—a device that was adopted in radio sets in a massive wave of innovation across the interwar United Kingdom.[81] But having no such positive response to his research from the Royal Society in 1873, Guthrie generally shunned its journals thereafter. Instead, he immediately began canvassing support at the British Association for the Advancement of Science summer meeting at Bradford for a new society to address the needs of experimental physicists like himself that were evidently not met by the conservative policies of the Royal Society toward research publication:

> I wish to try to form a Society for Physical Research: for showing new physical facts and new means for showing old ones: for making better known new home and foreign physical discoveries, and for the better knowledge of one another of those given to physical work. You who care for the being of such a Society, and are willing to help in its making, are hereby asked to write to me to that purpose before the first of October next. Whereupon you will be asked to meet so as to talk over the means [*sic*].[82]

At Bradford, Guthrie was immediately able to secure the support of a dozen fellows of the Royal Society, as well as his colleague T. H. Huxley, for his breakaway endeavor.[83] Recognizing the pedagogical orientation of the likely membership constituency, Guthrie dropped the word "Research" from the new organization's title, and the Physical Society of London was thus born in October 1873, with its first meeting in March 1874. Within the next three years, only the most prominant British physicists declined to join the Physical Society, both of whom were Cambridge professors at the heart of the Royal Society establishment. While George Stokes offered no response,[84] James Clerk Maxwell clearly saw the connection between this new society and Guthrie's recent rejection by the *Philosophical Transactions*. Maxwell wrote in December 1873 to Professor William G. Adams of King's College, London, who had been helping Guthrie to recruit members at the older universities:

I got Professor Guthrie's circular some time ago. I do not approve of the plan of a Physical Society as an instrument for the improvement of natural knowledge. If it is to publish Papers on physical subjects which would not find their place in the transactions of existing societies or in scientific journals, I think its progress towards dissolution will be very rapid. But if there is sufficient liveliness and leisure among persons interested in experiments to maintain a series of stated meetings to show experiments and talk about them, as the Ray Club do, then I wish them all joy. . . .[85]

The Physical Society in fact rapidly flourished without recourse to the cozy informality of the natural history gatherings in Cambridge to which Maxwell archly alluded. More strategically important was contrasting its role with that of the Royal Society to avoid any implication of competition. As the Physical Society's first president, John Hall Gladstone, reiterated at the close of its first year, it fulfilled a very different function from that of its senior institutional sibling:

The Royal Society readily receives important physical papers; but it is difficult to exhibit experiments or discuss them in detail, at Burlington House, and minor or unfinished papers are obviously unsuited for communication to this, the chief of the learned Societies.[86]

Guthrie's close FRS associate, George Carey Foster of University College London, offered a similar story in writing Guthrie's obituary for the Physical Society in 1887. Without interfering "with the field of action of the Royal Society," there was ample room for a society that took cognizance of "smaller matters, points of technical detail, useful laboratory contrivances, experimental methods of illustrating physical principles, questions connected with methods of teaching," and other important matters in physics that would "nevertheless be out of place before the Royal Society." Foster especially highlighted Guthrie's role as its first demonstrator, since the "actual exhibition of physical phenomena to the members" was a central feature of the Physical Society's meetings.[87] Some early presentations to the Physical Society that appeared in the first volume of its *Proceedings* (1874–75) bear out this pedagogical focus: in addition to Barrett's enhanced trombone, and Foster's didactic paper on graphical (nonalgebraic) methods for solving "electrical problems," there were papers on glass cells, microscopical techniques for demonstrating colored rings in crystals, and apparatus to illustrate the "Formation of Volcanic Cones."[88] As already indicated, however, this was not the whole story.

In fact, substantial theoretically informed empirical research was presented at the Physical Society's early meetings. One such example was the first paper presented at the society's inaugural meeting, in March 1874, by Guthrie's student Ambrose Fleming: "On the New Contact Theory of the Galvanic Cell." Unlike most of the pedagogical papers in the first year, this and John Rae's paper "On Some Physical Properties of Ice"[89] were swiftly published in the *Philosophical Magazine* dated June-July that year, and appeared later in the first volume of the Physical Society's *Proceedings*. New technological developments were also given both theoretical analysis as well as practical demonstration: with the arrival of the telephone and new forms of dynamo in 1876–77, papers on these topics by William Preece and David Hughes appeared in the second volume of the *Proceedings* in 1876–78.[90] Significantly, the society soon benefited from the regular attendance at its meetings of *Nature* journalists, who reported experimental demonstrations that were not necessarily recorded in the society's *Proceedings*, such as William Crookes' demonstration of his radiometer in December 1877.[91] This brought such good publicity to the society that, soon after Maxwell's death in 1879, virtually all Cavendish Laboratory staff in Cambridge joined. Thus, as the Physical Society flourished on the national scene, and sometimes held meetings at laboratories outside London, its British membership reached more than four hundred by 1900.[92]

Nevertheless, the Physical Society's early years were marked by other challenges that constrained its capacity to publish its own independent journal. Indeed, at its jubilee in 1924, the chemist Henry E. Armstrong commented that "in early days, you were subject to Royal Society influences, which sought to keep you in the nursery."[93] Thus, to avoid diplomatic difficulties, the Physical Society's founders at first deliberately avoided setting up any rival journal of its own. Instead, an agreement was reached with the proprietors of the *Philosophical Magazine* that any papers read before the Physical Society that its council, acting as quasi-editor, wanted to publish would in the first instance be printed in the *Philosophical Magazine*—instances of which we have seen above. Only afterwards were the papers collected and issued at irregular intervals in a new informal *Proceedings* that was at first circulated purely privately to the Physical Society's members, starting with the first volume in 1875. This arrangement continued for the first two decades of the society's existence, until 1895, by which time any concern that the Physical Society was a rival to *Philosophical Transactions A* appear to have faded away. The hitherto annual *Proceedings of the Physical Society* moved at this point to monthly publication, thus adopting at last a standardized model of an independent society journal.[94]

THE INTERNATIONALIZATION OF
BRITISH PHYSICS JOURNALS

Another significant move in 1895 that underpinned the new authority of the Physical Society was a decision to embark on an internationally oriented journals project. This was the publication in English translation of the abstracts of recent important papers on physical sciences printed in both British and "foreign" journals. As noted above, selected papers from outside British scientific societies had long been acknowledged and abstracted in the *Philosophical Magazine* and then *Nature*, among other publications. But the importance of a more systematic approach to such abstracting was becoming eminently clear by the late 1870s. As Elizabeth Crawford has shown, a movement for internationalism in science was growing, to overcome the increased nationalism in the imperialist politics of Europe and North America. Crawford highlights three initiatives in this period: the International Committee on Weights and Measures, the Institut international de physique Solvay, and the International Catalogue of Scientific Literature.[95]

For the purposes of this chapter, however, we can note that by the 1890s the communities of physics and electrical engineering were increasingly aware that much valuable research was being conducted in other countries in non-British journals. The sheer extent of it was impracticable for even the most polyglot practitioner to cover in the course of perusing journal outputs every month. Hence, the cross-communal demand arose for systematic abstracts of other non-UK periodicals in order to counter the dangers of a parochial reliance on the British-based journals outlined above. The Physical Society began this venture on its own, publishing with Taylor and Francis three annual volumes of *Abstracts of Physical Papers from Foreign Sources*, starting in 1895 with an international staff of thirty-five abstracters, led by James Swinburne. The fifty-three periodicals abstracted from France, Germany, Italy, Russia, the Austro-Hungarian Empire, and the United States epitomized the breadth with which the "physical" science periodical was conceived, including physics journals such as the American *Physical Review*; the French *Journal de Physique* and *Annalen der Physik und Chemie*; and numerous specialist chemical periodicals and electrical engineering journals such as the French *L'Électricien*, the German *Elektrotechnische Zeitschrift*, and the *Transactions of the American Institute of Electrical Engineers*.[96]

News of this globally oriented venture soon spread across the Atlantic. As an editorial in the US journal *Science* commented in July 1897: "It is hoped that these Abstracts will be of great use in facilitating a knowledge by English-speaking physicists of the work which is being done by their colleagues in other countries." But so great was the cost of the abstracting process that the annual subscription charged to the four hundred members of the Physical Society had to be increased from one pound to two pounds and two shillings.[97] To broaden the scope of this abstracting initiative, the Physical Society joined forces with the Institution of Electrical Engineers (IEE), the successor to the Society of Telegraph Engineers and Electricians from 1888. Thus, the first annual volume of *Science Abstracts: Physics and Electrical Engineering* appeared in 1898, published by the engineering specialists E. & F. N. Spon, Limited, in London, and Spon & Chamberlain in New York. The number of staff involved increased to fifty-three, including the Cavendish Laboratory Demonstrator, G. F. C. Searle, and both the international correspondent of the *Electrician*, E. E. Fournier d'Albe, and its recent editor, A. P. Trotter. The now monthly issues of the *Science Abstracts* covered more than one hundred physical science and electrical engineering periodicals, including publications from Australia, Canada, Ireland, Netherlands, Poland, and Switzerland.[98]

This metajournal publication of 1,400 or so annual abstracts in a new collaborative cross-institutional journal neatly complemented the Royal Society's *Catalogue of Scientific Papers*, produced from 1867 onward. For those who did not have access to the elite libraries containing the physics journals listed in the *Catalogue*, nor educational access to the polyglot culture required to read them in the original language, the Physical Society and IEE's *Science Abstracts* supplied invaluably succinct journal content to a broad English-language readership.[99] Given that the production costs of this new quasijournal publication had been underwritten by the Physical Society's membership, they all received the *Science Abstracts* free of charge. As the annual number of abstracts published almost doubled by the early twentieth century, the Royal Society joined in the financing of this project, along with the Institution of Civil Engineers and the British Association for the Advancement of Science. By 1903 the volume had become so large that its contents were split into two parts, A (Physics) and B (Electrical Engineering), though abstracting remained speedy, typically taking no more than two months. While the length of an abstract was not initially standardized, and sometimes ran up to several pages, the exigencies of the Great War, including paper shortages, led to the length being standardized to a single paragraph, which remains the standard format to this day.[100]

Overall, then, despite intermittent fears of competition between the journals discussed in this chapter and at other times starkly different approaches to engaging industrial and educational audiences, what brought the journals together was the huge enterprise of bringing to their readers a broader literature than just homegrown knowledge of physics. It is thus fitting to close this chapter with a recognition that, while British physics was exported around the world through its journals, we must not forget that what those journals produced on their own was by no means the sole constituent of physics in Britain, the communities of which had a firmly internationalist outlook.

CONCLUSIONS

In this chapter I have shown how the advent of the Physical Society's *Proceedings* and the *Electrician* marked an expansion of interest in the field of physics, reflecting the rise of new professional constituencies at the interface of what had been known as "natural philosophy" and engineering. While laboratory workers and school teachers joined the Physical Society of London and published in its *Proceedings* (initially via the *Philosophical Magazine*), the *Electrician* was especially useful to technical specialists and academic consultants who embraced electrical communications and power technologies as major sites of disciplinary innovation. A study of these initially overlapping but latterly divergent (except for the internationalization outlined above) journal publishing activities subverts the suggestion by Buchwald and Hong that this period was marked by a determinate disciplinary transformation from natural philosophy to physics.[101] Instead, the professional basis of these journals' readership, and the divergent editorial practices of the *Philosophical Transactions* and the *Electrician* with respect to speed and refereeing processes, are more striking.

Both new journals capitalized upon disaffection among these new professional audiences with the older elite publications and practices. Whereas publication in a Royal Society journal required a prospective author to persuade an existing fellow of the society to read the paper at a Royal Society meeting, publication in the *Proceedings of the Physical Society* or the *Electrician* was both far less exclusive and accomplished by very different routes. For the former, attendance at a Physical Society meeting to read a paper echoed the practice of the Royal and other scientific societies—typically with the requirement for experimental demonstration, which was absent at the Royal Society. Yet the Physical Society's meetings were much more inclusive of multiple approaches—ground was regularly given to incomplete researches, to demonstration of new apparatus or teaching equipment, and to technological explorations in ways alien to the Royal Society environment. Thus, dozens of school and college practitioners effectively excluded from the inner sanctum of the Royal Society could present papers unconstrained by the more formal expectations of the older society for theoretical or empirical rigor. Although not all papers were published, the Physical Society's committee on publications took a much broader view about what was appropriate publishable material in the *Physical Society's Proceedings* than did Stokes and his successors. By that means as well as by the movement of the younger society's meetings around the country, a new community of physics specialists could flourish with a journal and institution that have survived —albeit both retitled— to the present day.

As for the *Electrician*, we can see that this was a journal run on very different operational principles than those of a society's proceedings. Its contents were based more upon what the editor commissioned from house journalists or freelance writers concerning new industrial developments than upon any work spontaneously submitted by electrical practitioners. While the latter constituency could write letters to discuss matters raised in editorials or in commissioned series of articles, this journal was not aiming to reflect the activities and interests of the rising professional group of readers, so much as to lead them by highlighting what they *ought* to know in order to practice their professions more effectively. Thus, the crucible of critical appreciation was not the refereeing process that simultaneously improved the rigor and slowed the progress of society *Proceedings*; rather, it was the extent to which, as a commercial journal operating weekly, it could supply readers with what they needed even before they knew they needed it, and thus bring commercial success from satisfied readers renewing their subscriptions.

We can thus see how the *Electrician* survived in the longer term whereas Sturgeon's *Annals of Electricity* did not. The *Electrician* rooted its operations in the needs of rising new reading communities that could give a technical periodical sustainability, and thus longevity. A future project going beyond the scope of this chapter would be to develop a deeper understanding of the various readerships of the journals discussed above. From that we would be able see how far their participation shaped those journals, and how far the journals, in turn, shaped the readerships' identity as practitioners of the newly ubiquitous discipline of "physics."

NOTES

1. For a study of the *Lancet* in its broader publishing context, see chapter 9 in this volume.

2. Aileen Fyfe, "Journals, Learned Societies and Money: 'Philosophical Transactions,' ca. 1750–1900," *Notes and Records: The Royal Society Journal of the History of Science* 69 (2015).

DOI: 10.1098/rsnr.2015.0032. A fuller discussion of related matters is anticipated in Alex Csiszar, *The Scientific Journal: Authorship and the Politics of Knowledge in the Nineteenth Century* (Chicago: University of Chicago Press, 2018).

3. Alex Csiszar, "Peer Review: Troubled from the Start," *Nature* 532 (2016): 306–8. http://www.nature.com/news/peer-review-troubled-from-the-start-1.19763.

4. John Heilbron, "Benjamin Franklin in Europe: Electrician, Academician, Politician," *Notes and Records of the Royal Society* 61 (2007): 353–73. For further discussion of the narrowing and shifting reference of the term "electrician" in the century after Franklin's death, see Stathis Arapostathis and Graeme Gooday, "Electrical Technoscience and Physics in Transition, 1880–1920," *Studies in History and Philosophy of Science* 44 (2013): 202–11.

5. James Clerk Maxwell, *Treatise on Electricity and Magnetism* (Oxford, UK: Clarendon Press, 1873); James Clerk Maxwell, "A Dynamical Theory of the Electromagnetic Field," *Philosophical Transactions of the Royal Society of London* 155 (1865): 459–512.

6. Aileen Fyfe and Noah Moxham, "Making Public ahead of Print: Meetings and Publications at the Royal Society, 1752–1892," *Notes and Records of the Royal Society* 70 (2016): 361–79, especially 371.

7. Jonathan Topham, "Anthologizing the Book of Nature: The Circulation of Knowledge and the Origins of the Scientific Journal in Late Georgian Britain," in *The Circulation of Knowledge between Britain, India and China: The Early-Modern World to the Twentieth Century*, ed. B. Lightman, L. Stewart and G. McOuat (Leiden, Netherlands: Brill Academic Publishers, 2013), 119–52, especially 138–42.

8. Fyfe and Moxham, "Making Public ahead of Print," 371.

9. Imogen Clarke and James Mussell, "Conservative Attitudes to Old-Established Organs: Oliver Lodge and the 'Philosophical Magazine,'" *Notes and Records of the Royal Society* 69 (2015): 321–36. I thank Bernard Lightman for pointing out that Taylor and Francis relied on John Tyndall early in his career as a reader for the *Philosophical Magazine*. See letters in Ruth Barton, Jeremiah Rankin, and Michael Reidy, eds., *The Tyndall Correspondence Volume 3 (January 1850–December 1852)* (Pittsburgh: Pittsburgh University Press, 2017). For discussion of the refereeing process in the early-twentieth-century *Philosophical Magazine*, see Imogen Clarke, "The Gatekeepers of Modern Physics: Periodicals and Peer Review in 1920s Britain," *Isis* 106 (2015): 70–93, especially 74.

10. Iwan Morus notes that the contemporary London Electrical Society, in which Sturgeon was initially involved, had a similarly short life. Iwan Morus, *Frankenstein's Children: Electricity, Exhibition, and Experiment in Early-Nineteenth-Century London* (Princeton, NJ: Princeton University Press, 1998), 46.

11. Philip Strange, "Two Electrical Periodicals: The 'Electrician' and the 'Electrical Review,' 1880–1890," *IEE Proceedings,* 132A (1985): 574–81, especially 574.

12. Susan Sheets-Pyenson, "Popular Science Periodicals in Paris and London: The Emergence of a Low Scientific Culture, 1820–1875," *Annals of Science* 42 (1985): 549–72. For reminiscences of early years of the magazines' practices, see Lt.-Col. D. J. Smith, "A History of English Mechanics," *English Mechanic*, issue 3069 (18 January 1924): 5–6; and "K. C. A. J.," "Notes on the 'English Mechanic and Mirror of Science and Art,'" *English Mechanic*, issue 3066 (28 December 1923): 326–27.

13. Melinda Baldwin, *Making "Nature": The History of a Scientific Journal* (Chicago: University of Chicago Press, 2015). For the time scale of publication, see Fyfe and Moxham, "Making Public ahead of Print," 371.

14. See discussion of Macmillan's support for *Nature* in Graeme Gooday, "Sun-Spots, Weather and the Unseen Universe: Balfour Stewart's Anti-Materialist Representations of Energy in British Periodicals," in *Science Serialized: Representations of the Sciences in Nineteenth-Century Periodicals*, ed. Geoffrey Cantor and Sally Shuttleworth (Cambridge, MA: MIT Press, 2004), 111–47.

15. The Society of Telegraph Engineers changed its organizational name a number of times, becoming the Institution of Electrical Engineers in 1888. Rollo Appleyard, *The History of the Institution of the Electrical Engineers (1871–1931)* (London: Institution of the Electrical Engineers, 1939); William J. Reader with Rachel Lawrence, Sheila Nemet, and Geoffrey Tweedale, *A History of the Institution of Electrical Engineers, 1871–1971* (London: Peregrinus, 1987).

16. Simon Schaffer, "Late Victorian Metrology and Its Instrumentation: A Manufactory of Ohms," in *Invisible Connections: Instruments, Institutions, and Science,* ed. R. Bud and S. E. Cozzens (Bellingham, WA: SPIE Press, 1991), 23–56. For a recent study of how precise measurement techniques were developed in the global techniques of cable mending, see John Trelawny Brooks Moyle, "The Telegraphic Life: Maintenance of the System 1850–1914," PhD diss., University of Leeds, 2015.

17. Oliver Heaviside, "On Electromagnets," *Journal of the Society of Telegraph Engineers* 7 (1878): 303–23. Correlatively, a review of Maxwell's *Treatise* was published in the commercial monthly *Telegraphic Journal and Monthly Illustrated Review of Electrical Science,* launched in November 1872; this was rebranded as the weekly *Telegraphic Journal and Electrical Review* in 1881, and retitled again as the *Electrical Review* in 1891.

18. For a broad-ranging survey of chemical periodicals, see William H. Brock, *William Crookes (1832–1919) and the Commercialization of Science* (Aldershot, UK: Ashgate, 2008). For the Chemical Society, see *The Jubilee of the Chemical Society of London: Record of the Proceedings Together with an Account of the History and Development of the Society, 1841–1891* (London: Harrison, 1896; reprinted Hardpress, 2013).

19. For the intersecting boundaries of physics with other disciplines, see Robert Fox and Graeme Gooday, *Physics in Oxford 1839–1939: Laboratories, Learning and College Life* (Oxford, UK: Oxford University Press, 2005). For the First World War context of change, see Arapostathis and Gooday, "Electrical Technoscience." For the stably configured discipline of "modern physics" that emerged only in the mid-twentieth century, see Graeme Gooday and Daniel Mitchell, "Rethinking Classical Physics," in *The Oxford Handbook of the History of Physics*, ed. Jed Z. Buchwald and Robert Fox (Oxford, UK: Oxford University Press, 2013), 721–64.

20. Charles Babbage, *Reflections on the Decline of Science in England: And on Some of Its Causes* (London: Fellowes, 1830); Noah Moxham and Aileen Fyfe, "The Royal Society and the Prehistory of Peer Review, 1665–1965," *Historical Journal* 61, no. 4 (2018): 1–27.

21. Henry George Lyons, *The Royal Society 1660–1940, A history of its Administration under its Charters,* (Cambridge: Cambridge University Press, 1944), 254, 324.

22. G. G. Stokes to Mary Stokes (née Robinson), 30 January 1857, in *Memoir and Scientific*

Correspondence of the Late Sir George Gabriel Stokes, ed. Joseph Larmor, 2 vols. (Cambridge: Cambridge University Press, 1907), I:54.

23. Michael Foster recalled that, as Huxley's successor in the role, he continued the same division of labor with Stokes. See Larmor, ed., *Memoir and Scientific Correspondence* I: 97–99.

24. Clarke and Mussell, "Conservative Attitudes," 321–36.

25. For a more traditional view of the disciplinary emergence of physics in the late nineteenth century, see Jed Z. Buchwald and Sungook Hong, "Physics," in *From Natural Philosophy to the Sciences: Writing the History of Nineteenth-Century Science*, ed. David Cahan (Chicago: University of Chicago Press, 2003), 163–95.

26. Strange, "Two Electrical Periodicals," 574.

27. Graeme Gooday, "The Questionable Matter of Electricity: The Reception of J. J. Thomson's 'Corpuscle' among Electrical Theorists and Technologists," in *Histories of the Electron: The Birth of Microphysics*, ed. J. Z. Buchwald and A. Warwick (Cambridge, MA: MIT Press, 2001).

28. Arapostathis and Gooday, "Electrical Technoscience," 202–11.

29. For discussion of the commercial and commissioning operations of the *Electrician* by one of Biggs's editorial successors (from 1890), see the manuscript version of Alexander Pelham Trotter, "Reminiscences" (undated, c. 1920), in IET Archives, 610–28. A greatly shortened version of these reminiscences were published as Alexander P. Trotter and F. W. Hewitt, eds., *Early Days of the Electrical Industry, and Other Reminiscences of Alexander P. Trotter* (London: Institution of Electrical Engineers, 1948).

30. As Philip Strange notes, up to that point most articles in the *Electrical Review* (as it was latterly known) had been short unsigned pieces or translations of papers delivered abroad. In April 1885 it secured from Italy a series of papers by Galileo Ferraris on the efficiency of alternate current transformers, followed in 1886 by a series from Anthony Reckenzaun on the applications of electricity to transportation, and another in 1887 on the design of storage batteries. See Strange, "Two Electrical Periodicals," 580.

31. Moxham and Fyfe, "The Royal Society and the Prehistory of Peer Review," 15.

32. Fyfe and Moxham, "Making Public ahead of Print," 363.

33. David B. Wilson, *Kelvin and Stokes: A Comparative Study in Victorian Physics* (Bristol, UK: Adam Hilger, 1987). The scale of Stokes's correspondence is thus comparable to that of Charles Darwin.

34. Lamor, ed., *Memoir and Scientific Correspondence*, I: 239, 406.

35. Lamor, ed., *Memoir and Scientific Correspondence*, I: 34–35.

36. Lamor, ed., *Memoir and Scientific Correspondence*, I: 223.

37. The letters that follow are taken from David B. Wilson, ed., *The Stokes-Kelvin Correspondence, Vol. 1, 1846–1869* (Cambridge: Cambridge University Press, 1990).

38. Roland Jackson, "John Tyndall and the Early History of Diamagnetism," *Annals of Science* 72 (2015): 435–89, especially 465–67. Thomson to Stokes, 12 February 1855, reproduced in Wilson, *Stokes-Kelvin Correspondence*, 186; Thomson to Stokes, 14 March 1855, *Stokes-Kelvin Correspondence*, 189 (Royal Society of London archives, RR.2.253).

39. John Tyndall, "On the Nature of the Force by Which Bodies Are Repelled from the Poles of a Magnet; to Which Is Prefixed an Account of Some Experiments on Molecular In-

fluences," *Philosophical Transactions of the Royal Society* 145 (1855): 1–52, and *Philosophical Magazine* 10 (1855): 65, 66, 153–79, and 257–90.

40. Reuben Philips, "On the Aurora," *Proceedings of the Royal Society* 8 (1856–57): 214–15. See Thomson to Stokes, 16 May 1856, in Wilson, *Stokes-Kelvin Correspondence*, 212–15.

41. Frederick Guthrie, "Note of Experiments upon the Conduction of Heat by Liquids," *Philosophical Magazine* 35 (1868): 283–87.

42. Guthrie to Stokes, 15 October 1868, Royal Society Archives, MC.8.255.

43. Frederick Guthrie, "On the Thermal Resistance of Liquids," *Proceedings of the Royal Society* 17 (1869): 234–36.

44. Frederick Guthrie to George Stokes, 23 July 1869, Cambridge University Library ULC AD 7656 0833.

45. James Clerk Maxwell to William Thomson, 16 November 1869, Cambridge University Library ULC ADD 7342 M107.

46. Frederick Guthrie, "On the Thermal Resistance of Liquids," *Philosophical Transactions of the Royal Society* 159 (1869): 637–60.

47. Frederick Guthrie, "On Approach Caused by Vibration," *Proceedings of the Royal Society* 18 (1870): 93–94.

48. Thomson to Guthrie, 14 November 1870 in *Proceedings of the Royal Society* 41 (1870): 423–25. For the full correspondence, see *Proceedings of the Royal Society* 41 (1870): 420–29; and see Guthrie to Stokes, 17 November 1870, Royal Society Archives, R.S.MC.8.255.

49. Guthrie to Stokes, 15 June 1871, Royal Society Archives, R.S. MC.9.215.

50. Stokes's daughter, Isabella, recalled his uncomfortable encounter with a visitor who called to complain about the Royal Society's rejection of a friend's work. Mrs. Laurence Humphry, "Notes and Recollections," in *Memoir & Correspondence* I: 37.

51. See Derek Jackson and Brian Launder, "Osborne Reynolds and the Publication of His Papers on Turbulent Flow," *Annual Review of Fluid Mechanics* 39 (2007): 19–35. In mapping the long correspondence between Stokes and Thomson, David Wilson characterizes the former as representing the orthodoxy beyond which the latter was aiming to move. See Wilson, *Stokes-Kelvin Correspondence*, xxxviii–xl. From 1887 to 1892 Stokes served as a Conservative member of Parliament for the Cambridge University constituency; he was elevated to a baronetcy in 1889.

52. Paul J. Nahin, *Oliver Heaviside: The Life, Work, and Times of an Electrical Genius of the Victorian Age* (Baltimore: Johns Hopkins University Press, 2002), 104–5; "C. H. W. Biggs," in *Dictionary of Nineteenth-Century Journalism in Great Britain and Ireland*, eds. Laurel Brake and Marysa Demoor (London: Academia Press and British Library, 2009), 53–54; Robert Sharp, "Biggs, Charles Henry Walker (1845–1923), Technical Journalist and Journal Editor," Oxford Dictionary of National Biography. Retrieved 13 December 2017 from http://0-www.oxforddnb.com.wam.leeds.ac.uk/view/10.1093/ref:odnb/9780198614128.001.0001/odnb-9780198614128-e-58119.

53. "Note," 13 July, *Electrician* 1 (1878): 85; Strange, "Two Periodicals," 575.

54. F. De Chaumont, "On the Anatomy of the Organ of Hearing in Relation to the Discovery of the Principle of the Microphone of Prof. D. E. Hughes, and the Magnophone of Mr. W. L. Scott, A.S.T.E," *Nature* 18 (1878): 285–86.

55. John Joseph Fahie, *A History of Electric Telegraphy to the Year 1837* (London and New York: E. & F. N. Spon, 1884), x–xi; Strange, "Two Electrical Periodicals," 579.

56. Following Biggs's template, Trotter commissioned a series of articles in 1890 from, for example, John Ambrose Fleming ("Electrical Distribution by Transformers") and Gisbert Kapp ("Capacity and Self-Induction in Alternate Current Working."). A. P. Trotter, *Reminiscences* (undated. c. 1920) in IET Archives, 610–28.

57. Charles Biggs to Oliver Heaviside, 5 December 1882, UCL Oliver Heaviside collection.

58. All in these series of papers are reproduced in Oliver Heaviside, *Electrical Papers*, 2 vols. (London: Macmillan, 1891), vol. 1.

59. Bruce Hunt, "Oliver Heaviside: A First-Rate Oddity," *Physics Today* 65 (2012): 48–54, doi: http://dx.doi.org/10.1063/PT.3.1788; letter by O. Heaviside to G. F. FitzGerald (13 February 1894), in Bruce Hunt, *The Maxwellians* (Ithaca, NY: Cornell University Press, 1991), 71.

60. Nahin, *Oliver Heaviside,* 103. For the introduction of "impedance" in 1887, see Oliver Heaviside, "Electromagnetic Induction and its Propagation," *Electrician* (1886–87), volumes 18–19, specifically vol. 19, 27 May 1887. "Section XXIX," 50, was reproduced in Heaviside, *Electrical Papers* (1892) vol. 2, p. 64.

61. Letter from Heaviside recorded by A. P. Trotter, unpublished memoirs, IET Archives. Cited in Nahin, *Oliver Heaviside,* 105.

62. Hunt, *The Maxwellians,* and Nahin, *Oliver Heaviside.*

63. Charles Biggs to Oliver Heaviside, 20 September 1887, IET Archives UK0108 SC MSS 005/I/3. Nahin discusses this episode in detail, noting that fear of libel proceedings prompted the *Philosophical Magazine* to decline a paper by Heaviside that attacked Preece. Nahin, *Oliver Heaviside,* 152–53. The problematic articles that Heaviside could not publish in the *Electrician* nevertheless appeared in his *Electrical Papers*, vol. 2.

64. William Henry Snell to Oliver Heaviside, 30 November 1887, IET Archives UK0108 SC MSS 005/I/3; discussed in Nahin, *Oliver Heaviside,* 153–55.

65. Oliver Heaviside, "On Resistance and Conductance Operators, and Their Derivatives, Inductance and Permittance, Especially in Connection with Electric and Magnetic Energy," *Philosophical Magazine* (December 1887): 470; reproduced in Heaviside, *Electrical Papers*, vol. 2, 355–74. For Maxwell's Cambridge style of theorizing, see Warwick, *Masters of Theory: Cambridge and the Rise of Mathematical Physics* (Chicago: University of Chicago Press, 2003). On quaternions, see Nahin, *Oliver Heaviside,* 187–215.

66. Trotter, *Reminscences,* 631; W. Thomson, "Aether, Electricity, and Ponderable Matter," *Journal of the IEE* 18 (1889): 4–43.

67. Strange, "Two Electrical Periodicals," 579. "Mr. A. P. Trotter," *Nature* 160 (1947): 390.

68. Nevertheless, the misanthropic and severely deaf Heaviside never attended public events, so he never saw his subsequent papers read at any of the Royal Society's meetings.

69. Oliver Heaviside, "On the Forces, Stresses, and Fluxes of Energy in the Electromagnetic Field," *Philosophical Transactions of the Royal Society A* 183 (1892): 423–80.

70. Trotter, *Reminiscences,* 618, 632–38, especially 632, on the allegations from "orthodox Royal Society mathematicians" on Heaviside's alleged "want of rigour."

71. The Royal Society's assistant secretary, Herbert Rix, wrote to Heaviside on 21 December 1891 acknowledging that Heaviside objected "strongly" to the use in his paper of the typeface

customarily used in the *Philosophical Transactions*. In a postscript, Rix agreed to cast a new type for Heaviside, even though this would delay the publication of his paper. See letter from Herbert Rix to Oliver Heaviside, 21 December 1891, NLB/5/1052 GB 117, the Royal Society. In fact, discussions between the Royal Society and Messrs. Harrison & Sons on the cutting of the new typeface, and Heaviside's objections to the Royal Society's management of it in proofs of his paper, extended well into the summer of 1892.

72. Nahin notes that this outright rejection of an FRS-authored paper from the *Proceedings* had only happened once before: J. J. Waterston's paper on the kinetic theory of gases, rescued from the society's archives forty years later. Nahin, *Oliver Heaviside*, 222–27.

73. James Swinburne to Oliver Heaviside, 21 June 1894, IET Archives UK0108 SC MSS 005/I/6. Discussed in Nahin, *Oliver Heaviside*, 225–26, which reads "divergence" rather than "divulgence."

74. Graeme Gooday, "Precision Measurement and the Genesis of Physics Teaching Laboratories in Victorian Britain," *British Journal for the History of Science* 23 (1990): 25–51.

75. *The Physical Society of London 1874–1924, Proceedings of the Jubilee Meetings, March 20–22nd, Special Number* (London, 1924), copy in the Institute of Physics library and archives, 15.

76. *Physical Society of London 1874–1924*, 14–15. William Fletcher Barrett, "Modification of the Usual Trombone Apparatus for Showing the Interference of Sound-Bearing Waves," *Proceedings of the Physical Society of London* 1 (1874–75): 51–52.

77. Frederick Guthrie, "On New Relation between Heat and Electricity," *Proceedings of the Royal Society* 21 (1873): 168–69.

78. Fleeming Jenkin to George Stokes, 27 May 1873, R.S. R.R.7.244, Royal Society Archives.

79. Maxwell correspondence with Stokes, Maxwell to Stokes, Royal Society Archives, LS. R.R.7.245, undated (presumably May 1873).

80. Frederick Guthrie, "On a Relation between Heat and Electricity," *Philosophical Magazine* October 46 (1873) [4th series]: 257–66.

81. At the fifty-year-jubilee meeting of the Physical Society in 1924, Ambrose Fleming looked back to Guthrie's experiments of 1873 and emphasized that his observations were not fully explained "until the researches of Sir J. J. Thomson had made us acquainted with the electron and electron emission from incandescent bodies." *Physical Society of London 1874–1924*, 19–20.

82. Cited in G. C. Foster [Obituary of Frederick Guthrie], *Proceedings of the Royal Society* 8 (1887): 9–13.

83. Barrett recalled the following earlier joiners: Balfour Stewart (Manchester), Edward Frankland (London), John Hall Gladstone (London), John Tyndall (Royal Institution), William Crookes (editor of *Chemical News*), James Glaisher (president of the Royal Photographical Society), George Carey Foster (University College London), Alexander Strange, and Thomas Archer Hirst. Some fellows of the Chemical Society were also keen to sign up, notably Herbert McLeod and William Fletcher Barrett. *Physical Society Jubilee Proceedings* (1924), 15.

84. Guthrie approached Stokes a second time in 1885, when the latter became president of the Royal Society, but with no apparent response. Guthrie to Stokes, 12 August 1885, ULC ADD 7656 G834, Stokes Collection, University of Cambridge Library.

85. Maxwell to W. G. Adams, December 1873, cited in *Physical Society Jubilee Proceedings*

(1924), 17. In interpreting Maxwell's disapprobation of the Physical Society's essential activities, Russell Moseley claims that the society was populated by "second-order" experimentalists who engaged in merely "elementary" activities. Russell Moseley, "Tadpoles and Frogs: Some Aspects of the Professionalization of British Physics, 1870–1939," *Social Studies of Science* 7 (1977): 423–46, esp. 426–27. In citing the 1898 presidential address of Shelford Bidwell, with reference to the "lax regime" in early meetings that allowed the reading of papers that were "sometimes blemished by serious errors," Moseley disingenuously omits Bidwell's comment that "the demolition of the authors added much to the interest and liveliness of the discussions." *Proceedings of the Physical Society* 16 (1898): 12. This is discussed in detail in chapter 8 of Graeme Gooday, "Precision Measurement and the Genesis of Physics Teaching Laboratories in Victorian Britain," PhD diss., University of Kent, 1989.

86. John Hall Gladstone, [annual report], *Proceedings of the Physical Society* 1 (1874–75), appendix 5.

87. Foster, [obituary of Frederick Guthrie], 9–10.

88. F. Clowes, "Glass Cell with Parallel Sides," *Proceedings of the Physical Society of London* 1 (1874–5): 26; W. H. Stone, "On a Simple Arrangement by Which the Coloured Rings of Uniaxial and Biaxial Crystals May Be Shown in a Common Microscope," ibid., 34; G. Carey Foster, "On Graphical Methods of Solving Certain Simple Electrical Problems," ibid., 101; C. J. Woodward, "On an Apparatus to Illustrate the Formation of Volcanic Cones," ibid., 159.

89. John Ambrose Fleming, "On the New Contact Theory of the Galvanic Cell," *Proceedings of the Physical Society* 1 (1874–75): 1–12, reproduced in *The London, Edinburgh, and Dublin Philosophical Magazine and Journal of Science* Series 4 (48), 1874, issue 315, 401–11. John Rae, "On Some Physical Properties of Ice; on the Transposition of Boulders from Below to Above the Ice; and on Mammoth Remains," ibid., 14–18. Reproduced in *The London, Edinburgh, and Dublin Philosophical Magazine and Journal of Science,* series 4 (48), 1874, issue 315, 56–61.

90. William Henry Preece, "On Some Physical Points Connected with the Telephone," *Proceedings of the Physical Society of London* 2 (1876–78) 224–35; David Hughes, "On the Physical Action of the Microphone," ibid., 255–260; George Fuller, "Model of a Small Electrical Machine," ibid., 83–86.

91. "Societies and Academies: 'Physical Society, December 16, 1876,'" *Nature* 15 (1877) 210.

92. John Lewis et al., *125 Years of the Physical Society and the Institute of Physics* (London: Institute of Physics Publishing, 1999), 28–29.

93. *Physical Society Jubilee Proceedings* (1924), 3.

94. As president of the Physical Society in 1885, Guthrie encouraged members to think of the society's *Proceedings* as the first choice for communication. By 1894 the society had introduced formal refereeing processes to the *Proceedings*. Lewis et al., *125 Years of the Physical Society,* 22, 24.

95. Elisabeth Crawford, "Internationalism in Science as a Casualty of the First World War: Relations between German and Allied Scientists as Reflected in Nominations for the Nobel Prizes in Physics and Chemistry," *Social Science Information* 27 (1988): 163–201.

96. James Swinburne, *Abstracts of Physical Papers from Foreign Sources,* vol. 1 (London: Taylor and Francis, 1895). This enormous Taylor and Francis enterprise is not mentioned in

W. H. Brock and J. A. Meadows, *The Lamp of Learning: Two Centuries of Publishing at Taylor and Francis*, 2nd ed. (London: Taylor and Francis, 1998).

97. "The Physical Society of London," *Science,* new series 6 (1897): 53–54.

98. James Swinburne et al., *Science Abstracts: Physics and Electrical Engineering, Issued under the Direction of the Institution of Electrical Engineers and the Physical Society of London,* vol. 1 (London: E. & F. N. Spon; New York: Spon & Chamberlain, 1898).

99. For the origins of the *Royal Society's Catalogue of Scientific Papers,* see Alex Csiszar, "Seriality and the Search for Order: Scientific Print and Its Problems during the Late Nineteenth Century," *History of Science* 48 (2010): 399–434.

100. The launch of the *Science Abstracts* series in 1898 is mentioned in passing without specific explanation in Henry Lowood, "Journal," in *The Oxford Companion to the History of Modern Science,* ed. John Heilbron (Oxford, UK: Oxford University Press, 2003), 429–31, and Robert Mortimer Gascoigne, *A Historical Catalogue of Scientific Periodicals, 1665–1900, with a Survey of Their Development* (New York: Garland, 1985). For a full discussion of how the *Science Abstracts* projects have been maintained up to the present day, see http://www.theiet .org/resources/library/archives/inspec/1898–1914.cfm.

101. Buchwald and Hong, "Physics," 163–95.

CHAPTER 8

Late Victorian Astronomical Society Journals: Creating Scientific Communities on Paper

Bernard Lightman

In his letter to the editor of the *Journal of the British Astronomical Association* for 1895, Arthur Mee of Cardiff expressed concern about the correspondence section of the journal, which was then in its fifth volume. "Though our 'Journal' has a Correspondence Department," he pointed out, "few Members seem to avail themselves of it." Mee saw the correspondence section as vital to the journal. This was the section where members' letters to the editor were published and sometimes received public replies. Whereas papers, which he compared to "elaborate dishes," were in abundance, what was desperately needed was more correspondence to provide "a little more seasoning." Mee missed this feature "keenly," and "having suffered in silence through four volumes," he had decided to voice his dismay, confident that "there are more of my opinion." Mee was also convinced that the editor "would like to see the Correspondence Department more widely utilised, for but few Members comparatively can attend the Meetings, and to the great bulk the 'Journal *is itself the Association*,' and we ought to find in its pages the miscellaneous scraps of information that would soon be floating around were all the Members present in one big Parliament."[1] For Mee, correspondence was the key to the success of the journal, while the journal actually constituted the society for many members.

Mee was no ordinary correspondent. An amateur astronomer, he was in the process of founding a new society and journal at the same time that he wrote to the *Journal of the British Astronomical Association*. The first president of the Astronomical Society for Wales, Mee was also the editor of that society's journal. On the front cover of the preliminary number of the *Journal of the*

Astronomical Society of Wales, dated January 1895, Mee laid out his vision for the new organization. He welcomed all interested in astronomy to become members. "Let no one draw back from joining the Society," he declared. It was not necessary to own a telescope or to spend substantial sums of cash. He insisted, "There is no science less costly of pursuit than Astronomy." Using spectacle glasses and cardboard tubes, beginners could see "for themselves quite as much as the wonderful '£5 telescopes' reveal." He promised to explain how to construct these makeshift telescopes in the pages of the new journal. Clearly, in Mee's mind, the journal was just as central to the new society as the *Journal of the British Astronomical Association* was to the fledgling British Astronomical Association. "To a large number of members," Mee asserted, "our Journal must necessarily be the Society."[2]

Many of the astronomical journals published in Britain in the nineteenth century were either society journals or connected in some way to a society. Although William Brock argued in 1980 that scholars had paid too much attention to society-sponsored journals while overlooking the significance of commercial science journals, the scholarship on late-nineteenth-century astronomical society journals is nearly nonexistent.[3] Yet an analysis of these journals provides some fascinating answers to the questions raised by Sally Shuttleworth and Berris Charnley about the relationship between science periodicals and the construction of science. How, they ask, did various forms of science periodicals "help to create and maintain scientific communities and the production of scientific knowledge? Who was involved, and, in the era before the consolidation of professional science, how did elite and non-elite practitioners interact?"[4]

For the first six decades of the nineteenth century, the existing British astronomical journals were tied either to the Royal Astronomical Society or to government or university observatories.[5] These journals were initially geared toward the gentlemen of science, the predominantly Oxbridge-educated Anglicans who controlled all of British science during the first half of the century. Cambridge men such as John Herschel and George Airy dominated astronomy in this period. Herschel, who served as president of the Royal Astronomical Society three times, was renowned for his work on double stars and for mapping the sky of the Southern Hemisphere. Airy, who was elected president of the Royal Astronomical Society for four terms, was the seventh Astronomer Royal. They were part of an intellectual elite who used the astronomical journals of their time to communicate their astronomical discoveries, primarily to each other.

The appearance of a series of new astronomical journals created with new audiences in mind began with the founding of the *Astronomical Register* in 1863 and then the *Observatory* in 1877. Both of these journals were monthlies. Neither of them was formally affiliated with any astronomical society; they were commercial journals. The editors of these periodicals experimented with the format of the astronomical journal in order to appeal to an as yet undefined public. Often in conversation with their readers, the editors attempted to create ideal scientific communities through their journals. In the process they had to deal with the relationship of their periodicals to the elite Royal Astronomical Society. The *Astronomical Register* (1863–86), edited by Sandford Gorton, a printer and a member of the Royal Astronomical Society, was created for those interested in astronomy who found the papers in the Royal Astronomy Society journals to be too mathematical. Part of a growing body of amateur astronomers, the readers of the *Astronomical Register* were interested in observation rather than theory. Gorton included news about astronomical pieces in popular science journals, astronomical societies, and well-known popularizers of astronomy. The correspondence section received the most attention from readers, for this was the journal's most distinctive—and controversial—feature.[6] Neither of the Royal Astronomical Society publications had anything like it. The journal ended up serving as a site for criticism of the Royal Astronomical Society. This did not sit well with all its readers, but it established the *Register* as a publication independent of the senior society.

Like the *Astronomical Register*, the *Observatory* included reports of the meetings of the Royal Astronomical Society, though it did not have a formal relationship to the society. Founded by William Christie in 1877, it was edited by a series of elite astronomers, many of whom worked at Greenwich Observatory. The editors of the *Observatory* came up with a formula that proved to be more successful than that of the *Astronomical Register*. Not only did the *Observatory* best the *Register* in competing for the audience of British amateur astronomers in the late nineteenth century, it lasted into the twentieth century and continues to be published to this day. The *Observatory* borrowed some of the features of the *Astronomical Register*, such as making correspondence central to the publication and reviewing the important works of popularizers of astronomy, but it aimed for a higher tone. The controversies within the Royal Astronomical Society received far less attention. There is, then, some justification to Allan Chapman's assertion that the *Observatory* was "in some respects an unofficial organ of the Royal Astronomical Society."[7]

The six new society-based journals founded in the late Victorian period after the establishment of the *Observatory* were confronted by the challenge of differentiating themselves from the other astronomical periodicals without alienating their readers, who could be loyal to more than one journal. Should there be reports of the meetings of other societies, and if so, which ones? Should journals serve as a forum for criticism of any of the astronomical societies? Should they allow discussion of religious themes or unorthodox astronomical theories into their pages? Should women be permitted to participate fully in the life of a journal? Deciding on how to deal with these questions determined the way scientific findings were communicated, and established a community of readers and practitioners. With the example of previous astronomical journals in front of their eyes, society journal editors attempted to reimagine the scientific periodicals of their discipline. Editors were not always sure that their predecessors had gauged their audiences accurately. This led to experimentation with format to find niches for their publications. The British astronomical journals of the second half of the nineteenth century, then, vividly illustrate how, as Jonathan Topham has put it, "the history of scientific journals has been a history of continual reinvention, as both the project of natural enquiry and the culture of communication have undergone profound change."[8]

In this chapter I will focus on six society-based journals founded in the late Victorian period. "Letters to the editor" sections, the bread and butter of the *Astronomical Register* and the *Observatory*, are noticeably missing from all of these journals, and the sharing of observations and information is given the highest priority as a means for creating community. But issues of scale and identity could complicate forming communities of astronomers. Some of these societies were local, others national. Some started out as local, but then evolved into national societies. Two of the societies harbored international aspirations. The changing nature of these societies was reflected in their periodicals, and this played a major role in how long the respective journals survived.

The publication of the *Journal of the Liverpool Astronomical Society* in 1882 was the first of a series of new society-based astronomical journals that appealed to amateurs. Naturally, the members of the society were the target audience. Women were seen as making up a significant part of the readership. The journal format reflected the society's emphasis on organizing its members into observational teams, referred to as "sections." Reports on the observational work of the sections played a key role in the publication. The financial problems of the society and the difficulties in determining whether it was a local, national, or international organization led to the journal's early demise. In 1892 and 1893 the *Journal of the British Astronomical Association* and the *Memoirs* of the same society picked up where the *Journal of the Liverpool Astronomical Society* had left off. However, the British Astronomical Association (BAA) defined itself from the start as a London-based national society of amateurs. The BAA, in effect, aimed to "nationalize" astronomy. Drawing on the Liverpool innovation of dividing up the active observers into sections, the BAA *Journal* added an extensive set of abstracts of current astronomical periodicals to make the progress of the science more accessible. Like the Liverpool journal, the BAA publication included women as part of the reading audience, but it emphasized their role in astronomy even more. The founding of two new astronomical societies, one in Leeds in 1893 and a second in Wales in 1895, led to the creation of more society-based journals. These societies defined themselves primarily as provincial in nature. Their publications, titled the *Journal and Transactions of the Leeds Astronomical Association* and the *Journal of the Astronomical Society of Wales* (subsequently renamed the *Cambrian Natural Observer*), appealed largely to a local audience of middle-class professionals. They defined themselves in relation to the publications of the British Astronomical Association rather than those of the Royal Astronomical Society. Both made room for discussions of religious themes in astronomy. An examination of these new society-based journals illuminates how the attempt of astronomical periodical editors to create a community of readers and observers bound together by a shared identity and common scientific practices was complicated by issues of organization, scale, class, gender, and religion.

ORGANIZING OBSERVERS: *THE JOURNAL OF THE LIVERPOOL ASTRONOMICAL SOCIETY* (1882-91)

In his 1889 message "To the Readers of the *Journal* of the Liverpool Astronomical Society," editor Isaac H. Isaacs bragged about the success of the society and its journal. The society's membership was now the largest of its kind in the world, and it had produced original work of the highest quality. "To its *Journal* must be given the credit for keeping on permanent record such work," Isaacs claimed, "and for forming a bond of union and friendship between its widely distributed members and branches." Referring to the Royal Astronomical Society as "our elder sister society," Isaacs reported that at a recent meeting of that society its president asked, "Why it was that all of the best papers went to Liverpool?" Isaacs believed that the answer to the question was that "the aim of the Liverpool Society has always been the pure furtherance of Astronomical Science, and a furtherance of such a character as to make the Science appreciated by *all who are in any degree interested in the Science.*"[9] The last part of the sentence implied that Isaacs saw in the Liverpool Society and its journal a far more effective vehicle for engaging a broader audience than any of the other existing societies and journals. Whereas the *Astronomical Register* and the *Observatory*, journals not formally tied to any astronomical society, effectively used their correspondence columns to define their audience, the *Journal of the Liverpool Astronomical Society* capitalized on the society's organizational innovation: the establishment of sections. Amateurs were welcomed to join groups, referred to as "sections," that focused on specific fields of astronomical specialization. Each section—devoted to the moon, planets, meteors, double stars, variable stars, or colored stars—was led by an experienced observer. The results of the observations by members in each section were collected together and then communicated to the rest of the society through reports appearing in the journal. But, despite Isaacs's positive report on the state of the society and its chief publication, the journal ceased to publish two years later while the society struggled with financial difficulties.

The *Journal of the Liverpool Astronomical Society* was the first of the new society-based astronomical journals (fig. 8.1). The society was among the first big provincial astronomical societies, such as those also founded in Leeds (reorganized in 1892), Manchester (established in 1892), Ulster (established in 1893), Wales (established in 1894), and Newcastle (established in 1904). Chapman traces their lineage back to the broader-based nonspecialist literary and philosophical societies and mechanics' institutions. The founding of the provincial societies was a response to the growing interest in astronomy during the closing decades of the nineteenth century.[10] They came on the scene several decades after the new natural history field clubs, discussed by Samuel Alberti, that were founded in Great Britain in the 1860s and 1870s. Like the field clubs, the astronomical societies were populated by members of the middle classes eager to fill their leisure hours with a morally acceptable activity.[11] The Liverpool Astronomical Society was established in 1881. In the laws of the society, the stated objective was "to foster a liking for Astronomy, and to encourage Astronomical Observation, especially amongst the possessors of small Telescopes."[12] The number of members increased at a notable rate. Starting with ten members, by the end of the first session it had grown to seventeen.[13] In 1885 there were two hundred members, many of whom lived outside Liverpool.[14] On 8 July 1887, when the annual meeting was held, it was reported that the society now had 440 members.[15]

The membership of the Liverpool Astronomical Society included gentlemen of science as well as ordinary enthusiasts. The Reverend Thomas Henry Espinall Compton Espin, one of the society's founding members, was educated at Oxford. Espin was a regular contributor to the *English Mechanic* and

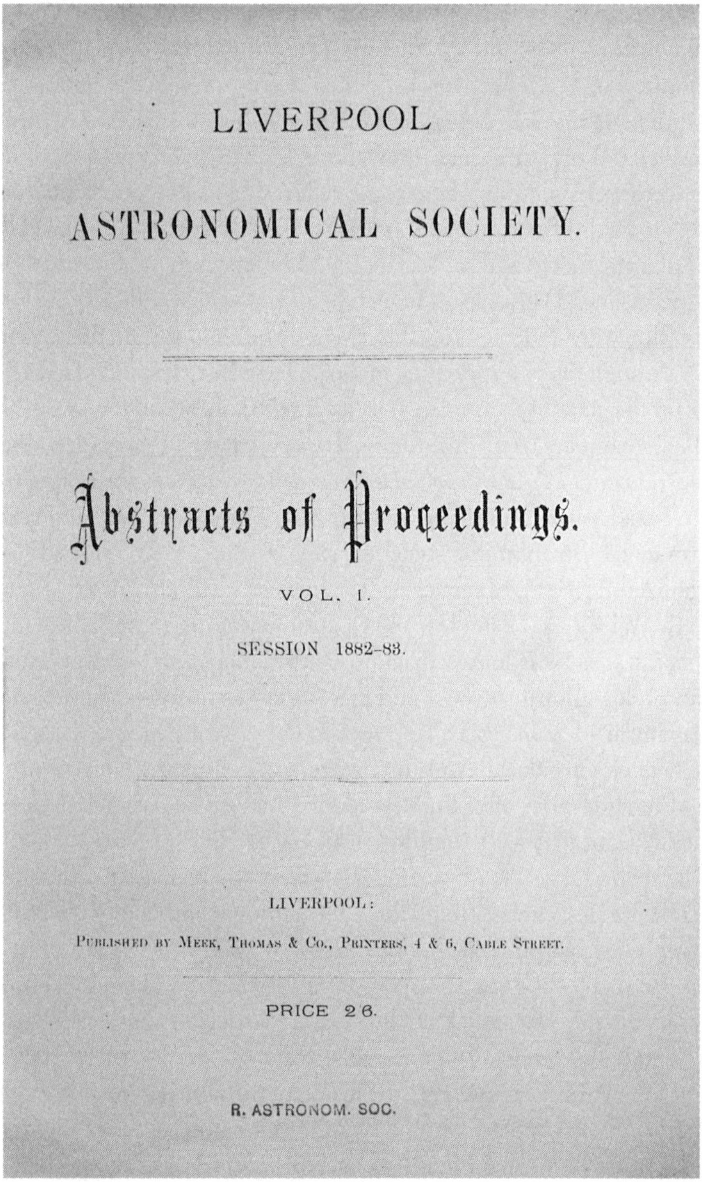

LIVERPOOL

ASTRONOMICAL SOCIETY.

Abstracts of Proceedings.

VOL. 1.

SESSION 1882-83.

LIVERPOOL:

PUBLISHED BY MEEK, THOMAS & CO., PRINTERS, 4 & 6, CABLE STREET.

PRICE 2 6.

R. ASTRONOM. SOC.

FIG. 8.1. Frontispiece of the *Journal of the Liverpool Astronomical Society*. Courtesy of the Royal Astronomical Society.

had worked with the well-known amateur astronomer Thomas Webb. His area of expertise was stellar astronomy. Early on, Espin was given a special, formal role: "Observer to the Society." The purpose of the post was to assist members who required help in observational astronomy.[16] Later, in 1884, he became the society's second president. Webb was also an active member, along with Isaac Roberts, William Denning, and Thomas Elger, well-known amateurs from the Liverpool area.[17] Ordinary and associate members included eminent practitioners such as Robert Ball, William Huggins, E. W. Maunder, Charles Piazzi Smyth, Giovanni Schiaparelli, and Otto Struve. Unlike the Royal Astronomical Society, the Liverpool Astronomical Society admitted women members. Its membership included Elizabeth Brown, who specialized in solar observation, and Agnes Clerke, a prominent popularizer of astronomy.[18]

When the Liverpool Astronomical Society was formed, there were already two astronomical journals in existence that catered to the audience of amateur astronomers: the *Astronomical Register* and the *Observatory*. The society's publications were geared primarily to its members, who, at least initially, were astronomers in the Liverpool area. In its early years the society actually had two publications, the *Abstracts of Proceedings* and the *Transactions*. The latter, four numbers of which appeared from 1883 to 1884, consisted primarily of reports from the solar, planetary, lunar, and variable star sections. Published by Meer, Thomas, and Company of Liverpool, the reports offered the collected results of the observations submitted by members participating in particular sections. In the opening remarks to the report by the solar section, E. W. Maunder argued that, though public observatories had taken up solar work "vigorously," it was not true that there was "little or nothing in this branch of Astronomy left for amateurs, and especially for those with but small telescopes, to do." In fact, "the work of the great observatories has still left abundant room for the labours of amateurs; indeed the one work would have been very incomplete without the other."[19] Amateurs could contribute to the progress of astronomy by regularly observing the appearance of both sunspots and faculae (bright spots) and recording their observations in drawings or photographs.

Meek, Thomas, and Company also published the other early publication, the *Abstracts of Proceedings*, which appeared in two volumes from 1882 to 1884. Volume 1 was priced at two shillings and six pence; volume 2 at three shillings. The *Abstracts and Proceedings* contained an account of the meetings, which included the papers read there as well as the remarks and questions that followed, short articles, observations of various astronomical phenomena, the laws of the society, and a list of the members. The third volume of the period-

ical, issued near the end of 1884, was retitled *Journal of the Liverpool Astronomical Society*, and it was now published by the society. The subscription
rate was fixed at the low rate of five shillings per year.[20] Since the *Transactions*
had ceased to be published, some material on the work of the sections was now
included in the *Journal*. This included occasional reports from the sections,
and the name and address of each section's director. The president's address
was also added to the *Journal*, along with a list of the officers and members
of the council. Several men tried their hand at editing the *Journal*. Starting in
volume 4 the editor was listed as J. W. Appleton of Liverpool, a fellow of the
Royal Astronomical Society (F.R.A.S.). In volume 6 the president announced
that Herbert Sadler was the new editor.[21] Sadler, also an F.R.A.S., was a Cambridge graduate and an expert on double stars and the moon. Isaac H. Isaacs,
whose name did not carry the F.R.A.S. designation, edited volume 7. Isaacs
conceived of the main object of the *Journal* as "the publication of papers
read before the Society." Although he said that "this will not be lost sight of,"
he believed that the *Journal* could do a better job of reaching those with an
interest in astronomy but little expertise. He pledged to devote a section of the
periodical "to the exclusive benefit of the beginner."[22]

But the society's publications had always attempted to reach out to the
astronomical novice. The first volume of the *Abstracts of Proceedings* included
an article by Isaac Roberts on the "Planet Indicator," a do-it-yourself paper di-
agram of the positions of the planets (fig. 8.2).[23] In the *Transactions*, Maunder
explained how beginners could do solar observations.[24] "Letters to the editor"
columns did not play a major role in the strategy of the society's publications
to engage readers, as they had in the earlier *Astronomical Register* and *Observatory*. A "letters to the editor" column was not instituted until volume 5,
and even then it was a relatively small part of the journal.[25] The role of the
sections was far more important for both the society and the *Journal*. In his
article describing the "Sectional work of the Society," W. S. Franks reminded
the readers, "The L.A.S. occupies a somewhat different position from that
of most other scientific societies. Constituted as it is on the broadest lines, it
seeks to popularise the study of Astronomy by offering such inducements as
will best meet the wants of a very large class of amateurs." To Franks, a member
of the Royal Astronomical Society who at that point worked on colored stars
out of his homemade observatory, the term "amateur" did not have negative
connotations.[26] As director of the Star Colour Section, he thought that amateurs provided the lifeblood of the sections, and therefore of the society. The
principle behind the sections was, as Franks put it, "co-operative," as they
depended on "the hearty support of its members for success." The sections

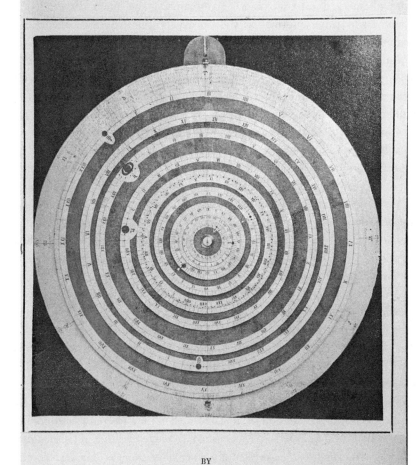

FIG. 8.2. Isaac Roberts's Planet Indicator. Isaac Roberts, "On a New Planet Indicator," *Abstracts and Proceedings of the Liverpool Astronomical Society* 1 (1882–83): 14–17, on 17. Courtesy of the Royal Astronomical Society.

were key to the society's success; in the past they had "rendered good service to the Society, in helping to establish its reputation." They would continue to play a crucial role in the future of the society and the journal, as they were in alignment with the move toward specialization in astronomy. By "appealing to a largely augmented constituency," Franks declared, the society would "considerably extend its sphere of usefulness" to the astronomical specialties that had emerged in recent years.[27] By organizing an army of disciplined astronomers to gather information, the journal and the society were redefining the identity of the amateur.[28]

Although the rapid growth in membership guaranteed an audience for the *Journal*, it also created problems that led to the end of the publication while weakening the society. Early on there were indications that the quality of the publications had not kept pace with the increase in society membership. At a meeting in late 1884, the society council was reluctant to choose between two equally unappealing choices: reduce the size of the publications or raise the subscription price. The latter, it was reported, "would be to take the Society out of the reach of the very class of amateurs it was intended to encourage." It was at this point that the two early publications were combined into the *Journal*, subsidized by scientific advertisements. Even though there was concern that advertisements would "lessen the dignity of a scientific society," those present at the meeting unanimously accepted this option, which transformed the publication into a commercial journal.[29]

Another problem associated with growth had to do with the identity of the society, which had begun as a local organization serving amateur astronomers in Liverpool and the surrounding area. The *Observatory* reported in 1887 that the society now "assumed a national character inasmuch as it included members from all parts of Great Britain and Ireland."[30] In recognition of the society's changing nature, the annual general meeting of 1886–87 was held in the Royal Astronomical Society rooms at Burlington House in London. But at that meeting it was announced that a branch of the Liverpool Astronomical Society had been founded at Pernambuco, Brazil, consisting of 111 members, and that associate branches in Australia were in the process of being formed.[31] Was the society a local, national, or international organization? By 1889 it was in serious financial difficulty. Substantial sums were owed to its printers, and it was clear that the small membership charge could not support a huge organization that had outgrown its regional status.[32] Although the society quietly survived in a much reduced form after the *Journal* ceased publication in 1891, the locus for a national amateur society shifted to a new association that proved to be far more enduring.

NATIONALIZING ASTRONOMY: *THE JOURNAL*
OF THE BRITISH ASTRONOMICAL ASSOCIATION
(1892-PRESENT) AND THE *MEMOIRS OF THE BRITISH*
ASTRONOMICAL ASSOCIATION (1893-1962)

Speaking at the annual meeting of the recently formed British Astronomical Association (BAA) on 28 October 1891, the president praised E. Walter Maunder's work as editor of the organization's journal (fig. 8.3). It was under Maunder's "most able and conscientious editorship" that the journal had "obtained a reputation and popularity to which the success of our Association is very largely attributable indeed." At the meeting, Maunder drew attention to the journal's distinctive feature, the abstract of current astronomical literature, and paid tribute to the fourteen gentlemen who had supplied him with the abstracts.[33] If the BAA's success depended on the journal, as the president asserted, Maunder claimed that the success of the journal, and therefore of the BAA, depended on the contributions of these men. The BAA actually produced two periodicals, both printed and published by Eyre and Spottiswoode. Besides the *Journal*, there was also the *Memoirs*, seemingly modeled on the Royal Astronomical Society's *Memoirs*. But the BAA's periodicals were quite different from the two Royal Society publications. While the *Journal* featured the abstracts, the *Memoirs* drew upon the work of the sections.

The emphasis on sections in the BAA points to the links between it and the Liverpool Astronomical Society. Peter Johnson has argued that the founding of the BAA in 1890, less than four years after the demise of the *Astronomical Register*, could not have been a coincidence. Having been deprived of the means of publishing their observations or staying in contact with each other, Johnson asserts, amateur astronomers must have felt the need for an organized body like the BAA.[34] The BAA was organized very closely on the model of the Liverpool Astronomical Society, which by 1890 was running into serious financial difficulty. The two BAA journals also drew upon the basic structure of the two publications originally produced by the Liverpool Astronomical Society. Whereas the *Astronomical Register* and the *Observatory* relied heavily on correspondence columns as a way of engaging their audiences, the periodicals of the BAA and the Liverpool Astronomical Society shared an emphasis on publishing the work of sections focused on distinct subfields of astronomy.

In a letter by E. Walter Maunder and T. F. Maunder in the first volume of the *Journal of the British Astronomical Association* concerning the reason for forming a new society, they made the connection explicit. "The success attained at one time by the Liverpool Astronomical Society, in spite of its local

F I G . 8 . 3 . E. Walter Maunder. Courtesy of the Royal Astronomical Society/Science Source.

title," they wrote, "appears to show that there is a real need for a Society beside the Royal Astronomical Society, and on somewhat different lines." The
Maunders conceived of the BAA as a solution to the need for a national organization that the Liverpool Astronomical Society was unable to supply. The
provisional committee, of which the Maunders were spokesmen, set down
two aims for the new association. First, it would cater to the needs of those
who found the subscription to the Royal Astronomical Society too high or
its papers too advanced, or "who are, as in the case of ladies, practically excluded from becoming Fellows."[35] Second, the BAA was to "afford a means
of direction and organisation in the work of observation to amateur Astronomers." The Royal Astronomical Society received papers and observations,
but did not undertake the work of direction. The BAA undertook to direct

research by instructing members of each section exactly how to record their observations. Since the provisional committee envisioned the new society as occupying "ground not covered" by the elite Royal Society, they deprecated "any idea of rivalry" between the two organizations.[36] To avoid any confusion between the two bodies, the original name of the society was changed at the first general meeting from "British Astronomical Society" to "British Astronomical Association."[37] Though the organization was thereafter referred to as an "association," it was organized on formal lines very similar to those of the Royal Astronomical Society. There was a council, a president, a vice president, a treasurer, and secretaries, whose roles were spelled out in the rules. But unlike in the Royal Astronomical Society, new members did not need to be part of the astronomical elite. Members were charged an entrance fee of five shillings, and a subscription rate set at half a guinea annually.[38]

The early history of the BAA resembles that of the Liverpool Astronomical Society. Both organizations experienced significant growth. But, as the Liverpool Astronomical Society ran into problems in the early 1890s, many of its members joined the BAA.[39] The steady increase in membership was reported extensively in the pages of the *Journal of the British Astronomical Association*. In December 1890 the total membership was 489, up by 169 since the last meeting. By October 1891 the count was at 584, and only a year later the association had increased to 700 members. However, the president, Captain Noble, warned that 36 of them were corresponding members, so there were actually 666 annual subscribers. "As 666 happens to be the number of the Beast," he joked, "the policy (not to say the urgent necessity) of adding to it becomes at once evident."[40] The numbers did indeed increase, to 1,144 in September 1898, 1,151 a year later, and 1,159 in 1900.[41] Originally seen as a London-based society, the BAA experienced an increase in membership due in part to the addition of branches including the North-Western branch (based in Manchester, established in 1892), the West of Scotland branch (based in Glasgow, established in 1894), the New South Wales branch (based in Sydney, established in 1894), the East of Scotland branch (based in Edinburgh, established in 1895), and the short-lived Victoria branch (based in Melbourne, established in 1897).[42] The membership was actually spread more widely around the world than the branches would suggest. A table showed the geographical distribution of the 1,151 members. Members came from England (716), Scotland (123), Australia (123), Europe (53), Ireland (43), North America (39), Africa (20), Wales (16), India (13), South America (3), and China and Japan (2).[43] The audience for the two periodicals was international in scope, though the vast majority of the BAA's members were located in Britain.

Like the Liverpool Astronomical Society, the BAA admitted women, which helped to increase the number of members. However, far more women joined the BAA. The *Journal* reported regularly on the candidates for election as members. Women frequently appeared in the lists, often each proposed by two other women members. For example, Annie Scott Dill Russell, a computer from Greenwich,[44] was proposed by Alice Everett and Edith M. Rix, who were also Greenwich computers.[45] Later, Ellen M. Clerke, a popularizer of astronomy, was proposed by her sister, Agnes Clerke, and by Alice Everett.[46] Of the fourteen new members proposed in 1896, five were women.[47] By 1898, 84 of the 1,151 members were women, or about 7.3 percent.[48] Several women played significant roles on the council and as officers of the BAA. The first elected council included Agnes Clerke and Margaret Huggins, while Elizabeth Brown served as director of the Solar Section.[49] Later, Alice Everett served as secretary from 1893 to 1896, and Annie Scott Dill Russell, now married to Maunder, was the vice president in 1896.[50] Annie Maunder was offered the presidency but turned it down, as she believed that her voice was not strong enough to carry over large meetings.[51] Women were welcomed as members and even officers of the BAA. But for some, their main purpose was to expand the membership, which ensured the financial health of the organization. One of the early presidents, Noble, declared that women were capable of using their feminine charms to recruit more members. "Lady Members," he affirmed, "with that irresistible power of persuasion which is so exclusively their own, might well bring in two apiece."[52] Though it could be patronizing, the BAA's policy toward women was touted in the *Journal* as a model to be emulated by other scientific societies. In 1893 the *Journal* proudly reported that the Royal Geographical Society had adopted it.[53] A year later, the *Journal* saw the issue of admission cards to Brown, Everett, Russell, the Clerkes, and Giberne to attend the Royal Astronomical Society meetings as a step forward.[54] Women were therefore seen as a significant part of the audience for the BAA's journals and a welcomed component of the astronomical community in general.

The *Memoirs of the British Astronomical Association* was dedicated to the reports of the observing sections (fig. 8.4). It was printed and published by Eyre and Spottiswoode and edited by E. Walter Maunder from 1893 to 1900, except for two years, 1895 and 1896, when Annie Russell took over the reins. Maunder had previously edited the *Observatory* for seven years. The work of the sections was considered to be one of the central features of the BAA. Noble acknowledged that the idea for the observing sections had come from the Liverpool Astronomical Society. In his presidential address of 1890 he discussed the advantage of "the division of labour," when research had become "so

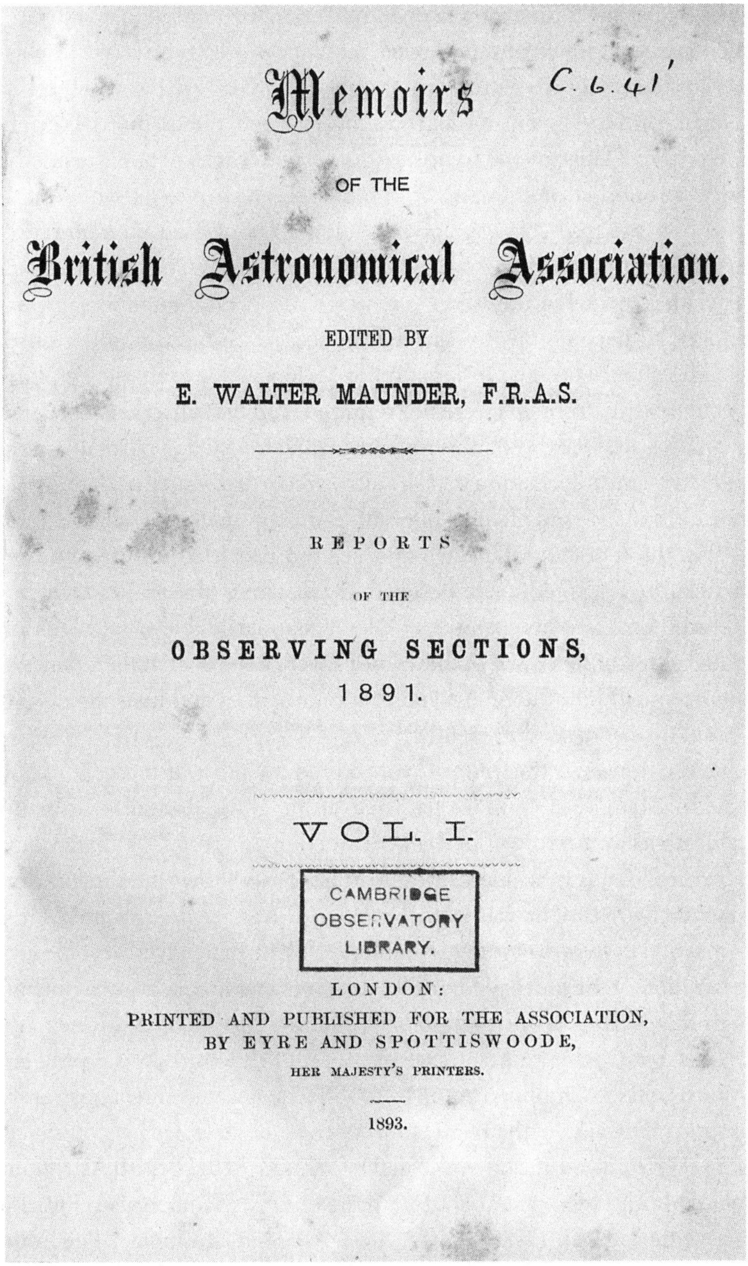

C. 6. 41'

Memoirs

OF THE

British Astronomical Association.

EDITED BY

E. WALTER MAUNDER, F.R.A.S.

REPORTS

OF THE

OBSERVING SECTIONS,

1891.

VOL. I.

CAMBRIDGE
OBSERVATORY
LIBRARY.

LONDON:

PRINTED AND PUBLISHED FOR THE ASSOCIATION,

BY EYRE AND SPOTTISWOODE,

HER MAJESTY'S PRINTERS.

1893.

FIG. 8.4. Title page of the first volume of the *Memoirs of the British Astronomical Association*. Reproduced with the permission of the Institute of Astronomy, University of Cambridge, and the *Journal of the British Astronomical Association*; www.britastro.org/journal.

illimitable that the individual observer must confine himself to a very circumscribed area of it if he is to do any useful work at all." Impressed by this approach to observational astronomy, Noble declared, "We have adopted from
our *confrères* of the Liverpool Astronomical Society the method of observing
in Sections or Departments of observation," with each section presided over
by an "Astronomer of eminence," often an elite astronomer who could su-
pervise how the research was done.[55] Volume 1 contained the reports of the
lunar, meteoric, star color, variable star, Jupiter, and solar sections. The reports
were written in such a way as to encourage association members, especially
beginners, to become involved in the sections by submitting their observations. The "Notes on the Observation of Meteors," for example, emphasized
that expensive instruments were not required, and that observers needed only
what "nature has bestowed." Those who were interested in participating were
instructed on the information that should be recorded—such as date, time, duration, and speed—and offered suggestions on how meteors could be drawn.[56]
Similarly, the report by the star color section provided information on the
nomenclature and method to be used in recording observations, including
the desired data and an example of what a standard table would look like.[57]
Potential contributors to all of the sections were assured that their data would
be incorporated into the report and that their names and instruments would
appear in the list of section members.

Like the *Memoirs*, the *Journal* was edited for most of the century's final
decade by Maunder. Maunder had help from Annie Russell, who had lent
valuable assistance in 1893 "in the general editorial work."[58] When Maunder
became president of the association in 1894, Russell became the interim editor for the two years he was in that office.[59] But since Russell and Maunder
were married in 1895, Maunder would have still been involved at some level in
the operation of the journal. Maunder drew on a wide variety of contributors,
though few were elite astronomers. William Henry Stanley Monck (1839–
1915) was among the frequent contributors. Professor of moral philosophy
at Trinity College, Dublin, from 1878 to 1882, he became chief registrar in the
Bankruptcy Division of the High Court of Ireland while working on stellar astronomy.[60] He played an active role in the founding of the British Astronomical
Association, and was a member of the first council.[61] Women also contributed
articles to the journal. The very first article of the initial volume, "The Rotation
Periods of Mercury and Venus," was by Agnes Clerke, while Elizabeth Brown
contributed a piece on sunspots to volume 4.[62] Besides short articles on sun-
dry astronomical topics, the usual format of a typical issue included a series of
regular columns, such as a fairly detailed report of the most recent meeting of

the British Astronomical Association, and brief reports by the directors of the observing sections. This was followed by a short, relatively little-used correspondence section, and reviews of new books and memoirs. A notes section came next, containing brief accounts of the meetings of the Royal Astronomical Society and other astronomical societies both inside and outside Britain, reports on the activities in observatories around the world, summaries of the content of recent astronomical periodicals, random news, and obituaries of important astronomers. Brief announcements from the association would ensue. The last few sections, taken together, filled a large portion of the *Journal*. They consisted of abstracts of astronomical periodicals, digests of articles in the general periodical press that focused on astronomy, and an index to recent astronomical publications noticed in the previous two columns.

One of the most distinctive features of the *Journal* was its attempt to bring to its readers information on the entire gamut of astronomical literature, especially the material appearing in periodicals. When the editor of the *Observatory* stepped down, the *Journal* reported it in the "Notes" section.[63] The *Journal* had earlier drawn the attention of readers to the "admirable little collection of ephemerides" in the *Observatory*, and expressed the hope that BAA members would subscribe to it.[64] When the *Journal of the Astronomical Society for Wales* began to be published, the *Journal* of the BAA referred to it as "a neatly got-up little publication."[65] Later, the *Journal* reported on the contents of the most recent issue, and reacted positively when the Welsh journal was revived under a new title, the *Cambrian Natural Observer*.[66] The *Journal of the British Astronomical Association* was also supportive of the publications of the Leeds Astronomical Society, praising the "interesting accounts" of lectures delivered at the society's meetings.[67] The *Journal* did not treat the other astronomical journals as rivals, but instead presented them as part of a larger community about which readers had to be informed. But reporting the important news contained in the other astronomical periodicals also made sound commercial sense. The *Journal of the British Astronomical Association* was the only periodical that readers needed to purchase to be kept abreast of all things astronomical.

In addition, the *Journal* provided an extensive set of summaries of important periodical articles in order to keep readers up-to-date on the latest discoveries. The editor announced this feature of the *Journal* in the first volume. Since the circulation of current astronomical information was one of the objects of the association, the editor planned to supply, on a monthly basis, "a complete summary of the periodical literature of Astronomy." Maunder then asked for volunteers to abstract each month the important papers of one periodical.[68] Twenty readers, including Captain Noble and Agnes Clerke, later volunteered to do the work.[69] The innovative review of periodical literature was a hit. The report of the annual meeting held on 28 October 1891 stated that the journal had been most highly praised for its "abstract of current astronomical literature." Maunder stated that about fourteen individuals had supplied the abstracts.[70] Now both the observing sections and an important feature of the journal were dependent on the participation of the association's members. By the fourth volume in 1895, the abstracts section had increased in size, as Maunder was reviewing not only the British astronomical journals, such as the *Monthly Notices, Nature*, the *Observatory*, the *English Mechanic*, and *Knowledge*; he was also covering select American, German, and French astronomical journals. When Annie Russell took over as editor of volume 5, she attempted to streamline this feature by organizing it according to topic rather than listing the articles in each periodical. The topics included the moon, planets and minor planets, variable stars, stellar photography, nebulae, and comets. The change in format was in part a response to objections that the précis of current astronomical literature took up too much of the journal. The issue was discussed at several meetings and in the journal's correspondence section.[71] This responsiveness to the desires of readers, along with the attempts to encourage participation in the sections, helped the journal to keep a loyal audience throughout the end of the century and beyond.

PROVINCIALIZING ASTRONOMY: *THE JOURNAL AND TRANSACTIONS OF THE LEEDS ASTRONOMICAL SOCIETY* (1893-1922), THE *JOURNAL OF THE ASTRONOMICAL SOCIETY OF WALES* (1895-97), AND THE *CAMBRIAN NATURAL OBSERVER* (1898-1910)

In the first volume of the new organ of the reconstituted Leeds Astronomical Society, William Barbour, the editor, reported that the British Astronomical Association had expressed "a willingness to receive and accredit" the Leeds organization as the Yorkshire Branch of the BAA. The editor pointed to one of the advantages of affiliating with the BAA. All of the members received "a Monthly Journal containing the latest astronomical information from all parts of the world." Moreover, many of the members of the Leeds group were already members of the BAA. But for the Leeds Society to become an affiliate, the number of its members had to be increased.[72] In the end, no formal affiliation was ever established, and the Leeds Astronomical Society has maintained its distinctive identity to this very day. The question for smaller provincial societies that formed in the 1890s was how to work out their relationship to the British Astronomical Association. The periodicals of two of the societies, the Leeds Astronomical Society and the Astronomical Society of Wales, will be the focus of this section of the chapter.

Since the BAA had established itself as the national society for amateurs, the provincial organizations conceived of their purpose as building a local community of observational astronomers, and this was reflected in their journals. Their readership was primarily society members, many of whom frequently interacted in person. There was no need for a letters to the editor column, the tool used by the *Astronomical Register* and the *Observatory*, to connect to their readers. Instead, these periodicals emphasized the sharing of observations published as articles or notes. In comparison to the *Observatory* and the BAA publications, the provincial journals were more permissive of unorthodox astronomical theories and of discussions of religious themes.

The Leeds Astronomical Society was actually the first distinct and specifically amateur astronomical society in Britain. Founded in 1859, it was short-lived, as was another incarnation in 1863. It took on a more stable form when it was reorganized in 1892.[73] Its stated object was "the Acquisition and Diffusion of Knowledge, connected with Astronomy in all its branches."[74] The members bemoaned the backward state of astronomy in Leeds, and looked forward to a time when the city would take its rightful place as one of the scientific centers of the kingdom with its own observatory. They envied the progress of one of the Canadian societies. "Even the comparatively small city of Toronto, in Canada," members complained at a meeting, "is possessed of a far more highly organised Astronomical Society than the city of Leeds," due to the generous support it received. The annual reports of the Toronto society rivaled in interest and size the *Journal of the British Astronomical Association*.[75]

Washington Teasdale, the first president of the Leeds Astronomical Society, wrote in a self-deprecating piece in the initial number of the journal that he had "little claim to be considered a practical astronomer," and that he was "constrained to assume" the position of president by a small group of astronomical observers who asked for his temporary assistance in organizing the society (fig. 8.5).[76] A native of Leeds, and a civil engineer by profession, he belonged to several local scientific and literary societies. He was elected to the Royal Astronomical Society in 1886, joined the British Astronomical Association when it began, and served as president of the Leeds Astronomical Society from 1893 to 1897.[77]

The majority of the members were from Leeds and the surrounding areas, and, unlike in the BAA, there were few elite astronomers. In its first year the society had forty-nine members, twenty-nine of whom were from Leeds.

FIG. 8.5. Washington Teasdale, president of the Leeds Astronomical Society 1893–97. *Journal and Transactions of the Leeds Astronomical Society* 5 (1897): frontispiece. Reproduced courtesy of the British Library.

There were no women, but eight men were members of the BAA. By 1897 the Leeds society had grown to eighty-one members, of whom fifty-four were from Leeds and five were women. Three years later, there was a modest increase in the Leeds society's total membership to ninety-eight, with fifty-six from Leeds. There were seven women (including antivivisectionist Frances Power Cobbe), three members of the Royal Astronomical Society, and ten members of the BAA.[78] Teasdale went out of his way to invite women into the Leeds society. At one meeting in 1895, he reminded attendees that women were "not only admissible as members, but will be welcomed by the members generally." He pointed out that the important and influential BAA "had wisely, from the first, admitted ladies to full and accredited membership."[79] In an 1896 meeting of the Leeds society, Florence Taylor, who had been admitted as the first female member in 1894, delivered a lecture on Caroline Herschel.[80] She also presented a paper on Mary Somerville's contributions to astronomy, after which Teasdale commented briefly on other women, such as Clerke, Brown, Maunder, and Everett, who had distinguished themselves in the realm of astronomy. Their accomplishments, Teasdale asserted, showed that "the mind of woman, when properly trained and cultivated in this and other scientific subjects, eminently fitted her to be the intellectual companion of man."[81]

The members constituted the primary readership for the Leeds society's publication. For the first two numbers, annuals for 1893 and 1894, the title was *Report and Transactions of the Leeds Astronomical Society*, and the publisher was William Brierly. The subscription for ordinary members was set at five shillings per annum. Starting in 1895 it was retitled the *Journal and Transactions of the Leeds Astronomical Society* and published by a local firm, Richard Jackson. It was still published annually. Teasdale referred to it in 1893 as a "pamphlet" that was "intended chiefly for local perusal."[82] The first issue was 48 pages in length. But it grew in size, and by 1900 it was 121 pages. In 1898 the price of a number for ordinary members remained at five shillings. Producing the journal was costly for the society. It absorbed almost all of the income generated by subscription and membership fees, but in the eyes of the society's committee the "favorable comments" on it "in accredited scientific journals" fully justified the expenditure.[83]

Up until his resignation as honorary secretary in 1903, William Donald Barbour was the editor of the journal. Originally from Glasgow, Barbour settled in Leeds in 1840 and set up a business as a colonial agent. He was fascinated by astronomy, especially the topic of the evolution of planetary life. Barbour brought a reverent frame of mind to his enthusiasm for the study of the heavens.[84] He offered his readers a varied format, though there were no correspondence columns. Since one of the journal's primary aims was to feature the society's work, Barbour reported on observations sent to him by members.[85] The journal also contained accounts of the society's meetings, the program of society lectures, a list of the officers and members, and the society's rules. In addition, Barbour included short articles on any and every astronomical topic, such as sunspots, colored stars, planets, meteors, and comets, as well as information on upcoming astronomical events and how to observe them.

Barbour allowed controversial topics into his journal, since they could be used for educational purposes. In the very first number a lecture on zetetic or flat earth theory was included. Barbour inserted an editorial note asserting that "our Zetetic friends" had based their scientific heresies on an antique book saturated with Eastern modes of thought.[86] When a defender of the geostatic system was asked and granted the opportunity to be heard during a meeting of the society, and failed to convince his hearers, Barbour again saw this as an ideal occasion for correcting erroneous beliefs circulating among the pubic. "If the science of astronomy were popularised," he declared, "and made easily accessible to the public in all our large towns, by institutions such as this Leeds Society, where error could be exposed, and misapprehension corrected, before being irremediably assimilated by the mind," then inaccurate theories of the heavens would be eliminated.[87] Barbour also did not hesitate to publish pieces that explored the religious significance of astronomy, though the tone of the journal was largely secular. Much of the religious language was contained in Barbour's numerous articles. In his 1894 article on the evolution of planetary life, Barbour discussed divine immanence in nature and how the evolutionary process was under "Divine sanctions."[88] In 1898 Barbour contributed an article on the ether as a revelation of the oneness of nature, concluding with the affirmation of science as an entrance into "the great temple of our Ethereal Universe," where we realized "in a far deeper sense the immanent, and everabiding presence of 'Our Father.'"[89] In the report on the annual meeting of January 1899, Barbour noted with satisfaction that the local press applauded the society's aim of replacing "trivial and non-elevating literature" with reading that encouraged a "taste for studying, at first, hand, the ways and methods of Creation."[90]

In 1894, two years after the reorganization of the Leeds Astronomical Society, another provincial society was established, this time in Wales. In his president's address, Arthur Mee laid out the history of amateur astronomy in Britain. First the Liverpool Astronomical Society had been formed, and then the British Astronomical Association; after that, Manchester, Glasgow, Chester, and Leeds had set up their own astronomy societies. At last, Mee declared, "we have an Astronomical Society of our own." But, he asked, what could the Astronomical Society of Wales do, given that they did not have a telescope like the one at Greenwich? Mee explained that observational astronomy could be pursued with a very moderate instrument, or even without a telescope.[91] He noted that the society was "intended in the first instance for Welshmen and Welshwomen, and lovers of astronomy resident in Wales," though it did not exclude others from membership.[92] As of January 1895 there were 59 members, mostly from Cardiff and the surrounding region.[93] By 1901 there were 114 full members, of whom 35 had some sort of academic or professional designation. Twenty-two were women (fig. 8.6). Like the Leeds Astronomical Society,

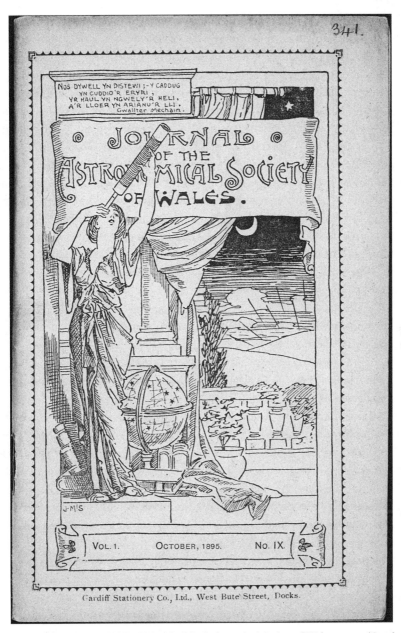

Nos DYWELL YN DISTEWI ;– Y CADDUG
YN CUDDIO 'R ERYRI ;
YR HAUL YN NGWELY'R HELI,
A'R LLOER YN ARIANU'R LLI.
Gwallter Mechain.

○ JOURNAL ○
OF THE
ASTRONOMICAL SOCIETY
OF WALES.

J·M·S

VOL. 1.   OCTOBER, 1895.   No. IX.

Cardiff Stationery Co., Ltd., West Bute Street, Docks.

F I G . 8 . 6 . Front cover of the *Journal of the Astronomical Society of Wales* 1, no. 9 (October 1895). Chapman points to this picture of a beautiful woman looking through a telescope as symbolic of the importance of female membership to the society. The woman may be a modern personification of the goddess Urania. See Chapman, *Victorian Astronomical Amateur*, 279. Reproduced courtesy of the British Library.

the Astronomical Society of Wales was dominated by middle-class professionals. Membership continued to increase, as by 1908 there were 242 members; however, at the beginning of World War I the society came to a sudden end.[94]

Mee not only served as the first president of the Welsh society but was also the editor of its journal. He was a journalist who had moved to Cardiff in 1892 to work for the *Western Mail* newspaper, serving many years as assistant editor. Mee was a popularizer of astronomy who lectured widely and wrote several books for beginners, and the *Journal of the Astronomical Society of Wales* was an extension of these activities.[95] The *Journal* was published monthly; each number was about ten pages in length. The subscription rate was set at the rather affordable cost of five shillings for men who lived within fifteen miles of Cardiff, and two shillings, six pence for women and those beyond Cardiff.[96] The cheaper price, fairly unusual, was intended to encourage women, as well as those not in close proximity to Cardiff and who could not participate as actively in society activities, to become members. The format included reports on society meetings, lists of officers and members, the objects of the society, short papers on various astronomical subjects, instructions for beginners on how to observe the heavens, information on upcoming astronomical events, and a "notes" section. As in the case of the *Journal and Transactions of the Leeds Astronomical Society*, there was no letters to the editor section.

Mee tried to keep his readers updated on important astronomical news, especially in the "Notes" section. In the first volume, for example, he informed them that the number of discovered planetoids stood at four hundred.[97] In the second volume he replaced the "Notes" column with one titled "Our Council Table," and began a series of reports on the activities of other astronomical societies and their publications. He told his readers that the president of the British Astronomical Association hoped that the BAA and the Wales society would be able to work together in the future, that the Leeds Astronomical Society published a valuable annual titled *Report and Transactions*, and that the current number of the BAA journal was "particularly rich."[98]

Keeping up with the most recent astronomical journals was important to Mee's readers, and the *Journal* facilitated a cooperative system first proposed in the "Editor's Notes" column. Mee reported that a member had suggested the formation of a "Magazine Club." Those who joined would buy one journal, read it, and pass it on.[99] A reader wanted to take this idea one step further, suggesting how the society could benefit from it. "If a number of the members would each [. . .] take in a different astronomical journal," G. T. Davis wrote, "pass them round mutually, and at the end of the year, or when the volume was complete, present it to the Society, they would then soon become possessed of a valuable library without a great deal of self-sacrifice on their part."[100] The spirit of cooperation was not limited to sharing periodical literature. In the first volume, Mee began a section composed of observations contributed by members. Mee pleaded with his readers, "whether possessing instruments or not," to "do their individual utmost to make this department a success."[101] In the second volume, in "Practical Hints for the Coming Season," Mee again implored his readers to submit their observations. If they were in doubt as to the procedure, he had written a "Programme of Astronomical Work," which contained instructions for beginning observers. Then Mee expressed his regret that "some of our members appear to be under the impression that it suffices to subscribe and to read the *Journal*." Active participation in observational astronomy was a requirement, in his opinion.[102]

Like Barbour, Mee was not averse to including articles on the ether, or on the religious significance of astronomical issues, though no pieces on unorthodox theories appeared. J. E. Southall contributed an article titled "The Ether," in which he concluded that "secondary causes possess a legitimate use, and are gradually unravelled by painstaking research, but the nearer we approach to primary causes the more reverence we need rightly to treat of them."[103] The Reverend Arthur J. Jenkins wrote a piece called "Other Worlds Than Ours," borrowing the title from Richard Proctor's earlier book (1870) endorsing pluralism. The universe, Jenkins argued, was built to support life. Just as the microscope revealed the presence of teeming life everywhere on the earth, the telescope showed us countless globes inhabited by living beings. Matter was made "to support life, and this we may accept alike of our earth and of the worlds of space. To the scientist the Sun is a star, and every twinkling star a sun, and the universe the temple of the Omnipotent."[104]

After two volumes, Mee's monthly *Journal of the Astronomical Society for Wales* was replaced by a new journal, the *Cambrian Natural Observer*. The new journal still served as the main publication for the Astronomical Society of Wales. It was a quarterly instead of a monthly, presumably to cut down costs. In the opening article of the journal, Mee, who was still editor, complained that "for some reason or other science does not seem to flourish in Wales." The *Observer* was "extending the field" in two senses. First, the journal would include material on meteorology and other subjects related to astronomy. Second, though precedence would be given to members of the Welsh society when it came to publishing submissions, Mee also welcomed contributions by those who were not members.[105] The subscription rate remained the same as it had been for the *Journal of the Astronomical Society of Wales*, five shillings for members resident within fifteen miles of Cardiff, and two shillings, six pence for ladies and all other members.[106] Although columns on the weather and on rainfall now appeared, the format was similar to that of the earlier publication, including the society's rules, lists of members, editor's notes, reports on society meetings, and observations submitted by readers. If anything, the number of pages devoted to observations increased (fig. 8.7). In 1900 a lengthy section called "Observations" included various reports by observers on meteors, Venus, the occultation of Saturn, sunspots, waterspouts, thunderstorms, earthquakes, heat waves, sound, geological models, and the size of the sun and moon.[107] The increase in observations was due in part to Mee's plea for help placed on the back cover of volume 2 in 1899. He asked for donations, more observations, and more subscribers. Donations were received from many of the other astronomical societies, including the British Astronomical Association, the Leeds Astronomical Society, the Liverpool Astronomical Society, and even the Royal Astronomical Society.[108] The cooperative spirit that Mee had tried to instill within his society also existed between major astronomical societies at the end of the century.

CONCLUSION: REACHING FOR THE STARS

In 1898, in the editorial kicking off the new quarterly journal of the Astronomical Society of Wales, Mee declared, "We compete with no existing Journal, Society, or Organisation, and we appeal for a friendly reception from those in the Principality and beyond it who are lovers and students of the Book of Nature."[109] Mee's point about the *Cambrian Natural Observer*—that it did not compete with any other periodical—would have been echoed by the editors of many late-nineteenth-century British astronomical journals. Each journal attempted to create a niche for itself by engaging a distinct audience defined by its level of expertise, its geographical location, and sometimes even its class associations. Editors experimented with different formats in order to appeal to their target audiences. They helped to differentiate themselves from other astronomical journals through the way they defined their relationship to the Royal Astronomical Society and the British Astronomical Association. Unavoidably, however, there was competition. Finding a niche where conflict was nonexistent was wishful thinking. A number of new astronomical journals in the second half of the century faltered as the reading public for these specialized periodicals was not large enough to support all of them.

For the society-based journals, having a "built-in" audience did not automatically lead to success. Those journals were dependent on the health of the societies to which they were attached. Many of the British astronomical

FIG. 8.7. "Great Meteor. Seen at Cardiff by Norman Lattey, 4 January 1900, 9h 23m." *Cambrian Natural Observer* 3, no. 2 (October 1900): frontispiece. Reproduced courtesy of the British Library.

journals founded in the nineteenth century, society-based or otherwise, ceased publication either late in the nineteenth century or early in the twentieth, including the *Astronomical Register* (which closed in 1886), the *Journal of the Liverpool Astronomical Society* (1891), the *Cambrian Natural Observer* (1910), and the *Journal of the Leeds Astronomical Society* (1922). Two journals lingered on into the second half of the twentieth century: the *Memoirs of the British Astronomical Association* (1962) and the *Memoirs of the Royal Astronomical Society* (1978). Only three of the nineteenth-century astronomical journals survive today: the *Observatory*, the *Monthly Notices of the Royal Astronomical Society*, and the *Journal of the British Astronomical Association*.

Although a number of astronomical journals were not published past 1900, during the final decades of the nineteenth century society-based periodicals played an important role in the process whereby an audience for a specific scientific discipline, astronomy, was created. Astronomical communities were conceived of in the pages of these journals as existing at a variety of levels, including local or provincial, national, and even international. These communities existed somewhat independently of the circle of elite astronomers who read the publications of the Royal Astronomical Society or the official observatories, though some members of the Royal Astronomical Society were key figures in the amateur societies. The British Astronomical Association took care of organizing amateur astronomy at the national level. The local or provincial societies oriented themselves in relation to the British Astronomical Association even when they were not formally linked to it. To some extent, this confirms Simon Naylor's assertion that frequently "regional science was forced to assume a more limited epistemological position."[110] The new societies saw as part of their mission the organization of their members into groups of observers, whether it be into more formal divisions, as in the case of the Liverpool Astronomical Society and the British Astronomical Association, or more informally into one unified group, via the societies themselves. This provided amateur astronomers with a role. Often led by experienced observers, and even by elite astronomers who were members of the Royal Astronomical Society, the amateurs trained the members of their societies how to observe as part of a team.

Part of the story, then, is how elite astronomers were able to draw upon the observations and insights of amateurs. Indeed, the new journals were partially responsible for creating an identity for the amateur astronomer, one that was celebrated in the pages of the periodicals. This identity proved to be attractive to male professionals, their wives, their daughters, and, in the case of the local societies, to those interested in the religious dimensions of astronomy. Examining the publications of the new astronomical societies of the second half of the nineteenth century provides us with new insights into how communities of scientific practice were constructed through the periodical literature.

NOTES

The author would like to thank Simon Schaffer, Lee Macdonald, and Robert Smith for sharing their knowledge of nineteenth-century British astronomical journals with me at an early stage of the research. Later, Robert Smith also read a draft of this chapter and provided helpful suggestions for revision. The author is indebted to the British libraries and archives that allowed me to examine their holdings of astronomical journals, including the University of Cambridge Library, the Institute of Astronomy at the University of Cambridge, the Royal Society, the Royal Astronomical Society, and the Radcliffe Science Library at the University of Oxford. I am particularly appreciative of the assistance provided by Mark Hurn of the Institute of Astronomy, University of Cambridge, and Sian Prosser of the Royal Astronomical Society Library.

1. Arthur Mee, "Correspondence: Our Correspondence Column," *Journal of the British Astronomical Association* 4 (1895): 309–10, on 309.

2. *Journal of the Astronomical Society for Wales* 1 (January 1895): front cover.

3. W. H. Brock, "The Development of Commercial Science Journals in Victorian Britain," in *Development of Science Publishing Europe*, ed. A. J. Meadows (Amsterdam, New York, Oxford: Elsevier Science Publishers, 1980), 95–122, on 96.

4. Sally Shuttleworth and Berris Charnley, "Science Periodicals in the Nineteenth and Twenty-First Centuries," *Notes and Records* 70 (2016): 297–304, on 298.

5. This would include the *Memoirs of the Royal Astronomical Society of London* (founded in 1822) and the *Monthly Notices of the Astronomical Society of London* (founded in 1830), the two journals of the Royal Astronomical Society; and *Astronomical Observations Made at the Observatory at Cambridge* (founded in 1828) and *Results of the Astronomical Observations Made at the Royal Observatory, Greenwich* (founded in 1750), examples of journals connected to the British government or to universities.

6. Bernard Lightman, "The Mid-Victorian Period and the *Astronomical Register* (1863–1886): 'A Medium of Communication for Amateurs and Others,'" *Public Understanding of Science* 27, no. 5 (2018): 629–36, on 633–34.

7. Allan Chapman, *The Victorian Amateur Astronomer: Independent Astronomical Research in Britain 1820–1920* (Chichester, New York, Brisbane, Toronto, Singapore: John Willey & Sons, 1998), 249.

8. Jonathan R. Topham, "The Scientific, the Literary and the Popular: Commerce and the Reimagining of the Scientific Journal in Britain, 1813–1825," *Notes and Records* 70 (2016): 305–24, on 308.

9. "To the Readers of the *Journal* of the Liverpool Astronomical Society," *Journal of the Liverpool Astronomical Society* 7 (August 1888): 1–3, on 1.

10. Chapman, *Victorian Amateur Astronomer*, 243, 251.

11. Samuel J. M. M. Alberti, "Amateurs and Professionals in One County: Biology and Natural History in Late Victorian Yorkshire," *Journal of the History of Biology* 34 (2001): 115–47, on 119–20.

12. "Laws of the Liverpool Astronomical Society," *Abstracts of Proceedings of the Liverpool Astronomical Society* 1 (1882–83): i–iii, on ii.

13. Gerard Gilligan, *The History of the Liverpool Astronomical Society* (Liverpool, UK: Liverpool Astronomical Society, 1996), 1.

14. "The President's Address," *Journal of the Liverpool Astronomical Society* 4 (October 1885): 3.

15. Gilligan, *History of the Liverpool Astronomical Society*, 2.

16. "Election of an Observer to the Society," *Abstracts of Proceedings of the Liverpool Astronomical Society* 1 (1882–83): 25–26.

17. Thomas R. Williams, "Espin, Thomas Henry Espinall Compton," in *Biographical Encyclopedia of Astronomers*, ed. K. Brachen et al. (New York, London: Springer, 2006), 343–44.

18. Chapman, *Victorian Amateur Astronomer*, 279. For a full list of the members in September 1884, see "List of Members of the Liverpool Astronomical Society," *Abstracts of Proceedings of the Liverpool Astronomical Society* 2 (1883–84): 1–8 (separate pagination).

19. E. W. Maunder, "Solar Observations for Amateurs," *Transactions of the Liverpool Astronomical Society* no. 2 (1883–84): 1–8, on 1–2.

20. Chapman reports this as the figure for 1886. See Chapman, *Victorian Amateur Astronomer*, 248.

21. "President's Address," *Journal of the Liverpool Astronomical Society* 6 (1887): 1–3, on 2.

22. "To the Readers of the *Journal* of the Liverpool Astronomical Society," *Journal of the Liverpool Astronomical Society* 7 (1888): 1–3, on p. 2.

23. Isaac Roberts, "On a New Planet Indicator," *Abstracts of Proceedings of the Liverpool Astronomical Society* 1 (1882–83): 14–16.

24. Maunder, "Solar Observations for Amateurs," 1–8.

25. "Correspondence," *Journal of the Liverpool Astronomical Society* 5 (1886–87): 134–36.

26. This is similar to how the naturalist Margaret Gatty embraced her role as a naturalist who disseminated scientific ideas as a member of a marine biology network. See Alberti, "Amateurs and Professionals in One County," 121.

27. W. S. Franks, "The Division of Astronomical Work," *Journal of the Liverpool Astronomical Society* 5 (1886–87): 13–16, on 13–14.

28. The situation in astronomy parallels Alberti's account of how the new field clubs and natural history societies of the later Victorian period organized their members into an army of disciplined naturalists. See Alberti, "Amateurs and Professionals in One County," 136.

29. "The Second Meeting of the Fourth Session," *Journal of the Liverpool Astronomical Society* 3 (1884–85): 29–30.

30. "Liverpool Astronomical Society," *Observatory* 10 (1887): 154–57, on 154.

31. "Annual Meeting of the Liverpool Astronomical Society," *Observatory* 10 (1887): 281–83, on 281.

32. Chapman, *Victorian Amateur Astronomer*, 251.

33. "Report of the Annual Meeting of the Association Held October 28, 1891," *Journal of the British Astronomical Association* 2 (1893): 1–5, on 2, 5.

34. Peter Johnson, "The *Astronomical Register* 1863–86," *Journal of the British Astronomical Association* 100 (1990): 62–66, on 66.

35. "Circulars Issued by the Provisional Committee," *Journal of the British Astronomical Association* 1 (1892): 17–19, on 19.

36. Ibid.

37. "British Astronomical Association," *Journal of the British Astronomical Association* 1 (1892): 1–14, on 5.

38. Ibid, 9–14.

39. Chapman, *Victorian Amateur Astronomer*, 252.

40. "Report of the Meeting of the Association Held December 31, 1890," *Journal of the British Astronomical Association* 1 (1892): 109–14, on 109; "Report of the Annual Meeting of the Association Held October 28, 1891," *Journal of the British Astronomical Association* 2 (1893): 1–5, on 2; "Report of the Annual Meeting of the Association, Held on October 26, 1892," *Journal of the British Astronomical Association* 3 (1894): 1–8, on 4.

41. "Report of the Council on the Work of the Eighth Session, October 1897 to October 1898," *Journal of the British Astronomical Association*, 8 (1899): 386–98, on 385; "Report of the Council on the Work of the Ninth Session, October 1898 to October 1899," *Journal of the British Astronomical Association* 9 (1900): 413–23, on 413; Howard L. Kelly, ed., *The British Astronomical Association: The First Fifty Years* (London: British Astronomical Association, 1989), 15.

42. Kelly, *British Astronomical Association*, 11.

43. "Report of the Council on the Work of the Ninth Session, October 1898 to October 1899," *Journal of the British Astronomical Association* 9 (1900): 413–23, on 414.

44. "Computers" was the term given to observatory assistants who could carry out routine calculations to turn raw observations into usable data. They also frequently were trained in the use of telescopes.

45. "Candidates for Election as Members of the Association," *Journal of the British Astronomical Association* 2 (1893): 75–76, on 76.

46. "Candidates for Election as Members of the Association, December 27, 1893," *Journal of the British Astronomical Association* 4 (1895): 36.

47. "New Members of the Association Elected March 25, 1896," *Journal of the British Astronomical Association* 6 (1897): 305.

48. *List of Members of the British Astronomical Association, September 30, 1898* (London: Eyre and Spottiswoode, 1898).

49. "[General Meeting of the British Astronomical Association,]" *Journal of the British Astronomical Association* 1 (1892): 1–8, on 2–3.

50. Kelly, *British Astronomical Association*, 131; "Report of the Annual Meeting of the Association, Held on October 28, 1896, at University College, Gower Street," *Journal of the British Astronomical Association* 8 (1898): 1–8, on 2.

51. Chapman, *Victorian Amateur Astronomer*, 288.

52. "Report of the Annual Meeting of the Association Held October 28, 1891," *Journal of the British Astronomical Association* 2 (1893): 1–5, on 2.

53. "Notes: Women as Members of Scientific Societies," *Journal of the British Astronomical Association* 2 (1893): 479.

54. "Notes: Ladies at the Royal Astronomical Society," *Journal of the British Astronomical Association* 3 (1894): 97–98, on 98.

55. "Report of the Annual Meeting of the Association Held November 26, 1890," *Journal of the British Astronomical Association* 1 (1892): 49–58, on 50.

56. "Notes on the Observation of Meteors," *Memoirs of the British Astronomical Association* 1 (1893): 17–20, on 17–18.

57. "Section for the Observation of the Colours of Stars," *Memoirs of the British Astronomical Association* 1 (1893): 33–52, on 33–35.

58. "Report of the Council on the Work of the Third Session, October 1892, to October 1893," *Journal of the British Astronomical Association* 3 (1894): 465–74, on 467.

59. "Officers and Council," *Journal of the British Astronomical Association* 5 (1896): 10.

60. Monck's career illustrates that at the end of the nineteenth century, British astronomy was still some way from being fully professionalized. Devorkin argues that Monck established a reputation among professionals because he "compared large samples of proper motion data of stars selected by spectral type in order to determine relative distances to these groups of stars." Monck's type of research, usually seen as being beyond the amateur, was done independently and prior to the work of the leading professional in the field, J. C. Kapteyn at Groningen. See David DeVorkin, "Stellar Evolution," *Astrophyiscs and Twentieth-Century Astronomy to 1950*, ed. Owen Gingerich (Cambridge: Cambridge University Press, 1984), 90–108, on 96. My thanks to Robert Smith for drawing this to my attention.

61. A. A. R., "William Henry Stanley Monck," *Monthly Notices of the Royal Astronomical Society* 76 (1916): 264–66.

62. A. M. Clerke, "The Rotation Periods of Mercury and Venus," *Journal of the British Astronomical Association* 1 (1892): 20–25; Miss E. Brown, "Papers Communicated to the Association: Notes on Peculiar Feature of the Large Sun-Spot of February 1894," *Journal of the British Astronomical Association* 4 (1895): 301.

63. "Notes: 'The Observatory,'" *Journal of the British Astronomical Association* 8 (1899): 98.

64. "Note: The Companion to the Observatory," *Journal of the British Astronomical Association* 4 (1895): 75–76, on 76.

65. "Notes: Astronomical Society for Wales," *Journal of the British Astronomical Association* 5 (1896): 177.

66. "The Journal of the Astronomical Society of Wales," *Journal of the British Astronomical Association* 5 (1896): 368; "Notes: The Cambrian Natural Observer," *Journal of the British Astronomical Association* 8 (1899): 320; "Notes: The Cambrian Natural Observer," *Journal of the British Astronomical Association* 9 (1900): 394–95, on 394.

67. "Notes: Leeds Astronomical Society," *Journal of the British Astronomical Association* 7 (1898): 437–38, on 437.

68. "Notices of the Association," *Journal of the British Astronomical Association* 1 (1892): 47–48, on 48.

69. "Report of the Council on the Work of the First Session. October 1890 to October 1891," *Journal of the British Astronomical Association* 1 (1892): 535–46, on 537.

70. "Report of the Annual Meeting of the Association Held October 28, 1891," *Journal of the British Astronomical Association* 2 (1893): 1–5, on 5.

71. "Report of the Meeting of the Association Held April 27, 1892," *Journal of the British Astronomical Association* 2 (1893): 319–23, on 322; O'N. F. Kelly, "Correspondence: Abstracts

of Current Astronomical Literature in the 'Journal,'" *Journal of the British Astronomical Association* 2 (1893): 414–15; "Abstracts of Current Astronomical Literature in the 'Journal,'" *Journal of the British Astronomical Association* 2 (1893): 468–72; "Correspondence: Abstracts of Current Astronomical Literature in the 'Journal,'" *Journal of the British Astronomical Association* 2 (1893): 537–41; "Report of the Annual Meeting of the Association, Held on October 26, 1892," *Journal of the British Astronomical Association* 3 (1894): 1–8, on 2–3.

72. "Leeds Astronomical Society," *Report and Transactions of the Leeds Astronomical Society*, no. 1 (1893): 6–7, on 7.

73. Chapman, *Victorian Amateur Astronomer*, 244, 247, 254–55.

74. "Rules," *Report and Transactions of the Leeds Astronomical Society* 1 (1893): 4.

75. Work of the Society, Oct. 4, 1893," *Report and Transactions of the Leeds Astronomical Society* 1 (1893): 24–25, on 24.

76. Washington Teasdale, "Personal," *Report and Transactions of the Leeds Astronomical Society* 1 (1893): 48.

77. E. W. M., "Washington Teasdale," *Monthly Notices of the Royal Astronomical Society* 64 (1904): 293–94.

78. "List of Members," *Report and Transactions of the Leeds Astronomical Society* 1 (1893): 46–47; "List of Members," *Journal and Transactions of the Leeds Astronomical Society* 5 (1897): 68–70; "List of Members," *Journal and Transactions of the Leeds Astronomical Society* 8 (1900): 117–20.

79. "Conversational Meeting August 14th, 1895," *Journal and Transactions of the Leeds Astronomical Society* 3 (1895): 34.

80. Florence Taylor, "Miss Caroline Herschel, the Astronomer," *Journal and Transactions of the Leeds Astronomical Society* 4 (1896): 33–36.

81. Florence Taylor, "Mary Somerville, the Great Woman Astronomer and Mathematician," *Journal and Transactions of the Leeds Astronomical Society* 5 (1897): 33–37, on 37.

82. Teasdale, "Personal," 48.

83. "Leeds Astronomical Society," *Journal and Transactions of the Leeds Astronomical Society* 3 (1895): 8–9, on 9.

84. C. T. Whitmell, "Obituary: Mr. W. D. Barbour," *Journal of the British Astronomical Association* 13 (1903): 286–87.

85. See, for example, "Work of the Society during 1893," *Report and Transactions of the Leeds Astronomical Society* 1 (1893): 35–38.

86. John Roberts, "Zetetic Astronomy, March 1, 1893," *Report and Transactions of the Leeds Astronomical Society* 1 (1893): 14–16, on 16.

87. "Work of the Society, Oct. 4, 1893," *Report and Transactions of the Leeds Astronomical Society* 1 (1893): 24–25, on 25.

88. W. D. Barbour, "Evolution of Planetary Life," *Report and Transactions of the Leeds Astronomical Society* 2 (1894): 41–45, on 45.

89. W. D. Barbour, "Light and Ether: Their Relation to the Universe," *Journal and Transactions of the Leeds Astronomical Society* 6 (1898): 62–77, on 77.

90. "The Leeds Astronomical Society: Annual Meeting of Members January 18, 1899," *Journal and Transactions of the Leeds Astronomical Society* 6 (1898): 15–16, on 16.

91. "President's Address," *Journal of the Astronomical Society of Wales* 1, no. 1 (1895): 2–6, on 4–5.

92. "Editor's Notes," *Journal of the Astronomical Society of Wales* 1, no. 6 (1895): 58–61, on 58.

93. The list of members included six women, three fellows of the Royal Astronomical Society, five members of the BAA, five reverends, two lawyers, a banker, a chemist, a draper, a school principal, and a professor. "List of Members Up to and Including January 1, 1895," *Journal of the Astronomical Society of Wales* 1, preliminary number (1895): v–vi.

94. Chapman, *Victorian Amateur Astronomer*, 256.

95. J. Bryn Jones, "Mee, Arthur Butler Phillips," in Brachen et al., eds., *Biographical Encyclopedia of Astronomers*, 764–65.

96. Chapman, *Victorian Amateur Astronomer*, 255.

97. "Notes," *Journal of the Astronomical Society of Wales* 1, no. 2 (1895): 17–18, on 17.

98. "Our Council Table," *Journal of the Astronomical Society of Wales* 2, no. 4 (1896): 65–66, on. 66; 2, no. 5 (1896): 72; 2, no. 3 (1896): 53–54, on 54. Mee was particularly full of praise for the BAA publications, extolling the virtues of the *Journal* and the *Memoirs* on several additional occasions. See "Current Notes," *Journal of the Astronomical Society of Wales* 2, no. 6 (1896): 77–78.

99. "Editor's Notes," *Journal of the Astronomical Society of Wales* 1, no. 9 (1895): 88–89, on 89.

100. "G. T. Davis, "A Few Thoughts on Perusing the Journal," *Journal of the Astronomical Society of Wales* 2, no. 8 (1896): 85–86, on 86.

101. "Observations Contributed by Members," *Journal of the Astronomical Society of Wales* 1, no. 9 (1895): 87–88, on 87.

102. "Practical Hints for the Coming Season," *Journal of the Astronomical Society of Wales* 2, no. 9 (1896): 99–102, on 100.

103. J. E. Southall, "The Ether," *Journal of the Astronomical Society of Wales* 1, no. 3 (1895): 22–23, on 23.

104. Rev. Arthur J. Jenkins, "Other Worlds Than Ours," *Journal of the Astronomical Society of Wales* 1, no. 3 (1895): 20–22, on 22.

105. "The Cambrian Natural Observer," *Cambrian Natural Observer* 1, no. 1 (1898): 1–3, on 1.

106. "Society Rules," *Cambrian Natural Observer* 1, no. 1 (1898): inside front cover.

107. "Observation," *Cambrian Natural Observer* 3, no. 2 (1900): 21–30.

108. "List of Donors," *Cambrian Natural Observer* 3, no. 2 (1900): [7–8], separate pagination.

109. "The Cambrian Natural Observer," *Cambrian Natural Observer* 1, no. 1 (1898): 1–3, on 1.

110. Simon Naylor, *Regionalizing Science: Placing Knowledge in Victorian England* (London: Pickering and Chatto, 2010), 180.

* 3 *

Managing the Boundaries of Medicine

"A Borderland in Ethics": Medical Journals, the Public, and the Medical Profession in Nineteenth-Century Britain

Sally Frampton

In 1881, luminaries of medicine from Joseph Lister to Louis Pasteur thronged to the International Medical Congress in London for what was the largest meeting of the profession ever to be staged. Held over seven days, with fifteen different sections and three thousand attendees, the congress was remarkable for both its size and its impact, as doctors debated the most pressing issues of the day, from the role of microorganisms in disease causation to state intervention in medical practice. In an opening address to the congress, the surgeon James Paget predicted that the event would see "a larger interchange and diffusion of information than in any equal time and space in the whole past history of medicine."[1] It was in this apt setting that the American surgeon John Shaw Billings addressed a large crowd on the state of medical literature, a subject that was rarely elaborated upon by doctors at their professional gatherings, but which was starting to receive increased attention. Billings, who at the Surgeon General's Office in Washington had helped amass the United States' largest medical library, drew attention to the vast and ever-expanding volume of medical periodicals being published. Billings calculated that in 1879, 655 different medical periodicals were published worldwide.[2] Between 1880 and 1899 that number would rise exponentially; in Britain alone the number of titles doubled from 60 to 120, marking an unprecedented era of growth.[3] The increase in medical journals, Billings speculated, brought both advantages and pitfalls. It was part of the great progress of modern thought, "that wonderful kaleidoscopic pattern which is unrolling before us," he wrote.[4] But he also cautioned that the propagation of journals, and the growing emphasis on their importance for professional advancement, promoted an undue haste in doctors who were eager to publish and see their name in print.[5]

Billings's speech reflected the prolific industry in journalism—medical and otherwise—that characterized British print culture in the late nineteenth century, as improved printing technologies and the abolition of stamp and paper duties gave rise to a rapid increase in the number of periodicals available, lowered their prices and increased their frequency. Contemporary commentators and historians have also aligned the increasing number of medical journals with a broader trend occurring in the organization and epistemology of medicine: the establishment of specialist practice, most notably in dermatology, ophthalmology, obstetrics, and gynaecology.[6] Certainly this trend was consolidated by the journals that were emerging at the end of the century; titles like the *British Gynaecological Journal*, begun in 1885, and the *British Journal of Dermatology*, which started in 1888, were part and parcel of the changes within medicine at this period. These were changes not just in content but in literary style, as medical journals underwent a gradual shift from the personable, gossipy tone of the weekly medical press in the 1820s—most famously characterized by the *Lancet*—to the more austere model of medical writing that was to predominate a hundred years later. The pages of medical journals were increasingly given over to specialized terminology emanating from physiology and other medical sciences, restricting the audience for professional medical journals to those cognizant of such scientific language.[7]

These changes signaled a move on the part of doctors to create a more closeted professional world. But the growth of specialist journals and their use of increasingly esoteric phraseology was accompanied by a no less significant trend in medical and health literature aimed toward a more diverse audience. During the 1880s, the ways in which different groups of people other than medical professionals might participate in healthcare was being reimagined. Through the public health movement, there was a growing emphasis on the place of personal hygiene and domestic management in containing and preventing disease. As is explored more fully in chapter 10 in this volume, public health reformers recognized the active role citizens needed to play for change to be satisfactorily accomplished. Preventive medicine relied not only on legalizing acts of compulsion like vaccination, but on communicating sanitary science to the public and encouraging citizens to manage their own health and the health of their community.[8] Meanwhile, the campaign to professionalize nursing by means of registration was also gaining ground, and elsewhere first aid training was becoming a force among the British population, as the St. John Ambulance organization began to instruct members of the public in how to treat illnesses and injuries. Following the introduction of more systematic management practices in British hospitals in the last decades of the century, the hospital manager was also being established as a professional role. These developments were reflected in the numerous publications that emerged in the 1880s and 1890s relating to nursing, public health, first aid, and domestic hygiene. Some titles, like *The Hospital*, were aimed toward both doctors and allied professionals working in medicine and health care; others were aimed toward one professional subset or another, such as the nursing community (*Nursing Notes*, *Nursing Record*, and *Nursing Mirror*). A number of new journals, including *Health*, *Baby*, and *First Aid*, were primarily aimed toward the public. Whether they anticipated it or not, many of these journals acknowledged finding an audience of both lay people and health care professionals. In this respect, they differed from the professional weeklies such as the *Lancet* and the *British Medical Journal*, which by the late nineteenth century were being marketed as strictly for doctors.

This new wave of journals built upon a longer tradition of medical and health journalism aimed to appeal to audiences outside the medical profession. Its emergence reignited perennial concerns among doctors about the diffusion of medical knowledge into newspapers and magazines, and the risks apparent in the public interpreting health advice without professional intermediaries. But the ensuing controversies surrounding the journals heralded a much more rigorous clampdown by the profession on the circulation of medical literature among the public and the emergence of a new ethical framework for medical journalism. These "quasi-medical" journals, as the *Lancet* labeled them,[9] were seen to be ethically dubious because they blurred the boundaries between lay and medical audiences and between professional and nonprofessional literature, thus bringing doctors into an unwelcome proximity to the worlds of commerce and advertising.

This chapter first seeks to contextualize the medical press within the broader framework of nineteenth-century print culture and to show how lay and professional audiences, broadly categorized, frequently intertwined. It then focuses on two journals, *The Hospital* and *Baby*, established in 1886 and 1887 respectively, which together brought to a head doctors' concerns about the public's interaction with medical journals. The audiences these journals solicited, the objectives they advanced, and the conversations they generated heralded a transformative period in British journalism during which the notion of what exactly constituted a medical journal was contested. As the number of titles relating to medicine and health expanded and diversified, these changes in print culture reshaped modes of communication between medical practitioners and the wider public. Doctors' responses to these changes revealed a professional community that sought to manage and regulate the dissemination of knowledge—knowledge to which they claimed authoritative ownership.

A CLOSED SHOP? THE MEDICAL
PRESS AND ITS AUDIENCES

In 1823, the surgeon Thomas Wakley began a weekly publication that quickly became the country's most popular and most controversial medical journal. In a reference to the sharp scalpel surgeons used to cut open the body, the journal was named the *Lancet*—a title reflecting Wakley's intention to expose and dissect the workings of the profession.[10] The *Lancet* bulldozed its way into a marketplace for medical journals previously dominated by monthly and quarterly titles, the most successful of which was the *Medico-Chirurgical Review*, edited by the former naval surgeon James Johnson, which in 1821 had attained a respectable circulation of around 1,300 copies.[11] In the space of just a year, the *Review* was outstripped by the *Lancet*, which was soon selling more than four thousand copies a week, an extraordinary achievement for a medical title.[12] Wakley used the weekly frequency of his journal to maximum effect, providing up-to-the minute news from the metropolitan hospitals, including details of recent cases, professional gossip, and tantalizing editorials that promised more exciting content in the next issue. The *Lancet* quickly became known for its savage takedowns of the nepotistic and closed world of the metropolitan hospital system, which led to Wakley attracting the wrath of London's medical elite. At the heart of Wakley's journalistic enterprise was a desire to make available to the medical profession en masse—the provincial practitioners, medical students, and countless doctors who worked overseas— intelligence and information usually only available to those able to access London's great teaching hospitals. Wakley used various strategies to undermine the dominance of those institutions and the individuals who worked at them, exposing corruption among hospital staff, and publishing without consent the lectures of well-known hospital surgeons.[13] The latter endeavor soon led to Wakley being accused of intellectual piracy, following his unauthorized publication of the lectures of the eminent surgeons Astley Cooper and John Abernethy, of Guy's and St. Bartholomew's hospitals respectively. The profession's indignation over the matter was not merely moralizing. By making their orations available in print for the weekly price of sixpence, Wakley was undermining the potential for hospital doctors to make a lucrative income through their lectures.

The *Lancet* and other weekly titles that followed in its stead—most notably the *Medical Times* in 1839, and the *Provincial Medical and Surgical Journal* in 1840 (retitled the *British Medical Journal* in 1857)—championed the rights and concerns of those working outside the elite rankings of the metropolitan hospitals. Their editors drew analogies between the closed shop of the corrupt hospital elite and the clandestine malfeasance of quack doctors and patent-medicine peddlers, and in contrast pitched their own journals as tonics of transparency and truth, working for the good of the profession and the public to rid the country of both. "SHALL THE PROFESSION BE REFORMED OR NOT?" asked the *Medical Times* in 1840. "We boldly commenced the work by unmasking chicanery; public support has been instant and hearty. It was all the encouragement we sought, all the guerdon we required."[14] While all three journals purported to share a reformist vision, Wakley was the most energetic in executing it. In the opening statement to the first issue of the *Lancet*, he proclaimed the journal to be not just for medical and surgical practitioners, but also for "every individual in these realms."[15] Wakley hoped the journal would furnish the wider public with enough knowledge that they might avoid the overtures of quackish practitioners. Thus, the public were recognized not just as a potential audience for the journal, but as active agents in the battle for medical reform that Wakley envisioned, which would see a dismantling of the elitist hierarchy within medicine and the erasure of unregulated quackery.

Toward this objective, Wakley emulated contemporaneous publications that educated and advised the general populace on medical and health issues.[16] He was especially influenced by the *Monthly Gazette of Health*, edited by the physician Richard Reece between 1816 and 1832, which was aimed toward both doctors and laypeople. The *Gazette* allowed readers to build up a repository of information and, through its correspondence columns, a chance to contribute queries, opinions, and medical recipes to the publication. Other medical journals that were also unabashedly aimed toward a public audience sprang up through the 1820s and 1830s. The *Medical Adviser*, which ran between 1823 and 1825, and *The Doctor*, which lasted from 1832 to 1837, both sold at a cheap price and targeted working-class readers, offering a mix of medical advice and treatment plans, lessons in anatomy and physiology, and articles on social and medical reform.[17] The market for medical literature was augmented by the considerable number of titles relating to nonorthodox systems of medicine, including homoeopathy and mesmerism, mostly established in the middle of the century, which attained varying degrees of success. Many nonorthodox titles attempted to straddle the boundary between the medical profession and the public, pursuing a lay audience while also trying to placate and even convert critics from within the profession.[18]

In the context of the literary marketplace, the potential for interaction that serial publications offered to readers arguably put them at a strategic advantage over books and pamphlets, which had been the more common modes of health literature in the seventeenth and eighteenth centuries.[19] It allowed journal proprietors and editors to create a sense of dialogue with their audiences. Richard Reece frequently published letters allegedly written by enthusiastic readers who praised the editor for making space for lay queries and contributions. "I am so great an admirer of your truly philanthropic undertaking that I have recommended it to all my friends, and I think I may say I have succeeded in procuring thirty subscribers in the course of last month," wrote one, a widow named Mary Ann Fletcher, in 1816.[20] Whether true or not—there is always the possibility that editors had a hand in inventing correspondence of this kind—the publication of such endorsements fostered the idea that a community building of sorts was occurring around the journal.

And yet the relatively short runs of titles like the *Medical Adviser* and *The Doctor* suggest that few orthodox medical journals aimed toward lay people were able to sustain a regular audience of readers. While claiming to denounce quackery in the same manner as the more professionally oriented medical weeklies, they themselves were vulnerable to accusations of impropriety. Critics bemoaned the unsuitability of the penny press in disseminating medical advice, and its potential for spreading inaccurate information. "It is to be feared that, if this penny Doctor's prescriptions were to be taken, a considerable reduction in the amount of population would be the inevitable result," wrote the *Mechanic's Magazine* about *The Doctor* in 1832.[21] The *Mechanic's Magazine* was a cheap weekly which, like its medical counterparts, was designed to cater to a working-class audience. But as a journal broadly focused on technology, it lay on rather less ethically shaky ground than did medical journals, where, it seemed, there was a real risk that the information they contained could lead to bodily harm. By the 1830s, reformists like Wakley were finding quackery in all places, from the highest echelons of the profession to the cheapest periodicals and magazines, all of which constituted a threat to their desire to see the establishment of a democratic but highly regulated profession. The rhetoric of reform within medical weeklies like the *Lancet* and *Medical Times* increasingly centered upon professional fraternalism, consolidated by a loosely consensual opposition to "quackish" systems of medicine such as homeopathy.[22] Despite opposition from the profession, nonorthodox medicines remained inexorably popular among a general populace that was often distrustful of the interventionist and at times brutal work of bloodletting, amputating "regular" doctors. For all Wakley's initial zeal for a readership that included autodidactic laymen, the public's compliance and cooperation in the erasure of quackery looked far from guaranteed. By the middle decades of the century, titles like the *Lancet* began to close ranks, distancing themselves from the rhetoric of their early days, where members of the public were positioned as participants in medical reform, and instead asserted themselves as publications limited to professional readers.

Connections between the medical press and what doctors termed the "lay" press, which included all types of print media beyond medical titles, nonetheless remained present. A reciprocal relationship based upon the reproduction of information enmeshed medical journals within wider print culture. Local medical news and gossip reported in the regional press was reproduced in the medical weeklies, as were satirical sketches from publications like *Punch*, particularly when they lampooned quacks and purveyors of nonorthodox medicine.[23] Conversely, the content and campaigns of medical journals provided good copy for the daily newspapers, who capitalized upon an ever-present curiosity among the public for the latest news on medicine, health, and disease. Popular titles like the *Morning Post* and the *Standard* were attentive to the discussions of the medical press, especially on public health. Foreshadowing the still-present fascination of the popular media with scientific research on food and diet, the issue that perhaps garnered the most interest among the press in the middle decades was the findings of the *Lancet*'s analytical sanitary commission, which ran between 1851 and 1854 as a response to concerns about food adulteration, and which attracted widescale attention from newspapers, women's magazines, and the penny weeklies. Thomas Wakley had employed the doctor and microscopist Arthur Hill Hassall to conduct a series of tests on common household foodstuffs including flour, bread, coffee, and chicory (the latter often used to adulterate coffee). Hassall wrote up his findings in a series of reports duly published in the *Lancet*. Hassall's study detected extremely high levels of impurity in samples of coffee and chicory, which were found to contain everything from sawdust to oak bark.[24] By sponsoring Hassall's revelatory work and connecting it to his journal, Wakley shrewdly spotted a way to market the *Lancet* as a trailblazing organ in public health innovation. He also took the opportunity to republish articles from the general press that praised the *Lancet*'s work in exposing the nefarious practices of grocers and merchants who were adulterating goods.[25]

On an individual level too, many prominent figures forged a living that relied upon a range of journalistic work in both medical and nonmedical publications. This included figures like Frederick Knight Hunt, the originator of the *Medical Times*, who spent much of his life vacillating between medical practice and journalism. Hunt, who had been brought up in straitened financial circumstances, started his career as a printing room boy at the *Morning Herald* before turning to medicine, in which he qualified in 1840. As a medical student he had begun the *Medical Times*, which would go on to achieve considerable success—though only after he had sold it on, anxious to avoid the threats of libel from medical men that had come his way since starting the publication—and in which he had emulated the caustic, controversial tone of the *Lancet*. While continuing to practice medicine, Hunt built up a reputation as a journalist of some importance and eventually became the editor of the *Daily News* before his untimely death at the age of just forty in 1854.[26] The career trajectory of the London physician Andrew Wynter was similarly peripatetic. Wynter edited the *British Medical Journal* between 1855 and 1860. After resigning its editorship, he went on to forge a successful career as a journalist and author writing on medical topics as well as on the most pressing subjects of the day, from railways to photography and telegraphy, contributing mainly to middlebrow and highbrow publications like *Once a Week* and the *Edinburgh Review*.[27] All the while he continued to practice medicine, primarily focusing on the treatment of insanity. However, Wynter's reputation would predominantly rest upon his writing rather than either his practice or medical editorship.[28] Both medicine and journalism could be financially precarious careers. For men like Hunt and Wynter, who displayed skill in both areas, harnessing the remunerative powers of both was a necessary step toward economic security.

While the medical press continued to share both content and personnel with other genres of print, it also attempted to forge a greater distinction between itself and the lay press. The attitude of the medical press toward medical content that featured in other types of periodical publications, particularly those aimed toward the working classes, was increasingly couched in the language of surveillance. In particular, popular titles that allowed readers to seek medical advice through their correspondence columns were policed. As Claire Furlong has detailed, the *Family Herald*, established in 1843, and *Reynolds's Miscellany*, begun in 1846, two hugely popular weeklies aimed at working-class families, both made this a feature of their publications. In the 1850s *Reynolds's* even began a dedicated "Medical Corner," suggesting, Furlong argues, that there was "a strong demand for health advice and that *Reynolds's* was more than willing to provide it."[29] In 1860 the *Lancet* publicly lambasted *Reynolds's* for providing dangerously inaccurate medical advice on its pages. In an article somewhat inflammatorily titled "How to Poison Correspondents," the *Lan-*

cet admonished *Reynolds's* for mistakenly advising a reader to consume ten drachms of tincture of foxglove to treat palpitations of the heart, a measurement that contained "enough digitalis . . . to poison six people."[30] The editor and namesake of *Reynolds's*, George W. M. Reynolds, responded in the *Lancet* the following week, conceding the mistake: The recipe in fact should have prescribed ten drops. He denied, however, that the error negated the journal's work in providing medical advice to readers, claiming to quote only from "those works which I consider to be reliable authorities, and that it is merely in the simplest cases I even undertake to furnish these receipts at all."[31] Although inquirers were at times referred to a doctor to receive advice, Reynolds seemed to be implying that he himself, and presumably his fellow editorial staff, took it upon themselves to diagnose and suggest treatment plans for illnesses.[32] Thus the journal assumed the role of medical adviser at a price that undercut that of the professional doctor. As we shall see, the *Lancet*'s reproach of Reynolds anticipated a more concerted effort among the profession to regulate and manage the circulation of medical knowledge among the general public.

The boundaries doctors constructed between "their" press and the lay press in the middle decades of the nineteenth century were erected both to mark the especial, authoritative status of the profession and to control the flow of information to the public. Medical journals increasingly identified themselves as sequestered spaces for discussion among doctors. But these boundaries were nonetheless very frequently transgressed as the medical press continued to influence, and be influenced by, other forms of media. As the century progressed, the medical press became more alert to the popular diffusion of medical information, particularly within the cheap periodicals. Culminating in a controversial episode involving the *Lancet* and a set of new publications in the 1880s, changing trends in journalism threatened to dissolve the boundaries between the medical and lay press altogether.

MEDICINE AND THE ETHICS OF PRINT CULTURE: THE *LANCET*, *THE HOSPITAL*, AND *BABY*

Near the end of 1887, the *Lancet* published a series of four editorials that criticized a new wave of "popular journals of a quasi-medical character,"[33] which they described as inhabiting a "borderland in ethics."[34] The *Lancet* anonymously implicated three recently established journals at the center of the trend, two of which were described in enough detail that they could be identified as *The Hospital*, a journal focused on hospitals and health care, which targeted an inclusive audience of medical professionals, hospital administrators,

FIG. 9.1. The first issues of *The Hospital* (1886) and *Baby* (1887). The opening edition of *The Hospital*, which pleaded for contributions to a number of different hospitals, attested to the charitable mission of its editor, Henry Burdett, while *Baby* quoted material from the *Lancet* and *British Medical Journal*, positioning itself as a periodical that was medical in nature. *The Hospital*: Institute of Healthcare Management; *Baby*: Bodleian Library, University of Oxford, shelfmark per. 2623 d12.

nurses, and members of the public, and *Baby*, a journal devoted to children's health (figs. 9.1a and 9.1b). *The Hospital* and *Baby* were markedly different publications. But both were sufficiently medical in their content yet populist in their styling that they aroused the *Lancet's* consternation. *The Hospital* was, it seems, the primary cause for concern. It had been established in 1886 by Henry Burdett, a businessman, philanthropist, and advocate for hospital reform. Having been a medical student at Guy's without qualifying, Burdett had gone on to forge a successful career at the periphery of the profession, working as a hospital administrator, during which time he garnered an impressive reputation for fund-raising. In his mid-thirties he had switched careers and begun working for the London Stock Exchange, though his links to the medical world remained strong, most notably through his championing of and fund-raising for the voluntary hospitals.[35] Burdett's personal experience, working at the borders of the lay and medical worlds, was reflected in his ambitious periodical project. He envisioned that hospital managers would be among *The Hospital*'s key audiences, using the journal as a way of exchanging information, strengthening their sense of professional community, and linking them to philanthropically minded members of the public in a bid to raise charitable contributions to the voluntary hospitals. As Burdett reminded readers, he used the word "hospital" in the widest sense, and also sought to make the journal accessible to a readership of doctors, nurses, and medical students as well as convalescents and their networks of friends, family, and clergy, thus bringing together all those who would have an interest in hospital life.[36] His polemical editorials on hospital work and charity sat side by side with news stories taken from the *Lancet*, original communications from doctors, and lists of operating times, as well as puzzles, quizzes, and fictional stories, including those for children. The latter three features, which were highly unusual for a medical journal, were indicative of Burdett's open commitment to achieving popular appeal. Astutely tapping into a sizeable audience that was only beginning to be catered to by publishers, the journal also included a supplement, *Nursing Mirror*, that provided nurses with job listings—crucial in an occupation where employment was often short-term. The *Nursing Mirror* additionally contained features useful to the everyday work of the nurse, including medical lectures and passages of prose to help comfort sick patients. Indeed, Burdett was perhaps overly ambitious in the expansive range of content he initially wished to include; items promised in the journal's opening editorial failed to materialize, such as the "madman's column." Burdett never specified what this was to be, although one might speculate that it was intended to provide a voice to the institutionalized and mentally ill.

This heterogeneous jumble of articles—in which medical and nonmedical content was presented side by side, each providing a smattering of knowledge, information, or trivia—troubled the *Lancet*.[37] One editorial complained that *The Hospital* consisted of "sermons, discourses on physiology, palmistry, bacteriology, enigmas, fever, food, Ruskin, vaccination, the action of remedies in disease—that hunting ground of adventurous empirics,—serial stories for young people, puzzles, and acrostics," which were "blended week by week in one incongruous whole."[38] This style of journalism, offering snippets of information on a wide range of subjects, was typical of the New Journalism, which had provoked anxiety in cultural spheres beyond medicine, with critics believing it encouraged a less intellectually rigorous "skim-reading" among audiences.[39] For the *Lancet*, the appropriation of this mode of communication by journals claiming to be medical was particularly harmful, encouraging lay people to diagnose and treat illnesses while having only minimal knowledge. "Have not those who have been brought into contact with the public by service at hospitals, and as lecturers on science applied to public needs, yet learned that 'a little knowledge is dangerous in everything, and in nothing else so much as in the science and art of healing?'" the journal asked.[40] Despite almost immediate criticism from the *Lancet*, Burdett's innovative vision quickly paid dividends; by the end of 1888, *The Hospital* had accrued an impressive weekly circulation of nine thousand copies, in comparison to the *Lancet*'s seven thousand copies.[41] Given the sales figures its competitor was generating, the *Lancet* had reason to be alarmed.

Baby, the other publication cited, had started up only a month prior to the first *Lancet* editorial on the matter. Edited by twenty-five-year-old Ada Ballin, the journal was intended to provide information and advice on the management of children, in illness and health, and was aimed toward a predominantly lay and female audience, eliciting and attracting mainly middle-class readers. Among its features were advice on ailments, nutritional recipes, music, poems, and specially commissioned dress designs for babies, as well as a range of lectures on health and medicine, either written for the journal or reproduced from other medical publications. Articles reflected upon controversial and popular topics of the day, from the developmental psychology of children to the nutritional value of infant formula food.[42] Like Burdett, Ballin had received a smattering of medical education but had not qualified or practiced. The Medical Act of 1876 had overturned previous legal barriers to women obtaining medical qualifications, allowing institutions to award them regardless of gender. But obstacles to women's entry into the profession remained, as organizations like the Royal College of Surgeons of England and the British Medical Association continued to refuse them admission, making it difficult for women to build professional careers. Entering University College in 1878, Ballin was, however, able to study with William Corfield, the university's professor of hygiene and public health, piquing her own interest in medicine.[43] A precocious teenager, she had been an ardent proponent of rational dress reform, largely on the grounds that practices such as tight-lacing were detrimental to women's health, and she began lecturing at the tender age of eighteen for the National Health Society, an organization that gave public lectures on preventive health measures. The society was primarily run both by and for women, and Ballin's motivations for becoming involved in health journalism were shaped by her perceptions of gender; she worried deeply about the way in which girls were brought up to be "mere conglomerations of frivolity."[44] Recognizing that it was wives rather than husbands who were engaged in the business of childrearing, she believed women were the most obvious audience for medical and scientific literature on child health. The opening editorial of *Baby* boasted a selection of high-profile contributors, among them the journalist and health lecturer Florence Fenwick-Miller (who had qualified in medicine), Robert Parker, surgeon at the East London Hospital for Children, and Edmund Owen and Catherine Wood, surgeon and matron respectively at the Great Ormond Street Hospital for Sick Children. Other high-profile names were promised for later issues. The dynamics of gender embedded in Ballin's journalistic work were complex and informed by the structural framework of male authority of which medicine and journalism were both part and parcel. In her 1885 book *The Science of Dress in Theory and Practice*, Ballin had blamed the lack of success of the rational dress movement on the ineffectiveness of male writers who lectured women on subjects they themselves were better qualified to address through their own experience.[45] But as Sally Shuttleworth has highlighted elsewhere, most contributions in the first issues of *Baby* came from male doctors and scientists; despite Ballin's wish for health literature created by and written for women, she was, to an extent, reliant on the authority of medical men to help legitimate *Baby*'s place in the medical journal market.[46] Nonetheless, Ballin also carved out space for her female readers, calling upon them to furnish the journal with tips on clothing, feeding, and the general management of children, as well as any new inventions they had crafted to help with child care.[47] Ballin anticipated that *Baby* would bring together "a mother's parliament," a community of readers who would prove to be a resource of expertise on child health, accrued through their collective experience. As the journal's success grew, correspondence both offering and asking for health advice increased, the latter of which saw Ballin increasingly taking on the role of health adviser to her readers.

The *Lancet*'s attack on both *The Hospital* and *Baby* had subtexts of professional vulnerability and pecuniary anxiety. Its claim that they were "quasi-medical" implied a deception, an intentional misappropriation on the latter's part of the medical journal genre.[48] Both publications emulated certain aspects of the content of professional medical journals in a way that was viewed as improper. In the case of *The Hospital*, the *Lancet* disapproved strongly of that journal's publication of details about hospital staff, which it believed would encourage the public to choose for private consultations those doctors written about "to the professional injury of all those whose names do not appear in the list."[49] With *Baby*, the *Lancet* criticized the journal's inclusion of original articles from doctors. It feared that by prominently featuring the names of doctors while soliciting a public audience, both journals would appear as little more than a surreptitious form of advertising on the part of the profession.

Medical journals marketed to the public were, as we have seen, not a new development; many had appeared over the century since Richard Reece had begun the *Monthly Gazette of Health* in 1816. The *People's Medical Journal and Family Physician*, for example, begun in 1850, which lasted two years, promised content on anatomy, physiology, surgery, chemistry, and pharmacy for those who were in the "industrious classes."[50] The *Medical Review and Invalid's Guide*, established in 1872, garnered considerably more success, surviving until 1887. Carefully targeting the chronically unwell, it printed tips on treatment for conditions like varicose veins and indigestion, and information about residences for invalids in Britain and abroad. *The Hospital* and *Baby* were arguably not that distinct from publications like the *Medical Review*, which had similarly brought information on doctors and disease to a general audience. Why then did *The Hospital* and *Baby* receive much greater criticism than their predecessors?[51] There was one obvious, significant factor: neither Burdett nor Ballin was a doctor.[52] This distinguished both *The Hospital* and *Baby* from earlier lay-oriented medical journals, most of which had been edited by doctors, and also from the contemporary sanitary journals, described in chapter 10 of this book, which were edited by high-profile medical men like Ernest Hart and Benjamin Ward Richardson.[53] The perceived lack of medical competence on the part of Burdett and Ballin no doubt crystallized the profession's fears regarding medical information being handled by those not adequately qualified to comprehend or promulgate it.

But the *Lancet*'s castigation of "popular" medical journals also revealed a profession situated uncomfortably within the increasingly tumultuous world of mass print, where in the 1880s commercialism, medicine, and literature were blending together in ever more visible ways, heightening tensions around the ethics of medical publishing. Facilitated by advertising space in newspapers and magazines, consumer culture was reaching into the papers—and pockets—of the literate working classes, and journal advertisements were becoming bigger and more extravagant.[54] Medicinal and health-promoting products were conspicuous in the age of advertising; Lori Loeb has described how an "unprecedented assortment of mass-produced and mass-marketed patent medicines flooded the market."[55] Gaudy advertisements proclaiming the benefits of a host of pills, tonics, lotions, and appliances found ample space in the press.[56] As Loeb has argued, there was a degree of cognitive dissonance in the profession's approach to patent medicines. Ostensibly they condemned them, claiming that the failure of patent medicine vendors to disclose the ingredients of their remedies posed a risk to the public's health. In actuality, many doctors used patent medicines in their practice, some invested in patent medicine companies, and professional journals including the *Lancet* and the *British Medical Journal* carried advertisements for them in order to sustain themselves financially.[57] In the face of the profession's complex relationship with commercialism, the *Lancet*'s interpretation of advertising was adapted to elide any mention of its own entanglement with it, and was instead used to criticize doctors who featured in and wrote for other journals intended for public audiences.

The *Lancet*'s editorial series on "quasi-medical" journals highlighted the complex set of expectations the profession put upon both itself and the public by the 1880s. In the view of the *Lancet* it was entirely permissible for the public to be cognizant of and indeed actively engaged in disease prevention and hygienic practices, and to possess a degree of practical knowledge such as the first aid movement was offering, in the event of sudden illness or injury. That was a form of good citizenship. What was not appropriate was when the public strayed into the territory of medical practitioners by consuming information that could lead them to try and diagnose and treat themselves. This was especially problematic given the audiences that *The Hospital* and *Baby* were targeting; *The Hospital*, priced at just a penny, was aimed at both men and women from the lower classes, while *Baby* was aimed at an almost exclusively female readership. Although not explicitly stated, questions of gender no doubt played into the *Lancet*'s worries. In February 1888 a correspondent to the journal raised concerns about female readers' consumption of medical knowledge through nursing journals. They criticized the recently established journal for nurses and midwives, *Nursing Notes*, for facilitating discussion of anatomical knowledge among nurses that it was not appropriate or necessary for them to learn.[58] This was a discussion pertinent to the era. Similar controversies were raging about the admittance of female medical students to anatomical lectures.[59] And yet, contemporaneously, the public health movement was placing responsibility upon women as the ideal conduits through which to establish hygienic practices within families. Thus, the extent to which women should be be privy to medical knowledge, and the reading matter available to them, remained a contentious issue. The boundaries between journals orientated toward women generally and those involved in health care on an occupational basis could be hazy and were indeed complicated by the periodicals themselves. The *Nursing Mirror* supplement in *The Hospital*, for example, featured a regular column called "The Book World for Women and Nurses," containing reading suggestions "likely to interest women and nurses," while *Nursing Notes* often featured articles relating to the progress of women within the world of paid employment, and operated a lending library which circulated both medical literature and other books and journals that focused on women's role in the workforce.[60] *Nursing Notes* was also published by the Women's Printing Society, which had begun as a direct result of women's exclusion from the printing industry, and which attracted business from women using publishing as a means of advancing a progressive agenda.[61] During a period in which the role of women as social, professional and economic actors was being hotly contested, the overt solicitation of female readers for professional medical literature, nursing or otherwise, implicitly played into the broader context of gender politics.

The *Lancet* cast its critique of popular medical journalism as drawn from a moral standpoint. But it coalesced with concerns about the changing demographics and financial fortunes of the medical profession at large. The economic boom of the mid-century was grinding to a halt. Doctors expressed fears that the profession was becoming overcrowded by young men unable to find positions in business and industry. *The Hospital* and *Baby* were perceived as both a symptom and a cause of increased competition in the medical marketplace. In a climate of commercial depression, the presence of "popular" journals was seen to encourage doctors in financial straits to resort to advertising.[62] But the journals themselves were also viewed as competitors. If the public chose a cheap periodical over a more costly—and potentially harrowing—face-to-face encounter with a professional doctor, they stood to rob doctors of their financial livelihood.[63] One *Lancet* editorial lamented that such journals administered "not medicines, indeed, but medical advice to the public, one and all, for one penny!"[64] The *Lancet*'s invocation of the growing power of the penny press echoed the sentiments of contemporary journalism. In 1886 W. T. Stead, editor of the socially conscious and widely circulated *Pall Mall Gazette*, had published his essay "Government by Journalism."[65] Reflecting on a new era of populist, anti-elitist journalism, Stead passionately evoked the power of the press in shaping society, likening readers to an electorate who "register their vote by a voluntary payment of the daily pence. There is no limitation of age or sex. Whosoever has a penny has a vote."[66] Stead, through his periodical endeavors like *Review of Reviews*, begun in 1890 following the end of his editorship at the *Pall Mall Gazette*, sought to undermine the grasp of the intellectual elite upon expert knowledge—of which science comprised a key component—in a lively and vivid manner.[67] In a not dissimilar way, *The Hospital* and *Baby* offered a significant challenge to the authority of doctors, allowing readers a more active role in shaping their medical experiences at a time when the profession as a whole was attempting to capitalize on its increased scientific and cultural authority.

The *Lancet*'s animosity toward *The Hospital* put its editor, Henry Burdett, in a delicate situation. His years of work in hospital management and nursing reform meant that he held something akin to an honorary position among the British profession, and he often mingled socially with medical men. Their patronage was essential for his hospital charity work.[68] In private correspondence, his friend and confidant across the Atlantic, John Shaw Billings, advised him to rethink his foray into medical journalism. "If you go on with the 'Hospital' on your present plans you will have the great body of the Medical Profession either opposed to you or looking very much askance at you," Billings warned.[69] Burdett appears to have heeded his advice. The early 1890s saw him attempt to take *The Hospital* in a different direction, as he began to market the journal primarily toward doctors and medical students, though he continued to solicit a general audience too, perhaps to ensure its continued commercial viability. The subtitle of the journal underwent a transition. Initially *The Hospital* billed itself as "an institutional, family and congregational journal of Hospitals, Asylums, and all Agencies for the Care of the Sick"; but this changed to "a weekly journal of science, medicine, nursing, and philanthropy" with "Medical Practitioners, Students, Nurses, and the Charitable Public" mooted as the intended audience. This reordering, with doctors' readership prioritized, marked a concerted effort on Burdett's part to establish the journal's identity as a truly medical paper. Regular content such as reviews of medical books, information on new medical devices, and an opinion column called "The Doctor's Armchair" were all designed with a professional audience in mind. Especially striking was *The Hospital*'s attempt to corner the market for medical students and newly qualified doctors, with large portions of its pages given over to job advertisements as well as news of scholarships, prizes, and pass lists.

The *British Medical Journal*, now by far the most read medical journal in Britain, on account of being sent out to members of the British Medical Association as part of their subscription, refuted the notion that changes in style and content meant that *The Hospital* had become a medical journal in any real sense. To Burdett's chagrin, the *Journal* refused to include *The Hospital* in its summary of the circulation numbers of the leading medical journals, ignoring the estimated weekly circulation of more than 9,500 copies that Burdett now claimed for *The Hospital*.[70] In private correspondence with Thomas Henry Wakley (son of Thomas Wakley, who had died in 1862) and Ernest Hart, editors of the *Lancet* and *British Medical Journal* respectively, both of whom cast doubt on the credibility of *The Hospital*, Burdett had attempted to argue his case, but he made little ground with either man.[71] In 1897, the issue still evidently perturbing him, Burdett changed tack and publicly addressed the matter of *The Hospital*'s audience and identity in one of its editorials. In it, he freely admitted to the populist, conversational tone of *The Hospital* compared to those of the *Lancet* and the *British Medical Journal* and, in an effort to style the other two as behind the literary mores of the day, criticized the "fat" volumes of his competitors, replete with "ponderous communications" that the busy doctor had little time or inclination to wade through. Burdett claimed that all three journals were read by a mixed audience of medical professionals and the public—the only difference being that Burdett was prepared to acknowledge this fact. Of the *Lancet* and *British Medical Journal*, he wrote:

> In our view, both these are admirable papers. But is it true that they are read by the medical profession alone? Are they not sold on bookstalls? Are they not to be found on club and hotel tables? Are they not taken at free libraries? Are they not subscribed for by mechanics' institutes, and gloated over by the educated but unwashed youth of the period?[72]

The accusation would have hit a nerve with those in the medical profession. As Sarah Bull has detailed, the middle decades of the century had seen a clampdown upon the selling of obscene literature, in which texts containing sexual advice purporting to be medical were often implicated. Journals like the *Lancet* not only sought to distance themselves from vernacular medical texts that might be construed as obscene but, by the 1870s, were active in sponsoring prosecutions of irregular doctors on the grounds of obscenity.[73] As Bull observes, such actions on the part of the *Lancet* and similar titles were dependent on a claim of demarcation between readers of professional literature "disciplined enough to be unaffected by works that would incite antisocial behavior in others," and lay readers of nonprofessional literature who would be easily corrupted by sexually oriented reading matter.[74] In reality the lines were blurred, and Burdett capitalized on this, brazenly demolishing the idea that the *Lancet* was read only by professionals, instead evoking the possibility that young men read it for its salacious and possibly sexual content. Buoyed by the healthy circulation figures of *The Hospital*, Burdett's position on the periphery of the profession gave him latitude to play with the identity of the medical journal. He was less constrained by the performative rhetoric that compelled doctors to disavow any associations with profit and advertising. By emphasizing the commercial nature of *all* medical journals, Burdett cut through the rigid distinctions between the professional and popular press that the orthodox medical weeklies liked to project. Print culture complicated and challenged professional boundaries, blending audiences and belying any essential difference between doctors and lay people in their consumption of medical literature.

The original controversy involving *The Hospital* and *Baby* in 1887 aroused the attention of the Royal College of Physicians, who swiftly condemned those among their membership who had been involved with either journal. In a move indicative of the power the *Lancet* wielded in professional politics, members and fellows of the college who had contributed to either *The Hospital* or *Baby* were issued with a letter from the college registrar, Sir Henry Pitman, asking them to explain their actions. Many of the responses, particularly from those who contributed to *Baby*, are retained in the Royal College archive along with their correspondence with Ada Ballin. They offer a glimpse of the pressure those involved were under to extricate themselves from a potentially embarrassing situation. Most quickly moved to disassociate themselves from the publication; "As soon as I had seen the first number I at once sent a letter to the Editor informing my disapproval of certain of the contents of the magazine and also stating that the scope of the publication seems to extend further in the way of strictly 'medical' advice than I considered expedient," wrote an obstetrician and fellow of the college, William Graily Hewitt.[75] Some were more sympathetic to Ballin's endeavor, their correspondence revealing the competing demands of professional etiquette and public duty. Thomas Lauder Brunton, a high-profile physician at St. Bartholomew's, as well as one of the most popular consulting doctors among London's society elite, explained to Ballin the difficult position he had been put in with *Baby*. Drawing on his bleak experiences at St. Bartholomew's Hospital, where he frequently encountered children with severe diarrhea caused by unsanitary feeding practices, Brunton had agreed to write for the journal on the correct use of feeding bottles, believing that his contribution to the magazine could lead to a decrease in mortality from the condition. Brunton allowed publication to go ahead, but under pressure from the college he insisted that the article be published anonymously. "This step is necessary because the objectionable practice of advertising seems to be creeping into the Medical Profession in this country," he wrote to Ballin, "and as it is somewhat hard to draw the line between what is legitimate and what is not, extra care requires to be taken."[76] Here then, was the borderland in ethics that the *Lancet* referred to; journalism complicated the process of managing and authenticating medical knowledge. It placed doctors on ethically dubious ground, plunging them into a world of commercialism which contradicted efforts to project an image of the profession as motivated by altruistic concerns.

WHAT COUNTS AS A MEDICAL JOURNAL?

On 2 February 1888, less than three months after the *Lancet*'s first editorial on the subject, the Royal College of Physicians issued a resolution stating that it was "undesirable that any fellow, member, or Licentiate of the College should contribute articles on professional subjects to journals professing to supply medical knowledge to the general public." Notes on the resolution in the Royal College archives confirm that it was passed primarily in reference to *Baby* and *The Hospital*.[77] In a show of significant power, the *Lancet* had managed to shape the medical regulatory framework through the denunciation of its rivals. And yet, despite this, both journals that had come under fire endured—*Baby* until 1915 and *The Hospital* until 1924. The condemnation by the college appeared to have had little effect on diminishing the readership of either. Both journals continued to publish articles authored by doctors, though interestingly, in the case of *Baby*, a decline in male medical contributors was countered by an upsurge in female doctors writing for the magazine. In a literary marketplace where journals so frequently failed, both *The Hospital* and *Baby* can be construed as success stories. Their longevity in comparison to publicly oriented medical titles published earlier in the century attests to an increased demand from the public for cheap, accessible health information, that could be facilitated with relative ease in the expansive world of commercial publishing in the 1880s.

And yet neither *The Hospital* or *Baby* would be recognized today by any but a handful of medical or periodical historians. How then to understand the historical obscurity of these journals despite their obvious relevance? Much can be explained by an enduring association between the growing significance of the medical journal and doctors' professionalizing endeavors, one of these things seemingly reflecting the other over the course of the nineteenth century. It is the titles whose emergence appeared to signal the onset of professionalization in medicine—the *Lancet* and *British Medical Journal*, both of which remain today powerhouses of medical journalism—that continue to occupy center stage in historical investigations. This position is augmented, no doubt, by the digitization of the entire runs of both, which has made them reliable and accessible historical sources, more so than the many other titles for which full-scale digitization is yet to exist.

While no one can doubt the impact of the *Lancet* and *British Medical Journal*, publications like *The Hospital* and *Baby* complicate the condensed narrative of medical journalism that the former represent. Moreover, the existence of Henry Burdett and Ada Ballin's successful journalistic ventures necessitates an unpacking of the categories we use. How have we chosen to define "medical" journals? And how does that definition differ from that employed by our historical actors? In what ways were professional medical journals distinct from other health-related periodicals, and how were those boundaries erected, enforced, blurred, and contested? This chapter has tried to find a way through some of these complex issues, negotiating the vast swaths of medical periodical material available to doctors, nurses, and members of the public in the nineteenth century, of which historians have perhaps only begun to scratch the surface.

The connections and intersections between medical journals, the profession, and the public did not stop with the escalation of medical professionalization. As British medical journalism underwent a transformative period in the latter decades of the century, during which medical and health journals increasingly diversified in objective and scope, conceptualizations of the medical journal became significantly more complex and their relations to commerce and advertising more closely scrutinized. The controversies surrounding *The Hospital* and *Baby* reveal the anxieties of a profession struggling to acclimatize to the voluminous, vibrant and commercial culture of late nineteenth-century publishing, which provided doctors with both professional and economic opportunities but also exposed them to situations in which their propriety could very easily be called into question.

NOTES

1. James Paget, "International Medical Congress, 1881: Opening Address," *New England Journal of Medicine* 105 (1881): 145.

2. John Shaw Billings, "Address on our Medical Literature," *Lancet* 118 (1881): 265.

3. W. F. Bynum and Janice C. Wilson, "Periodical Knowledge: Medical Journals and Their Editors in Nineteenth-Century Britain," in *Medical Journals and Medical Knowledge: Historical Essays*, ed. W. F. Bynum, Stephen Lock, and Roy Porter (London and New York: Routledge, 1992) 30.

4. Billings, "Address on Our Medical Literature," 266.

5. Billings, "Address on Our Medical Literature," 270.

6. Bynum and Wilson, "Periodical Knowledge." In its first issue in 1888, the *British Journal of Dermatology* declared that "it will be allowed that the interests of the special branches of so

complex a science as that of medicine are best developed and maintained when those who are engaged in them have at their disposal separate literary channels for the interchange of ideas." "The Editors' Prologue," *British Journal of Dermatology* 1 (1888): 1.

7. M. Jeanne Petersen, "Medicine," in *Victorian Periodicals and Victorian Society*, eds. J. Donn Vann and Rosemary T. VanArsdel (Aldershot, UK: Scolar Press, 1994), 24.

8. Sally Shuttleworth, chapter 10 in this volume. See also Graham Mooney, *Intrusive Interventions: Public Health, Domestic Space, and Infectious Disease Surveillance in England, 1840–1914* (Rochester, NY: University of Rochester Press, 2015), 2.

9. "The Relation of Medical Ethics to Popular Journalism," *Lancet* 130 (1887): 971.

10. Wakley has been quoted as saying he named the journal the *Lancet* because "a lancet can be an arched window to let in the light or it can be a sharp surgical instrument to cut out the dross and I intend to use it in both senses." As Michael Brown has observed, the quote is apocryphal and its origin unknown. One imagines the latter description would have had more obvious meaning to its audience, given the medical content of the journal. As quoted in Mary Bostetter, "The Journalism of Thomas Wakley," in *Innovators and Preachers: The Role of the Editor in Victorian England*, ed. Joel H. Wiener (Westport, CT: Greenwood Press, 1985), 290. Michael Brown, "'Bats, Rats and Barristers': *The Lancet*, Libel and the Radical Stylistics of Early Nineteenth-Century English Medicine," *Social History* 39 (2014): 188.

11. Preface, *Medico-Chirurgical Review* 2 (1822): i.

12. This is the figure quoted by Samuel Squire Sprigge in his biography of Thomas Wakley. As Michael Brown has noted, Wakley himself claimed that the *Lancet* had a circulation of ten thousand by 1824. Samuel Squire Sprigge, *The Life and Times of Thomas Wakley* (London: Longman, Green and Co., 1899), 102; Brown, "'Bats, Rats and Barristers,'" 183.

13. For more on this, see Carin Berkowitz, *Charles Bell and the Anatomy of Reform* (Chicago and London: University of Chicago Press, 2016), 82–83.

14. "The Medical Times," *Medical Times* 1 (1839): 36.

15. Preface, *Lancet* 1 (1823): 2. The *Lancet* drew much of its antagonistic and populist style from contemporary political reformist periodicals, in particular William Cobbett's *Political Register*. Brown, "'Bats, Rats and Barristers,'" 184.

16. A biographical portrait of Thomas Wakley that appeared in 1839 cited the *Monthly Gazette of Health* as a key influence upon him. "Medical Portraits: Thomas Wakley M.P., Coroner for Middlesex," *Medical Times* 1 (1839): 17.

17. The *Medical Adviser*, for example, demanded an end to female prisoners being subject to treadmills, recently introduced to prisons, claiming that the women were physically unsuited to such arduous labor. "Important Medical Considerations upon the Treadmill," *Medical Adviser* 1 (1823): 1–4.

18. In the context of mesmerism, Jennifer Ruth has explored the somewhat contradictory tone of the journal *The Zoist*, established by the University College Hospital physician John Elliotson, an advocate of mesmerism, in 1843. Ruth notes, "The *Zoist* desperately wanted to be taken seriously by the professional community. Thus, despite its reliance on the words and presumably subscriptions of this reforming segment of the lay public, it strove to distinguish itself from the 'popular.'" Jennifer Ruth, "'Gross Humbug' or 'The Language of Truth'? The Case of the *Zoist*," *Victorian Periodicals Review* 32 (1999): 310. Britain's most successful ho-

meopathic journal, *The Homeopathic World* (established in 1866), while primarily read by the public, appears to have had doctors among its readers and contributors, and increasingly strove to reach out to the medical profession during the course of the century.

19. Roy Porter cites the popularity of William Buchan's *Domestic Medicine* and John Wesley's *Primitive Physick*. Roy Porter, "Lay Medical Knowledge in the Eighteenth Century: The Evidence of the *Gentleman's Magazine*," *Medical History* 29 (1985): 140.

20. "To the Editor of the Gazette of Health," *Monthly Gazette of Health* 1 (1816): 110.

21. "The Penny Press," *The Mechanic's Magazine* 18 (1832): 93.

22. However, titles differed in their treatment of the various 'alternative' systems of medicine. In its first years, for example, the *Medical Times and Gazette* openly and somewhat controversially supported John Elliotson in his practice of mesmerism.

23. "Hospital That Beats Bedlam," *Lancet* 57 (1851): 560; "'Punch on Quackery," *Lancet* 64 (1854): 496.

24. S. D. Smith, "Coffee, Microscopy, and the *Lancet*'s Analytical Sanitary Commission," *Social History of Medicine* 14 (2001): 183.

25. "The Analytical Sanitary Commission: Notices of the Press," *Lancet* 57 (1851): 228. Three similar reports on press coverage of the commission appeared in the *Lancet* that year.

26. Richard Garnett, "Hunt, Frederick Knight (1814–1854), journalist." Oxford Dictionary of National Biography, accessed 20 February 2018. http://ezproxy-prd.bodleian.ox.ac.uk:2167 /view/10.1093/ref:odnb/9780198614128.001.0001/odnb-9780198614128-e-14191.

27. These were some of the issues discussed in Wynter's popular collection of essays *Subtle Brains and Lissom Fingers*, originally published in 1863, which addressed a range of subjects, medical, scientific, and otherwise. Andrew Wynter, *Subtle Brains and Lissom Fingers: Being Some of the Chisel-Marks of Our Industrial and Scientific Progress* (London: Robert Hardwicke, 1863).

28. Peter Bartrip has noted that Wynter resigned his editorship of the *British Medical Journal* following criticism that the journal was out of touch and inaccessible to members of the British Medical Association, for whom the journal was intended. P. W. J. Bartrip, "Wynter, Andrew (1819–1876)," Oxford Dictionary of National Biography, Oct 2007, http://www.oxforddnb .com/view/article/30163, accessed 17 May 2017.

29. Claire Furlong, "Health Advice in Popular Periodicals: *Reynolds's Miscellany,* the *Family Herald*, and Their Correspondents," *Victorian Periodicals Review* 49 (2016): 35. Roy Porter identified this practice as one also found within eighteenth-century periodical culture, pointing to the prominent place of health advice within the highly successful *Gentleman's Magazine*. Porter, "Lay Medical Knowledge in the Eighteenth Century."

30. "How to Poison Correspondents," *Lancet* 75 (1860): 72.

31. "How to Poison Correspondents," 95.

32. Furlong, "Health Advice in Popular Periodicals," 36.

33. "The Relation of Medical Ethics to Popular Journalism," 971.

34. "Popular Medical Teaching," *Lancet* 130 (1887): 1124.

35. Frank Prochaska, "Burdett, Sir Henry Charles (1847–1920), philanthropist and hospital reformer." Oxford Dictionary of National Biography, http://ezproxy-prd.bodleian.ox.ac.uk :2167/view/article/38827, accessed 18 May 2017.

36. Steve Sturdy and Roger Cooter, "Science, Scientific Management, and the Transformation of Medicine in Britain c. 1870–1950," *History of Science* 36 (1998): 425.

37. Ironically, these were similar to criticisms leveled at the *Lancet* in its very early days when the journal, often carried nonmedical articles such as theater criticism and chess puzzles. Brittany Pladek, "'A Variety of Tastes': The *Lancet* in the Early-Nineteenth-Century Periodical Press," *Bulletin of the History of Medicine* 85 (2011): 573.

38. "Popular Medical Teaching," 1124.

39. Gowan Dawson, "The *Review of Reviews* and the New Journalism in Late-Victorian Britain," in *Science in the Nineteenth-Century Periodical*, eds. Geoffrey Cantor, G. Dawson, G. Gooday, R. Noakes, S. Shuttleworth, and J. Topham. (Cambridge: Cambridge University Press, 2004), 186.

40. "Teaching and Touting," *Lancet* 130 (1887): 1028.

41. "Tittle-Tattle for the Tea Table," *Pall Mall Gazette* 48 (1888): 7. Burdett cited these statistics in the headed paper he used for correspondence relating to *The Hospital*; correspondence of Henry Burdett, MS 5966, Bodleian Library, University of Oxford.

42. In the mid- to late nineteenth century, psychiatrists were beginning to develop theories about the mind of the child, in separation from more general psychiatric theories. Sally Shuttleworth has described the final decades of the century as a time when "'the mind of the child' became not only a scientific discipline but also almost a cultural obsession." Sally Shuttleworth, *The Mind of the Child* (Oxford, UK: Oxford University Press, 2010), 16. Similarly, with the advent of mass-scale production of artificial infant formula in Europe and America, doctors were becoming involved in debates about the relative merits of breastfeeding and formula milk, and the subject was often discussed in *Baby*. Rima D. Apple, "'Advertised by Our Loving Friends': The Infant Formula Industry and the Creation of New Pharmaceutical Markets, 1870–1910," *Journal of the History of Medicine and Allied Sciences* 41 (1986): 4–9.

43. Anne M. Sebba, "Ballin, Ada Sarah (1862–1906), Magazine Editor and Proprietor, and Writer on Health." Oxford Dictionary of National Biography, https://ezproxy-prd.bodleian.ox.ac.uk:4563/10.1093/ref:odnb/55732, accessed 20 June 2019.

44. "Editorial Address," *Baby* 1 (1887): 12.

45. Ada Ballin, *The Science of Dress in Theory and Practice* (London: Sampson Low, 1885), iii.

46. Shuttleworth, *The Mind of the Child*, 286.

47. "Mother's Parliament," *Baby* 1 (1887): 11.

48. "Mother's Parliament," 11.

49. "Popular Medical Teaching," 1124.

50. "Announcement," *The People's Medical Journal and Family Physician* 1 (1850): 1.

51. Despite its relatively long run, the *Medical Review and Invalid's Guide* only briefly caught the attention of the *Lancet* when, in the back pages of an 1883 issue, the latter complained about the inclusion in the *Medical Review* of a list of medical consultants, which included "amateur doctors." "A List of Consultants," *Lancet* 121 (1883): 760.

52. However, Henry Burdett did receive editorial assistance from a doctor, George Potter, MD.

53. Sally Shuttleworth, Chapter 10 in this volume.

54. Peter Gurney, *The Making of Consumer Culture in Modern Britain* (London: Bloomsbury Academic, 2017), 70–72.

55. Lori Loeb, "Doctors and Patent Medicines in Modern Britain: Professionalism and Consumerism," *Albion* 33 (2001): 409.

56. "Popular Medical Teaching," 1124.

57. Loeb, "Doctors and Patent Medicines in Modern Britain," 423.

58. W. H. Allchin, "The Teaching of our Nurses," *Lancet* 131 (1888): 345.

59. For more on the intersecting experiences of female doctors and nurses, see Vanessa Heggie, "Women Doctors and Lady Nurses: Class, Education and the Professional Woman," *Bulletin of the History of Medicine* 89 (2015).

60. Periodical titles the lending library offered included *Work and Leisure*—published by the social reformer Louisa Hubbard, who was an advocate of female employment—as well as a range of literary and women's magazines including *Literary World* and *Lady*.

61. Michelle Elizabeth Tusan, "Performing Work: Gender, Class, and the Printing Trade in Victorian Britain," *Journal of Women's History* (2004): 120.

62. "The Relation of Medical Ethics to Popular Journalism," 971.

63. "Popular Medical Teaching," 1125.

64. "The Relation of Medical Ethics to Popular Journalism," 971.

65. The medical profession had been touched by Stead's controversial brand of journalism when it became mixed up in one of his most notorious pieces of investigative work, "The Maiden Tribute of Modern Babylon," which purported to expose the trade in child prostitution in London. As part of his investigation, Stead "purchased" a thirteen-year-old girl, Eliza Armstrong, to show the ease with which girls could be bought and sold. To avoid being accused of sexually assaulting her, Stead brought in the obstetrician Heywood Smith to confirm the girl's virginity through a vaginal examination. Smith was heavily criticized by his fellow doctors for his role in the scandal, and only narrowly avoided being expelled from the Obstetrical Society. "The Royal College of Physicians and Dr. Heywood Smith," *Lancet* 126 (1885): 1209–10; "Obstetrical Society of London," *Lancet* 127 (1886): 255–56.

66. As quoted in Graham Law and Matthew Sterenberg, "Old v. New Journalism and the Public Sphere; or, Habermas Encounters Dallas and Stead," *19: Interdisciplinary Studies in the Nineteenth Century* 16 (2013): 9, accessed May 16, 2017, doi: http://www.19.bbk.ac.uk/articles/10.16995/ntn.657/; W. T. Stead, "Government by Journalism," *Contemporary Review* 49 (1886): 655.

67. Dawson, "*The Review of Reviews*," 187–88.

68. "Obituary: Sir Henry Burdett K. C. B., K. C. V. O.," *British Medical Journal* 1 (1920): 657.

69. Letter to Henry Burdett from John Shaw Billings, 4 March 1888, MS 5966, no. 10, Bodleian Library, University of Oxford.

70. *British Medical Journal* 1 (1896): 36; letter from Henry Burdett to Ernest Hart, 7 January 1895, no. 120, MS 5966, Bodleian Library, University of Oxford.

71. Letter to Thomas Henry Wakley, 12 November 1895, no. 117; correspondence between Henry Burdett and Ernest Hart, 7–9 January 1896, nos. 120–24, MS 5966, Bodleian Library, University of Oxford.

72. "'The Hospital to its Readers," *The Hospital* 22 (1897): 2.

73. Sarah Bull, "Managing the 'Obscene MD.': Medical Publishing, the Medical Profession, and the Changing Definition of Obscenity in Mid-Victorian England," *Bulletin of the History of Medicine* 91 (2017): 737.

74. Sarah Bull, "Managing the 'Obscene MD,'" 726.

75. Letter from William Graily Hewitt to Henry Pitman, 2 December 1887, MS 2412/40, Royal College of Physicians.

76. Letter from Thomas Lauder Brunton to Ada Ballin, 28 November 1887, MS 2412/35, Royal College of Physicians.

77. A medical article authored by a doctor which had appeared in *The Modern Review* was also cited as a secondary factor. Royal College of Physicians resolution, 1888. RCP/LEGAC/2410/50, Royal College of Physicians.

CHAPTER 10

"National Health Is National Wealth": Publics, Professions, and the Rise of the Public Health Journal

Sally Shuttleworth

The launch of a new journal frequently tempts its editor into hyperbole and overly ambitious statements of need and purpose. For the first number of the *Journal of Public Health and Sanitary Review* in 1855, its editor and proprietor, Dr. Benjamin Ward Richardson, went one step further, creating a motto, "National health is national wealth," which was emblazoned on the title page and went on to become a reforming cry for the sanitary and public health movements more generally, as they emerged in the latter part of the century (fig. 10.1). Although there has been considerable work on Victorian sanitary movements, little attention has been paid to the many sanitary and public health journals that sprang up in this period, and the role they played in the development of these campaigns and communities.[1] Following on from Sally Frampton's discussion of medical journals in chapter 9 of this book, this chapter looks at the rise of public health and sanitary journals, from the launch of the *Journal of Public Health and Sanitary Review* in the 1850s to the *Journal of the Sanitary Institute* in the 1890s, which carried detailed reporting of what by then had become huge annual congresses devoted to issues of public health. Such congresses brought together broad constituencies of both professional and lay participants, from engineers, chemists, and medical officers of health to teachers, local councillors, and ladies' sanitary associations. In this chapter I track the ways in which journals facilitated the growth of both the public health movements themselves and the associated professional groupings and wider communities involved in campaigns for the improvement of "national health," from the control of cholera epidemics to the monitoring of air quality and the introduction of horticulture for women.

THE

SANITARY REVIEW,

AND

Journal of Public Health.

EDITED BY

BENJAMIN W. RICHARDSON, M.D.,

LICENTIATE OF THE ROYAL COLLEGE OF PHYSICIANS,
PHYSICIAN TO THE ROYAL INFIRMARY FOR DISEASES OF THE CHEST, AND LECTURER
ON PUBLIC HYGIENE AT THE GROSVENOR PLACE MEDICAL SCHOOL.

NATIONAL HEALTH   IS NATIONAL WEALTH.

ὙΓΙΕΙΑ

πρεσβίστη μακάρων.

VOL. IV.

PRINTED AND PUBLISHED FOR THE PROPRIETOR BY

THOMAS RICHARDS, 37, GREAT QUEEN STREET.

1858.

FIG. 10.1. Title page of *The Sanitary Review and Journal of Public Health* 3 (1858). The title of the journal was reversed after the first two years. Courtesy of the Wellcome Collection.

Public health journals in the nineteenth century played a crucial role in helping to forge new academic disciplines like epidemiology. They also supported the development of new professional and occupational groupings, such as the medical officers of health and sanitary inspectors, giving them a sense of communal identity and purpose while also helping to define the appropriate language and subject matter of their callings. Richardson's *Journal of Public Health*, for example, was directly bound up with the development of the London Epidemiological Society, and with the emergence of the professional role of medical officer of health, while the later *Journal of the Sanitary Institute* promoted the professional training of sanitary officers.

In many of its forms, the public health journal also attempted to bridge the worlds of the professional and the lay reader, enabling concerned medical professionals and sanitarians alike to extend their messages of health reform to a broader public. Although these journals overlapped in many areas with the campaigning aspects of various medical titles, most notably the *Lancet* and the *British Medical Journal* under Ernest Hart, their remit was both narrower—solely public health issues—and also broader, in their capacious definitions of what might come within the purview of public health. As Tina Choi Young has commented, "It is difficult to overestimate the elasticity of sanitary reform as a set of concerns and practices."[2] Public health and sanitary journals enabled this multiplicity to flourish, offering a forum for discussion of issues as diverse as diet, education, and smoke and water pollution. Together with the local sanitary associations they often supported, they sowed the seeds for the environmental campaigning of the present day.

Given the medical opposition to the "quasimedical" journals, the *Hospital* and *Baby*, noted by Frampton in the previous chapter, it is perhaps surprising that the public health journals that engaged with a general readership did not attract the same forms of medical censure. In part this can be explained by the fact that the editors were usually medical men, such as Richardson, or Ernest Hart, editor of the *British Medical Journal*, who was also the first editor of the popular *Sanitary Record* (1874–). High-profile figures such as Sir Henry Acland, FRS (Regius Professor of Medicine at Oxford from 1857) gave their support, as well as eminent figures in other fields, from the meteorologist G. J. Symons, FRS (registrar of the Sanitary Institute from 1880 to 1895) to the sanitary engineer Sir Douglas Galton, FRS. The Sanitary Institute and the National Health Society (with which Ernest Hart was closely involved) also managed to acquire aristocratic patronage to ward off criticism and garner support for their campaigns, with the Queen herself becoming a patron of the

Institute in 1882.[3] Perhaps most significantly in this regard, public health and sanitary journals were usually not focused entirely on medicine, but ranged widely across other scientific fields and social concerns, and so were not construed necessarily as a direct challenge to medical authority (though many of the more unorthodox publications attracted criticism in their own right from sections of the general periodical press).

It is difficult to determine the exact numbers of public health and sanitary journals, given the uncertain boundaries with fringe medicine in the domain of "health"; but, taking a fairly inclusive approach, the number is in the region of seventy to one hundred, with the vast bulk emerging from the 1860s onwards.[4] As with other periodical forms, the history of these journals is marked by uncertain starts, amalgamations of titles, and high mortality rates, though some titles have survived as professional journals (albeit much transformed) into the present day, with accompanying alterations of title which themselves map changing social and cultural attitudes and professional identities. For example, *Public Health: The Journal of the Society of Medical Officers of Health* (1888–) is still published, but with the subtitle *Journal of the Society of Community Medicine.* The *Sanitary Record* (1874–) went through many titles until it ended up in the late twentieth century as *Municipal Engineering.*

Richardson's journal is generally seen as the first public health journal,[5] though it did have various short-lived predecessors, including most pertinently the *Journal of Public Health and Monthly Record of Sanitary Improvement* (1847–49). This journal was published, as it announced, "under the sanction of the Metropolitan Health of Towns Association," and was edited by John Sutherland, MD, a campaigning sanitarian.[6] The establishment of the Health of Towns Association in 1844, following the creation of the Health of Towns Commission chaired by Edwin Chadwick, was part of the context for the passing of the 1848 Public Health Act and the rise of public health journals and campaigns from the mid-century onwards.[7] When Richardson decided to start his journal, he was doing so in light of intense public interest in questions of public health, particularly in cities. The work of William Farr, at the General Register Office, on statistics of health and disease from the late 1830s onward, together with Edwin Chadwick's *Report on the Sanitary Condition of the Labouring Population of Great Britain* (1842), created an appetite, at least in some quarters, for further statistics and greater understanding of the relationship between locality, social conditions, and problems of health and disease.[8] It also introduced the term "sanitary" into the cultural lexicon.[9] On a more popular front, literary works like Gaskell's *Mary Barton* (1848) and

Dickens's *Bleak House* (1852–53), with their portrayals of life in the slums, and Henry Mayhew's blistering letters in the *Morning Chronicle* of 1849–50 (subsequently published as *London Labour and the London Poor*), all galvanized public interest in what came to be known as sanitary issues.[10] Such interest was intensified by the cholera epidemics of 1849 (a catalyst for Mayhew's reports) and 1853, as well as the reports of Florence Nightingale from the Crimean War (still ongoing when the *Journal of Public Health* was launched) on the colossal loss of life through lack of sanitary knowledge.

In introducing his new journal, Richardson laid stress on its social utility:

> The *Journal of Public Health* is issued under the earnest conviction that it is a work of which the English nation has much need. During the past ten or fifteen years a new fact has opened on the English mind. *The national wealth is in a great degree dependent on the national health.* This fact, which some less civilized nations have been acquainted with for many long ages past, but which has so lately been rightly understood and valued here, is already accomplishing results, which cannot in the end fail to add greatly to the happiness of our homes, to the health of all classes of the community, and to our prosperity as a great people.[11]

Interestingly, the motto of the title page is here reversed, and the radical implications somewhat tempered, although the marked irony of "less civilised nations" indicates a highly critical voice. Where "National Health is National Wealth" could imply that true riches lie in improving the health of the people for its own sake, the formulation here directly links health to wealth creation and "our prosperity as a great people." Reading through the journal, however, this particularly jingoistic formulation and sentiment is less in evidence; though there is, for instance, concern about the sanitary conditions for imperial troops, the overwhelming focus is on improving environmental and health conditions for city dwellers, without any overt reference to economic efficiency. In this introduction, Richardson celebrates the fact that science and the understanding of physical laws are at last entering both parliamentary legislation and the practice of medicine. "Preventive" medicine, he argues, is emerging as a principle, with its seven elements "Pure air—Proper nourishment—A regulated temperature—Bodily exercise—Cleanliness—Mental Education—Good morals."[12] The sanitary agenda is here encapsulated, but again the moral proselytizing strain—indicated by that final reference to "good morals," placed in such alarming proximity to "cleanliness"—does not generally emerge within the articles themselves. There is no doubt a level of secular fervor in the pursuit of the principles of preventive medicine, but the journal does not push any simplistic line associating dirt and poverty with low morals, as can be found in many of the religious sanitarian tracts.

Dickens, in creating those figures of "rapacious benevolence," Mrs. Jellyby and Mrs. Pardiggle in *Bleak House*, established an indelible cultural association between the distribution of sanitary and health tracts and monstrous, patronizing self-complacency, coupled with a complete disregard for the actual conditions and health and welfare of the poor.[13] Scholarly work on the sanitarians over the last thirty years in a largely Foucauldian vein has offered a similarly negative though very differently framed interpretation, assimilating Victorian work in sanitation and public health to technologies of surveillance and control. Peter Stallybrass and Allon White, for example, in their Bakhtinian analysis, explored how the hierarchy of the body was transcoded onto that of the city, with the sewers, slums, and their inhabitants linked to ideas of dirt, waste, and the lower bodily regions.[14] In *Making a Social Body* (1995), Mary Poovey drew on Chadwick's work to show how regulation of the individual body was linked to the "professionalized, bureaucratized apparatuses of inspection, regulation, and enforcement that we call the modern state."[15] More recent work in both veins can be seen in William Cohen and Ryan Johnson's edited collection, *Filth: Dirt, Disgust and Modern Life* (2004), or in Graham Mooney's *Intrusive Interventions* (2015).[16]

While such work has been of immense value, it has tended to lump all sanitary activities together, and hence to ignore the diverse constituencies and interests that came under this broad umbrella. In exploring the development of the public health journal, this chapter will help to restore some of that complexity and bring to the fore forgotten voices and elements of social and environmental campaigning that have been submerged under the overarching sanitary label. As with the other areas of science considered in this volume, it will explore the role played by the journals in bringing together, and indeed helping to create, both professional and amateur communities. At a time when public campaigns for improvements in social health and environments were just emerging, the journals offered a space both for the recording of data and for the development of new audiences and strategies. The new science of epidemiology, based on the large-scale collection of medical data, first came into being in the pages of these journals, while they also helped to articulate and refine the agendas of campaign groups and the new professional roles emerging in the field of public health.

THE JOURNAL OF PUBLIC HEALTH AND SANITARY REVIEW (1855-59): EPIDEMIOLOGY, DATA GATHERING, AND MEDICAL OFFICERS OF HEALTH

In concluding the introduction to his new journal, Richardson observed that improved knowledge was of "limited use, unless it is largely distributed." The aim of the journal was thus "to supply the reader with sound information on Health subjects, written in simple language, and published in such a form that all who are really anxious to possess it may obtain it without difficulty."[17] There is a hesitancy here as to the nature of the audience addressed: the seemingly admirable commitment to public dissemination, and hence to plain language, is checked by a wariness, perhaps an unwillingness to be associated with the hierarchy and condescension linked with the mass distribution of tracts. Those who wish to access the information may do so, but only if they are really anxious and determined. In format, the journal is neither a dry compilation of statistics nor a didactic form of advice literature. Rather, it is written for experts and interested lay people within the field, but with what might strike us now as a rather unusual blend of material—with poems and ancient history, for example, that reflected Richardson's own literary and historical interests,[18] scattered amid reports of the progress of epidemics, or forms of occupational disease.

Before looking in depth at some of the journal's content, it will be helpful to offer a little background on Benjamin Ward Richardson (1828–96) at this period. On his death he was acclaimed as one who was "better and more widely known than any other living medical man," but he has tended to be relegated in recent scholarship to a footnote to other peoples' lives.[19] Richardson was not from a privileged background, and had entered medicine by the traditional route of being apprenticed to a surgeon (subsequently obtaining his MD from St. Andrews in 1854). In 1849 he moved to Mortlake to work in partnership with Robert Willis, librarian of the Royal College of Surgeons, who quickly introduced him into the medical world of London. He subsequently moved to central London in 1854, combining over the next few years work at various dispensaries with appointments as lecturer at the Grosvenor School of Medicine, and as physician to the Royal Infirmary for Diseases of the Chest. From the early 1850s he was giving papers at meetings of the Medical Society of London, the Physiological Society, and the Epidemiological Society (which was founded in 1850), and was writing for the *Association Medical Journal* (the forerunner of the *British Medical Journal*), the *Medical Times and Gazette*, and the *Lancet*.[20] He was also introduced to more literary circles by Douglas Jerrold, joining a social group known as "Our Club," whose members included Dickens and Thackeray. Through this circle he became a lifelong friend of the illustrator George Cruikshank and also George Godwin, the architect and editor of *The Builder*, a journal which, as Ruth Richardson has rightly pointed out, played a major role in highlighting issues of public health.[21]

Richardson had also been directly caught up in epidemiological work in 1853 when cholera had broken out in Mortlake. Together with his neighbor, the anatomist Richard Owen ("a man full of energy in sanitation"), he set out to investigate its sources, founding the West Surrey Cholera Society.[22] Richardson's work on epidemiology, and also on forms of anesthetics, brought him into close association with John Snow, who often shared his laboratory, and he was to champion Snow's work and theories in the pages of the *Journal of Public Health*. Two further figures from this early period of immense influence on the journal were Edwin Chadwick and William Farr—both were to remain lifelong friends, and Richardson a devoted disciple. He was Chadwick's executor, and published a biographical memoir and condensed collection of his work, fittingly titled *The Health of Nations*, after his death. Richardson also dedicated one of his own books, *Diseases of Modern Life* (1876), to Farr.[23] The *Journal of Public Health* follows the agendas of both men in its belief in the power of statistics and evidence in the pursuit of sanitary goals.

With the abolition of the General Board of Health in 1854, Chadwick had stepped down from public office. Scholarship has tended to ignore his subsequent work, yet he continued to campaign vigorously on diverse areas of public health, from education to parks and gardens, and served as the first president of the Association of Public Sanitary Inspectors from 1883 to 1890, to be followed in office by Richardson himself.[24] Many of the key issues of the public health debates in the journals over the ensuing decades bear Chadwick's stamp. For Richardson, he was a man who "had not been properly understood, and, in fact, had some of his noblest ideas perverted."[25] Chadwick's reputation has inevitably been tainted by his role as architect of the 1834 Poor Law, and by some of the language and assumptions of the *Sanitary Report*, but in following through the discussions and debates of the public health journals from the 1850s to the 1890s, it is possible to construct a more complex picture of his legacy.

It was not unusual for ambitious young medics to undertake a spell of editing early on in their careers,[26] but when Richardson convened a meeting at the house of the founder and president of the Epidemiological Society, Dr. Benjamin Guy Babington, to establish the *Journal of Public Health and San-*

itary Review, he offered himself as both editor and proprietor. As a venture it was made less risky, however, by his association with the Epidemiological Society. The society agreed that he would publish their *Transactions* as an addendum to the *Journal*, and that each member would purchase a copy (thus supplying further evidence of Csizar's argument, in chapter 3 of this volume, of the rising interest of societies in making their papers and discussions open to a wider readership).[27] Richardson thus played a role in the establishment of epidemiology as a discipline in England. The *Journal* appeared monthly, moderately priced at one shilling.[28] After two years the title was reversed, and it became the *Sanitary Review and Journal of Public Health*. The title change was made apparently at the suggestion of the publisher and owner of the *Westminster Review*, John Chapman; this seems to tie Richardson and the journal further into the radical circles of literary London.[29] There was, however, no discernible change in mission or material, and one can only presume that, rather surprisingly to a modern reader, the newly coined term "sanitary" was felt to offer a more modern and attractive package.

The journal ran until 1859, when pressures of work led to Richardson's decision to cease publishing. The first volume of the *Journal of Public Health* set the pattern for the following volumes, with editorials and original articles intermixed with accounts of sanitary conditions in selected towns, details on the progress of epidemics drawn from the registrar general's quarterly reports, book reviews, and digests of relevant news and society meetings, together with various extracts from other journals such as the *Journal of the Society of Arts* and the *Builder*. It clearly attained quite a wide readership, as Richardson records in his autobiography his gratification at hearing that two copies had been taken by the Royal Library, and that Prince Albert had really liked his article "Health of the English Soldier."[30] The articles themselves ranged from discussions of hygienic rules for tropical climates to the value of gymnastics as an essential part of education, and from work on the diseases of colliery operatives to accounts of epidemics before Hippocrates and of the sanitary regulations of ancient Rome. The latter two articles were in line with Richardson's belief that the study of epidemics, and of medicine more generally, required detailed data from the past and also a sense of historical trajectory. The volume opened with a lengthy article called "Sanitary and Social Conditions of the English Poor," by a "Practicing Physician"; it addressed the same area of concerns as Chadwick, but with the emotive reporting style of Mayhew or Dickens. The focus was on overcrowding, both in agricultural dwellings and in city slums, ending with a rather melodramatic claim, following the writer's visit to a graveyard: "Were the choice offered me to be forced either to live in the den I had previously visited, or to lie in cold oblivion amongst the sleepers around me, I should prefer the last named alternative, with the deep quiet earth for my bed, and the waving boughs for my canopy."[31] Although the tone of other articles was more restrained, the romance of graveyards featured quite prominently in the journal. As in Chadwick's *Sanitary Report*, there was considerable interest in mortality rates, but in place of Chadwick's dry figures, the volume offered a series of articles by Dr. John Webster on his visits to graveyards across Britain, and the evidence of longevity gleaned from gravestones. These included an article titled "The Patriarchs of Pinner," in which Webster concluded that the numbers of centenarians demonstrated that Pinner in Middlesex had been a remarkably healthy place to live.[32]

The majority of the signed articles were by medical men, although there was a smattering of contributions by clergy, such as the Reverend Charles Girdleston, whose "On the Scientific Investigation of Sanitary Questions" contained the suggestion that households should acquire gauges to measure atmospheric impurities, thus enrolling the public in the monitoring of environmental pollution.[33] One of the key concerns of the journal was the role of the medical officer of health. Following the successful appointment of John Simon as the first medical officer of health (MOH) for the City of London in 1848, the Metropolis Local Management Act of 1855 called for the appointment of an MOH for every London district (with forty-seven appointed in 1856). Richardson's journal was thus part of the ferment of debate, and indeed jockeying for position, that accompanied this development (and Richardson himself did stand unsuccessfully for election).[34] The first issue carried the initial installment of "Short Notes on Some of the Details of Sanitary Police," by Robert Druitt (who was subsequently elected and later became president of the Metropolitan Association of Medical Officers of Health, as well as editor of the *Medical Times and Gazette*).[35] Druitt opened with what he termed a "burst of etymology," to demonstrate to his readers that the term police should be reclaimed from vulgar ideas of thief catching, to an understanding of a "liberal and ennobling" science, dedicated to the public good.

Richardson himself also contributed to discussions of the role and purposes of the MOH. In "Medical Police of London" he outlined the areas of expertise and personal skills the successful applicant should possess. The passage is worth quoting in full, since it also encapsulated the subject matter and range of the journal itself:

He should have a knowledge of practical chemistry; of the symptoms, causes, and treatment of disease; of physiology, or the laws of life; of pathology, or the laws of morbid action; of forensic medicine, or the connexion that exists between the science of medicine and the law; of various matters of a mechanical kind, bearing on sewerage, house-building, street-cleansing, and water supply; of meteorology, or the effects of climate, weather, and atmospheric influences on the body; of the physical characters or dynamics of the atmosphere; of ventilation; of statistics of life and mortality; of the literature of epidemics, and of all sanitary improvements. Lastly he should possess sound logical faculties, so that in dealing with facts and opinions he may neither mistake coincidences for causes, nor build up great theories on insignificant data, nor from great facts deduce absurd conclusions.[36]

The list was demanding indeed, but it offered a clear sense of the ambitions and agenda of the emerging sciences and professional bodies within the sphere of preventive medicine. It also rightly signaled necessary caution regarding the implementation of the new science of statistics. Richardson's prescriptions aroused considerable interest, and were taken up by the Board of Health in their instructions to those applying for the posts.[37] Richardson went on to suggest that when appointed, the medical officers of health would form a wonderful scientific council, which, together with the Epidemiological Society, would be well placed to advise the government. His vision was that of a close and growing interchange between this new field of experts and the making of government policy. In the event, the MOH met the following year and decided to form an independent organization under the presidency of John Simon. Tellingly, although they published annual reports, and *Transactions* from 1875, they did not start their own journal, *Public Health*, until 1888—a lag that speaks volumes about the difficulties of starting and maintaining a journal amid the crowded market that was emerging at this time.[38]

In concluding his article, Richardson offered a grandiose vision of the contribution the MOH could offer to medicine through the writing of reports: "A yearly report of the sanitary state of all London, drawn up by scientific and independent men, would in twenty years throw more light on the general causes of disease, and on the principles of prevention, than all the stray medical writings on these subjects which have appeared since the days of Hippocrates."[39] Allowing for the rhetorical flourish, the observation was well founded, giving voice to the excitement felt in these early days of epidemiology at the possibilities offered by the collective investigation of disease. It was duly decided that each MOH should individually produce an annual report, thus creating an extraordinary wealth of data that has since proved invaluable to historians and epidemiologists.[40]

Richardson shared with Druitt, Farr, Chadwick, and others an enthusiastic vision of the potential of coordinated data gathering in the field of health. He used the journal as a basis for setting up his own nationwide network of investigation of disease, to supplement the records of the registrar general, which focused solely on mortality. Richardson's article in the January 1858 issue of the journal, "The Registration of Disease in England," describes how the journal, when it commenced, "took as one of its leading objects the collection of the histories of epidemics from various observers."[41] Richardson used the journal as a means of recruitment of his "staff of observers" from the medical profession, and also as a conduit for publishing the results, thus paralleling practices in societies and journals in the natural sciences, explored elsewhere in this volume. By 1858, Richardson had created a network of forty medical observers who recorded not only the instances of human disease but also animal and plant diseases, along with meteorological conditions.[42] Ideally, he hoped to extend his system to include all 3,233 parochial medical officers in England and Wales, arguing that the weekly reports they already produced for their local boards could be easily transformed into "really scientific registers of diseases," if the information were collected more systematically and then collated at a national level. Richardson was working singlehandedly, however, with no organizational infrastructure to support him. Inevitably the labor proved too much, and his application to Benjamin Hall at the Board of Public Health for governmental support was rebuffed, with the observation that his plans, though laudable, were a century ahead of their time.[43]

Richardson nonetheless continued to champion the cause for the next forty years. In this early work he took his model from astronomy, which also involved what he termed the "fleeting" phenomena of nature: "If we would unravel the mysteries of either class of these phenomena, we must do so by the agency of a multitude of minds. . . . In the investigation of the fleeting phenomenon, the single individual becomes a nullity; and the difficulty of discovering its nature can only be overcome by bringing the whole competent mental strength of the time to bear upon it."[44] Richardson's vision was that of concerted large-scale observations conducted across time, considering the phenomena of disease both when they were absent as well as present: "We have often thought that if the eclipses of the sun or moon had always been observed in the same manner as physicians observe transient diseases, the nature of an eclipse would have remained as yet an unsolved problem."[45] Epidemiology was to follow the lead of astronomy, bringing together the individual records of multitudes of observers which, once amalgamated, could help to build understanding of the laws governing the progress of epidemics, both past and present. Although the experiment did not yield lasting results, given the volume of records which would have to be collated by one individual, the journal played a role in articulating the emerging principles of epidemiology, drawing attention to the need to think historically when considering the spread of epidemics. It was also well ahead of its time in calling for coordinated records of diseases across human, animal, and plant life in order to consider the potential intersections of disease across species barriers, and within wider environmental contexts.

In addition to obtaining and coordinating field observations from its readers, the journal also took on the role of public advocacy, championing in particular the work of John Snow, commonly labeled the "father of epidemiology," whose theories with reference to the causation of cholera were at that point still highly controversial. The first issue of the *Journal of Public Health* closely followed the outbreak of cholera in 1854, and Snow's now legendary intervention, with reference to the Broad Street pump, when he suggested that the local outbreak could be solved by removing the handle of the pump which supplied the street with water. Yet, at the time, scientific and public health specialists were by no means convinced of Snow's arguments regarding the waterborne nature of cholera. Richardson contributed to the June 1855 issue a signed review article, "Water Supply in Relation to Health and Disease," which focused primarily on Snow's *On the Mode of Communication of Cholera* (1855). It initially offers strong support of Snow's argument, while later slightly distancing itself, clearly trying to mollify unbelievers (of which Chadwick was one) by referring to Snow's "peculiar views" and suggesting that cholera could be both waterborne and also "wafted into the lungs by the medium of the air."[46] Numerous other supportive references occur through the early issues,[47] and in October 1856 Snow's own lengthy paper "Cholera and the Water Supply in the South Districts of London, in 1854" was published.[48] In addition, the *Transactions of the Epidemiological Society of London*, appended to the journal, also carried in its first issue Snow's paper "On the Comparative Mortality of Large Towns and Rural Districts."[49] Snow unfortunately died in 1858, and Richardson, who was his close friend, appended to his edition of Snow's papers on chloroform a memoir in which he noted that it was his duty as Snow's biographer "to claim for him not only the entire originality of the theory of the communication of cholera by the direct introduction of the excreted cholera poison into the alimentary system; but, independently of that theory, the entire originality of the discovery of a connection between impure water supply and choleraic disease."[50] The battles for acceptance of Snow's theory initially conducted in the pages of the *Journal of Public Health* are here continued in another form.

These early articles, together with Richardson's dramatic account in the *Memoir* of the "stranger" who entered the vestry meeting and offered the astounding suggestion to the "incredulous" committee that the Broad Street pump handle be removed, have played a significant role in constructing Snow's current status as a heroic figure in medical and epidemiological history.[51]

SOCIAL SCIENCE REVIEW (1864-66) AND THE RISE OF PUBLIC HEALTH JOURNALS IN THE 1860S

Despite his decision to terminate the *Journal of Public Health* in 1859 due to pressures of work, by 1864 Richardson was up and running again with a new journal, the *Social Science Review and Journal of the Sciences*, which despite its title was still largely focused on public health issues. In format it was fairly similar to its predecessor, but with less focus on reports of the progress of disease and local sanitary conditions, and more on discursive articles. The first issue reported on attempts by the National Association for the Promotion of Social Science (NAPSS, founded in 1857) to establish London premises and lectures, and it is clear that Richardson's choice of title was linked to the emergence of "social science" as a new and powerful framework for social reform. Richardson's agenda overlapped with that of the NAPSS (and he was subsequently to be president of its Health Section), but his focus was more medical, and also more idiosyncratic.[52] In part, the journal served as a vehicle for Richardson's own work: the first volume opened with the first of his series "Diseases of Overworked Men," the installments of which were to be collected subsequently in his 1876 volume *Diseases of Modern Life*. Covering both mental and physical strain, the essays contributed substantially to the growing discourses on occupational health, and also to the emerging concerns about mental stress and strain which were later popularized in America by George Beard and Silas Weir Mitchell, under the label "neurasthenia."

Like the NAPSS, the journal was supportive of a range of reforms, pressing, for example, for an end to both corporal and capital punishment, for more sensitive handling of prostitution, and for the development of female medical education.[53] The first issue also saw articles such as "Crime and Insanity" by Mackenzie Bacon, and an interesting article (by JTD) on the hardships in the lives of London cabmen, which argued: "Every grade in society is represented among them. Men drive cabs who have been physicians, clergymen of the Established Church, lawyers whose legal experience proved unfortunate, merchants and tradesmen, and artisans and mechanics of every kind. One or two baronets are stated to have driven cabs."[54] Allowing for street legends and the amplifying effects of rumor, the article nonetheless gives a fascinating insight into the perceived instabilities of the middle-class professional order at mid-century, and the threats of economic reversals.

In the positions it adopted, the *Social Science Review* was in general more radical in orientation than the NAPSS itself. Thus, unlike the NAPSS, it was openly critical of the middle-class bias in the vision of education offered by Matthew Arnold.[55] Rejecting calls for middle-class-only schools, it argued, in terms highly relevant for current education: "Our own impression is that the encouragement of middle-class schools will have a tendency to perpetuate and aggravate that class-feeling, so akin to caste-feeling, which is one of the greatest evils of our social economy. It adds the difference of education to difference of occupations in life, erecting social barriers which neither the instincts of humanity nor the authority of religion can break down, so that there is division and consequent weakness in our commonwealth."[56] The article was written by an anonymous author, so it is unclear whether it was by Richardson, but its powerful egalitarian thrust was at one with the overall stance of the journal, which undercuts any presumption that reforming public health journals were necessarily condescending or patronizing (at best) toward the working classes. Richardson's decision to close the *Social Science Review* in 1866 was probably linked to the launch that year of *Social Science: Transactions of the National Association for the Promotion of Social Science*, edited by George Hastings, which, although broader in focus, carried on much of the campaigning work highlighted in Richardson's journal.[57]

The Journal of Public Health and the *Social Science Review* were part of a wave of new public health–related journals that by the late 1860s became a veritable flood, with innumerable journals started in the field (many of them disappearing within an issue or two, or subsequently amalgamating with other titles). They attempted to tap into rising concerns with public health, but also into interest in the new industries and occupations related to the creation of sanitary infrastructures. Some, despite general titles, were very specific in their remit. Thus, the *Sanitary Reporter* (January 1863–February 1864) and subsequently the *Sanitary Review* (October 1864–January 1865), started ambitiously as "a weekly record relating to gas, water, sewerage and general science," was largely related to new gas installations, in the vein of the trade and technological journals discussed by Gooday in chapter 7 of this volume. But it could not sustain an audience. An even more short-lived periodical, *The Chloralum Review: A Sanitary Journal* (July 1871–January 1872), was directed at both medical and general readers with a particular focus on the use of antiseptics, but it overestimated public interest in this topic. Another evanescent journal was *The Sanitary Record and Social Observer*, which endured for a year in 1869 and in turn incorporated *Public Health: A Record and Review* (1868–69). Fairly radical in orientation, the first number of *The Sanitary Record* opened with an article by Charles Drysdale, a noted proponent of contraception, and also included an editorial highly critical of Burdett Coutts's dwellings for the poor, as well as a hard-hitting article by J. De Blaine on the unacceptable health conditions for workers in London printing offices.[58] The appeal in the second number for all who were interested in sanitary matters to take out annual subscriptions (six shillings, post-free) suggested, however, the inevitable fate of this title.

THE SANITARY RECORD (1874–): PUBLIC AND DOMESTIC HEALTH, AND "INFORMED LOCAL ACTION"

Far more successful was another, entirely separate journal of the same title, the *Sanitary Record: A Journal of Public Health* (1874–), edited for the publishers Smith, Elder by the indefatigable Ernest Hart, alongside his long and highly successful editorship of the *British Medical Journal*.[59] Hart, who was medically trained, had had a successful early career as a surgeon, and had been on the staff of the *Lancet* before taking up in 1867 the editorship of the *BMJ*, which he proceeded to transform into a flourishing title. Fairly astonishingly, Hart used his magic touch to produce the *Sanitary Record* on a weekly basis until July 1879, when a new series was initiated and it became a monthly. In his opening "L'Envoi," Hart drew on the statistics of Farr to present a formidable picture of disease and mortality which could be prevented by greater knowledge and "informed local action."[60] The object of the journal, he announced, was to be the "organ of intercommunication for the many thousand persons who are now officially employed in helping to advance the interests of health," a category that ranged from officers of sanitary boards and inspectors of mines and factories to medical officers of health, "members of health associations, visitors of the poor [and] managers of public institutions."[61] Moving beyond specific interest groups and occupational groupings, Hart tried to embrace the burgeoning new professions linked to areas of public health, and also those involved in local and voluntary initiatives, all of whom were to be brought together in a community of readers and activists fostered by the journal. The extensive list united professional and voluntary workers, the public sphere and the domestic, picking up on one of the primary messages of the sanitary movements: that women, as custodians of household hygiene and health, held the keys to successful reform. "The house is the unit of sanitary administration," Hart observed, "and household hygiene will be the theme which we shall treat as fully as the larger questions of public health."[62] With its explicit targeting of women and health within the home, the journal appeared to move into territory associated later with the journal *Baby*, discussed in the previous chapter of this book; but Hart's own medical credentials and standing in the community clearly deflected any potential criticism on this front.

The *Sanitary Record*'s emphasis on the domestic aspects of sanitary reform was reinforced by Hart's close association with the National Health Society, an organization set up in 1871 by a group of radical reforming women, which included Elizabeth Blackwell, with Fay Lankester (daughter of the sanitary reformer Edwin Lankester) as secretary and Ernest Hart as chair.[63] Although the balance of the *Sanitary Record* was undoubtedly toward the public and more professional aspects of reform, women were given a clear voice in its pages. The first number, for example, had an article by Dorothea Beale, principal of Cheltenham Ladies College, titled "Health and Education," in which she argued fiercely against suggestions that overwork and mental pressure were bad for women. It was mental idleness, rather, that would destroy them. Mrs. Edmund Maurice also contributed a series of articles titled "Playgrounds for Poor Children," suggesting, in a decidedly gentle attempt at social reformism, that London squares could be opened to the poor when the rich were at the seaside.[64] More radical in its social implications was the suggestion that there were possible career openings for women in analytical chemistry.[65]

Given the rapid demise of so many predecessors and emerging rivals in the field of public health, how did Hart manage to create and sustain a weekly audience for his new periodical? In part, the answer lay in the *Sanitary Record*'s constant shifting of tone and address, and its crossing of generic borders. The first number opened not with "L'Envoi," but with an article on arsenic poisoning, thus playing on associations with sensation fiction and police reports, though the subject matter was not deliberate murder but rather the more insidious form that could take place in homes with poisonous wallpapers and other arsenic-drenched household items.[66] The attractions of the periodical lay in the combination of the macabre and the mundane, as in the article "Putrid Cauliflowers," a title that, while playing on its associations with putrid corpses, almost compels reading by its seeming intensification of the triviality of the quotidian.[67] There was also the fascination of the defiantly local: those tired of the "Eastern Question" in their daily newspapers might take a vicarious thrill in reading "Goole Sewage Question"; or, even closer to home, "Tea Poisoning," an article that claimed that tea was even more of a poison than alcohol or tobacco.[68] The body, in all its needs, ailments, and excretions, was set at center stage in the journal's eclectic mix. There were lots of warnings about the adulteration of food, or hints about where to take holidays away from sanitary pollution. From the *Sanitary Record*'s relaunch as a monthly in July 1879, attractiveness was increased with the use of more pictorial material, including illustrations of rational dress or the use of "vegetable substances" such as "patent pulp" for domestic utensils or "life saving buoys" (fig. 10.2).[69]

From the second volume, Hart ran an invitation to secretaries of sanitary associations and kindred societies, which was to become a permanent feature: "The Editor will be glad to receive, with a view to publication, announcements

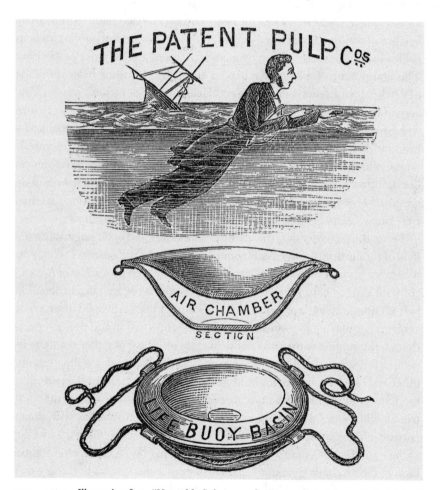

FIG. 10.2. Illustration from "Vegetable Substances for Domestic Articles, Life Saving, and Sanitary Purposes," *Sanitary Record*, n.s. 3 (15 June 1882): 528. Courtesy of the Wellcome Collection.

of meetings, reports of proceedings, abstracts of, or originals of papers read before the members of any sanitary or kindred association."[70] The net was cast deliberately wide, and from this point on Hart solved, at least temporarily, the problem of creating weekly copy by publishing both short abstracts and lengthy papers from a wide range of societies and meetings. Volume 2, for example, contained extensive coverage of the Birmingham conference "On the Sanitary Condition of Large Towns," including reports of papers and discussions, and the entire speech of the mayor of Birmingham, a youthful Joseph Chamberlain, who delivered an impassioned call to end the social inequalities that produced the problems of slum housing. It was no longer possible to ignore such an evil, he argued: "To do so would not only be a shameful dereliction of our duty, but a positive danger to the state: for there is danger in the continuance of this ever-widening contrast between the wealth and luxury of a few individuals and the deepening squalor and the wretched misery of a large class of the population."[71] In all the reports of society meetings and campaigns there were repeated calls to action, whether on housing, air, and water pollution or on social inequality. Concern about the material environment extended in multiple directions, including to the loss of open spaces. Thus, Francis George Heath from the Park Preservation Society entered a "Plea from East London" to save land "from the dread encroachment of bricks and mortar."[72] Although the aims and orientations of the societies and individuals brought together in the pages of the *Sanitary Record* could vary widely, there was an emphasis throughout on the need both to monitor social developments and to bring about social and environmental change.

JOURNALS, THE GROWTH OF HEALTH PROFESSIONS, AND THE SANITARY CONGRESS

By the time the *Sanitary Record* moved to a monthly format in 1879, it was competing with a plethora of popular health journals, as well as with the emergence of journals targeted specifically at emerging professional cohorts. Thus, the first volume of the *Transactions of the Sanitary Institute*, which aimed, among other goals, to provide training for sanitary inspectors, was published in 1879, while the medical officers of health finally published their own dedicated journal, *Public Health*, in 1888. As noted earlier, both journals are still in publication today, though the associated professions have undergone considerable change.[73] *Public Health* was very much one of the new breed of professional journals, designed to build solidarity in the ranks. As the first editor, Dr. A. Wynter Blyth, observed in the preface to the opening number, "Its object is to maintain and strengthen the existing bond of unity and sympathy between the various grades of officers of Sanitary Authorities."[74] Its aims were to keep its readers up to date with developments in legislation, sanitary architecture, and engineering, and with relevant research abstracted from foreign periodicals. As Margaret Pelling has noted, it was wide-ranging in content, but "never tried to reach or to proselytise among a wider public."[75] Its editors were unpaid, and generally active medical officers of health, drawn from the Council of the Society; they also tended to be members of other professional groupings, such as the Sanitary Institute and the Epidemiological Society.[76] My focus in this final section will be on the more eclectic *Transactions of the Sanitary Institute*, which offers detailed insight into the range and diversity of sanitary and public health movements and professions in the last decades of the nineteenth century.

The Sanitary Institute itself had been formed in 1876, following the passing of the Public Health Act in 1875. Its stated aims were to create a London base and national network of local branches, to monitor and push for further legislative changes in public health provision, and to create a system of examinations and qualifications for "Surveyors of Sanitary Districts and Inspectors of Nuisances."[77] It was a voluntary organization, with the backing of the Duke of Northumberland, who served as first president, and it financed itself through subscriptions. Benjamin Ward Richardson was involved from the start, and following his election as chairman of the council in 1877, he presided over the first congress, at Leamington Spa. The first publication of the *Transactions* followed the third congress, at Croydon in 1879, again under the presidency of Richardson (Edwin Chadwick had presided over the intervening conference at Stafford in 1878).[78] Richardson had evolved from the ambitious young medic who had started the *Journal of Public Health* to a leading medical health campaigner—the "prince of popular exponents," in the words of his obituary in *Public Health*.[79] His popularity following his lecture "Hygeia" in 1875, an outline of an ideal city of health that had received international acclaim, made him an invaluable asset for the new organization.[80]

Given his enthusiasm for journals, it is slightly surprising that Richardson did not edit the *Transactions* (though he did start yet another journal in 1884, the *Asclepiad*, which ran until 1895, with all content composed entirely by himself). The initial editors of the *Transactions* were the enterprising hospital administrator Henry Burdett (who from 1886 to 1920 would edit the weekly journal *The Hospital*, discussed in the chapter 9 of this book), and Francis De Chaumont, professor of military hygiene at the Army Medical School.[81] Both of them were also active contributors to the journal; the first issue, for example, had an article by De Chaumont on drinking water, while Burdett offered a piece called "The Unhealthiness of Public Institutions," which made the rather startling claim that since prisons had "come under the control of Government, they have gradually been made probably the healthiest residences in the United Kingdom."[82] The journal opened, however, with Richardson's presidential address, "Salutland: an Ideal of a Healthy People," which followed the popular vein he had pursued with "Hygeia." The lecture is an engaging vision of a future utopia in which people would live out their natural life spans of one hundred or more years in a city filled with parks and gardens; and when long-distance telephone calls would end feelings of separation, and polluting coal-powered locomotives would be replaced by aerial locomotion.[83]

The general tenor of the journal, however, was more directly practical. The Sanitary Institute existed in part to develop sanitary education, and particularly examinations for sanitary officers, but more broadly it set out to promote the diffusion of sanitary knowledge. The institute, its annual congresses, and the journal worked in consort, developing a public and readership that went well beyond the immediate audience of sanitary inspectors. By the time the congress met in Leeds in 1897, it attracted 1,380 attendees and encompassed individual conferences for medical officers of health, municipal representatives, municipal and county engineers, and ladies on domestic hygiene, as well as a conference on river pollution.[84] The growth of the congress was a measure of the increase of public health professions as well as voluntary organizations fostered by the journal, which took pains to include not merely the main papers but also the debates which followed, giving voice to many who might otherwise elude the domain of print. In their careful record of the contributions to the congress, the *Transactions* and *Journal* opened a neglected area of both social and scientific history, offering insight into the preoccupations and participants of these annual events which, while not quite rivaling the British Association for the Advancement of Science (BAAS) in size, were undoubtedly more socially diverse.[85]

The *Journal* by this stage was edited by a committee, with J. F. J. Sykes, medical officer of health for St. Pancras, in the chair. The 1897 issue, which I take as my final case study, contained papers delivered at sessional meetings during the year, and full and abstracted papers from the annual congress, as well as details of the range of participants, and associated activities. Readers were informed that more than three hundred local authorities had sent delegates to the congress, as had numerous school boards, and representatives from more than seventy societies including universities, the Royal Colleges, and organizations such as the Cremation Society and the St. John Ambulance Association.[86] The editors carefully recorded that the health exhibition attracted more than 75,790 visitors in the three weeks in which it was open, and they included lists of prize-winning sanitary wares and domestic goods on display (from Cadbury's Cocoa to the intriguing "Fin de Siècle Bath"). The level of community involvement was also emphasized, with accounts of 620 school children demonstrating physical drill, cooking competitions, and nursing demonstrations in a temporary hospital constructed for the purpose. Demonstrations of the very latest scientific developments, from Röntgen X-rays to the cinemetograph [*sic*], were also proudly recorded.[87] Both for attendees and for readers coming fresh to the material, the detailed publication of papers, discussions, and events was designed to build a sense of the social importance of the work, and to foster a sense of membership within an ongoing community that spread beyond the various local sanitary organizations, or the diverse campaigning or public health bodies and occupational groupings that came together for the congresses.

The voices captured in the *Journal* offer sharply variant tones and politics. The initial address by the president, Robert Farquharson (a doctor turned laird and member of Parliament) was uncompromisingly patrician. The working classes, he announced, "no less than their employers, are beginning to be connoisseurs in the quality of their atmospheric conditions [and] enjoy a good wash as much as a university graduate."[88] A more democratic and indeed defiant note was struck in discussion by Councillor B. Womersley (chairman of the Sanitary Committee of the Leeds City Council), who called for more local control in decision making, arguing that the representatives of the people on city councils "understand far better the conditions and requirements of their town than a few members of Parliament, promiscuously brought together."[89] The journal also gave a voice to women, albeit corralled within the Ladies' Conference on Domestic Hygiene, a title that suggests committees packed full of Mrs. Jellybys and Pardiggles. The reality proves more interesting, and subtly subversive. Thus, Alice Ravenhill, in a talk called "Women as Hygiene Teachers," argued that women, with their "love of detail and quick perceptions" were eminently suited to teach hygiene, and called for them to be given full-time remunerated employment as factory and sanitary inspectors, and made members of school boards, and urban and rural councils. The congress duly passed a resolution that Leeds City Council be urged to appoint women as sanitary inspectors.[90]

In his opening speech, Farquharson had foregrounded one of the dominant health concerns of the 1890s: that mind and body were being stretched too far in modern society. Ada Goodrich-Freer, honorary secretary of the Horticultural College, Swanley, seized upon the notion in order to make her case in a talk titled "The Sanitary Aspects of Gardening as a Profession for Women."[91] Gardening, she argued, afforded the "soothing and calming influences" that were much required at "the present stage of our social evolution." Such a beguiling prelude led not to traditional pieties regarding women's place within domestic flower gardens, however, but to an argument for a rigorous scientific training for women in horticulture, a profession to which they were peculiarly suited: "Horticulture, besides the art of gardening, includes some knowledge of chemistry, botany, entomology, geology, of surveying and mensuration, of light and heat, of horticultural building, of book-keeping and of many other applied sciences." Her college, Goodrich-Freer boasted, had trained more than one hundred women, who had taken the highest places in external examinations, including the Gold Medal of the Royal Horticultural Society, and were now head gardeners in private estates and market gardens, and were employed as teachers in various institutions, including the Royal Horticultural Gardens at Kew and Edinburgh.[92] She claimed for women a high degree of scientific mastery as well as the physical strength required to undertake all the demanding manual labor of a gardener, citing the support of both Sir Edward Sieveking and Elizabeth Garrett Anderson for her position regarding the "healthfulness of gardening as an occupation for women."[93] "Health" became the banner under which a radical call for women's scientific training, and wider entry into the employment market, could be packaged as a social good.

In bringing together the diverse aspects of the congress into the pages of the *Journal*, the editors created a wonderful cross-section of public health activity at the time, producing an in-depth picture which varied markedly from that created by subsequent scholarly accounts that have focused on parliamentary legislation, or on only one aspect of the closely interwoven sets of activities and concerns. Metropolitan bias was undercut, and significant local initiatives and scientific figures were given space and prominence. Papers selected for print in the "Chemistry, Meteorology and Geology" section, for example, included one by the naturalist and curator of the Leeds City Museum, Henry Crowther, in partnership with the eminent Leeds chemist and pioneer of X-ray photography Richard Reynolds, on ten years of meteorological observations of sunshine by the Leeds Philosophical and Literary Society, which had been undertaken to measure air pollution in the city.[94] Based on daily readings in the city center (at the museum) and at stations one and a half and four miles out respectively, the research found a considerable diminution of sunshine in the center, with figures improving during the weekends following reduction in manufacturing activity. These records, from more than ten thousand observations, the authors noted, would be of "prospective rather than retrospective" interest, pointing the way forward for action to be taken to diminish the amount of "unconsumed carbon" in the atmosphere from both industry and household coal fires.

The editors of the *JSI* devoted considerable space to the ensuing discussion on air pollution which was harshly critical of manufacturing interests. Walter Rowley of Leeds—who, interestingly, was probably a mining engineer—argued that manufacturers' outcries against legislation designed to prevent air and water pollution, on the grounds that "they were hampering and fettering trade," were a major obstacle to sanitarian activity. He questioned employers' rights to inflict "these great acts of injustice [on] the community at large. To plead the requirements of industry was no justification for polluting the air or the stream common to all alike."[95] Similarly, Dr. J. B. Cohen, a lecturer in organic chemistry at the Yorkshire College, argued that "persons who sent smoke into the atmosphere unnecessarily were really poaching on other people's preserves; an offence of this kind was quite unexcusable [*sic*]."[96] In his own paper "Air Pollution in Towns," Cohen also argued that it was not the people themselves but "air pollution" which was "largely responsible for the squalid conditions of urban habitations among the poorer classes."[97]

The papers and subsequent exchanges captured in the journal highlight both the interconnected nature of sanitarian concerns and the strong environmental consciousness that underpinned early work in public health. Issues of housing, physical health, and air and water pollution were all perceived to be interrelated, and members of the largely middle-class public health organizations spoke out in defense of the rights of all to unpolluted air and water. The Coal Smoke Abatement Society (1898) has been described as one of the world's oldest environmental organizations, but it traces its roots to the earlier Smoke Abatement Societies run in Leeds and Manchester and other northern cities, and to the labors of the individuals who painstakingly took daily recordings of sunlight or air pollution.[98] By printing the papers and discussion of the annual congresses, the *Journal of the Sanitary Institute* enabled local societies to keep abreast of what was happening in other towns while also fostering a sense both of community and common cause for those who were laboring, in diverse societies and multiple scientific and practical fields, to improve public health.

CONCLUSION

From the *Journal of Public Health and Sanitary Review*, and its involvement with the early days of the medical officers of health in the 1850s, to the *Jour-*

nal of the Sanitary Institute in the 1890s and the consolidation of the sanitary officer's role, one can track the interdependence of public health journals and the rise of associated professions. As we have seen, however, these journals also fostered growing public interest and engagement in the domains of public health, from local Ladies' Sanitary Associations, to Smoke Abatement Societies. Ironically, the very growth of professional groupings fostered by the journals led in turn to a lessening of public involvement, and to a decline in the twentieth century of journals focused on broad readerships. Instead, one finds in the early decades of the twentieth century a narrowing of focus as journals survive but turn inward to a captive professional audience. For some titles, such as *Public Health* itself, which had from the start been oriented to a professional audience, the story is one of triumphant survival. But it is also one of loss, with popular, large-scale involvement in issues of public health declining once journals and their associated domains came under the control of professional bodies.

The story of the rise of public health in the nineteenth century has been told many times; but if we place the various public health journals at center stage, a rather different picture emerges—one in which dominant figures such as John Simon and Joseph Bazalgette appear almost to drop from sight, and those with bit parts elsewhere assume far larger importance. The names also proliferate as the focus shifts from the metropolis to include influential campaigners in towns and cities across Britain, who often do not merit a footnote in more mainstream histories—or, alternatively, who do figure in scholarly historical accounts, but in very different areas.[99] Thus, George J. Symons, for example, remembered primarily for his work on meteorology with the British Rainfall Organisation emerges as an absolute stalwart of the Sanitary Institute; he was registrar for fifteen years between 1880 and 1895, with prime responsibility for devising and overseeing the new professional exams for sanitary inspectors.[100] His address as president of the Meteorology, Geology, and Geography Section at the third congress, held at Croydon in 1879, was a wonderful demonstration of his sense of the importance of engaging with a wide general audience. He anticipated the sceptics' questions: "What is the use of making three days of it, and having a lot of dry scientific papers of no use to anybody, and incomprehensible by any but dreadfully scientific people?"[101] His answer was a firm rebuttal of this view. There is no such thing as a separation between "science and practice," he argued, and he went on to show how meteorological understanding was fundamental to daily life, from ventilation in theaters to the positioning of houses. His vision was of a general public deeply engaged in issues of health and the environment, willing to give their own time to help record and improve their environmental surroundings. He depicted experiments in which householders were given ozone test papers to record air quality, and advocated a further development of such a system which could produce "a rough-and-ready test of the purity and healthiness of the air in different localities."[102] As Christopher Hamlin has argued in his excellent short introduction to *Sanitary Reform in the Provinces*, the history of such reforms is one of local action and empowerment, and also of equity, since the stories of local engagement are "about the extension of biomedical rights, which include rights to environmental determinants of health and amenity."[103]

Public health journals in all their diversity, from publications of local sanitary groups to professionally oriented titles, capture the depth and range of engagements in the fields associated with public health. They chronicle not just the major advances in science or legislation, but the forms of argument of their day. They also capture the closely interwoven networks of figures in diverse fields, from engineering to education, who worked together both locally and nationally to bring about change. The long lists of prize-winning exhibits at the Sanitary Congresses are testimony to the endless resourcefulness of the Victorians in their attempts to promote healthy living, while the detailed figures of air pollution observations conceal behind them the long hours devoted, often over decades, to making daily recordings that would be of value not just to contemporaries but to posterity, and thus to the climate scientists of today. One finds in these pages a level of ambition and optimism at odds with the general picture of dour, controlling sanitarians and the rise of state intervention. Instead, what emerges is a highly dynamic model and understanding of the interactions between mind, body, and environment.

NOTES

1. Even the superb six-volume *Sanitary Reform in Victorian Britain* ed. Michelle Allen Emerson et al. (London: Pickering and Chatto, 2012), has tended to publish articles in their stand-alone pamphlet form, thus losing the context of the sanitary journals in which they often first appeared. Tom Crook's excellent *Governing Systems: Modernity and the Making of Public Health in England, 1830–1910* (Oakland: University of California Press, 2016) makes good use of sanitary journals, but does not consider them in their own right.

2. Tina Young Choi, ed., introduction to *Medicine and Sanitary Science*, vol. 1 of Emerson, ed., *Sanitary Reform in Victorian Britain*, xiii.

3. Louis C. Parkes, *Jubilee Retrospect of the Royal Sanitary Institute, 1876–1926* (London: Sanitary Institute, 1926), 25.

4. Public health and sanitary journals are best tracked through W. R. Lefanu, *British Period-*

icals of Medicine: A Chronological List, 1640–1899, revised edition, ed. Jean Loudon (Oxford, UK: Wellcome Unit for the History of Medicine, 1984), which helpfully lists medical and some medically related titles in order of publication. A rough count suggests sixty-nine titles here, but this does not include more social science-oriented journals to be found in John North, ed., *Waterloo Directory of English Newspapers and Periodicals: 1800–1900*. A search here yields sixty journals under the rubric of sanitary journals, and forty-seven under public health. The best introduction to the overall field of medical periodicals remains W. F. Bynum, Stephen Lock, and Roy Porter, eds., *Medical Periodicals and Medical Knowledge: Historical Essays* (London: Routledge, 1992).

5. Gowan Dawson, "Richardson, Benjamin Ward (1828–1896)," in *Dictionary of Nineteenth-Century Journalism in Great Britain and Ireland*, ed. Laurel Brake and Marysa Demoor (Ghent and London: Academia Press and the British Library, 2009), 541.

6. *Journal of Public Health and Monthly Record of Sanitary Improvement*, ed., John Sutherland, MD. Sutherland had previously edited the Liverpool *Health of Towns Advocate*. The journal lasted from November 1847 to October 1849 (Lefanu, *British Periodicals*). There had previously been a *Journal of Public Health; or, Family Guide to Medicine*, edited by James Scott, which had a brief life in 1823, but it was more of a family medical compendium, as its subtitle implies, and not part of the sanitarian agenda that emerged from the 1840s.

7. See Christopher Hamlin, *Public Health and Social Justice in the Age of Chadwick. Britain, 1800–1854* (Cambridge: Cambridge University Press, 1998), 243–64. Hamlin offers the best overview and analysis of the early growth of public health campaigns. See also Norman Parkinson, "The Health of Towns Association and the Genesis of the Environmental Health Practitioner," *Journal of Environmental Health Research* 14 (2014): 5–16.

8. For details on the work of William Farr and the General Register Office, see John M. Eyler, *Victorian Social Medicine: The Ideas and Methods of William Farr* (Baltimore: Johns Hopkins University Press, 1979). M. W. Finn offers an excellent and extensive introduction to Chadwick's report in Edwin Chadwick, *Report on the Sanitary Condition of the Labouring Population of Great Britain* (Edinburgh: Edinburgh University Press, 1965 [1842]), 1–73.

9. The *Oxford English Dictionary* cites Chadwick's report as first usage. There were other uses beforehand, however, such as the *Annual Report of the Local Government Board for Ireland* (Dublin, 1834), which had sections on sanitary improvements.

10. See Henry Mayhew, *London Labour and the London Poor*, ed. Robert Douglas-Fairhurst (Oxford: Oxford University Press, 2010), xliii–xliv.

11. [B. W. Richardson], introduction to *Journal of Public Health and Sanitary Review* 1 (1855): 1.

12. Ibid., 2.

13. Charles Dickens, *Bleak House* (London: Bradbury and Evans, 1853), ch. 8, 71.

14. See, for example, Peter Stallybrass and Allon White, "The City: The Sewer, the Gaze and the Contaminating Touch," in *The Politics and Poetics of Transgression* (Ithaca, NY: Cornell University Press, 1986), 125–48.

15. Mary Poovey, *Making a Social Body: British Cultural Formation, 1830–1864* (Chicago: University of Chicago Press, 1995), 116.

16. William A. Cohen and Ryan Johnson, *Filth: Dirt, Disgust and Modern Life* (Minneap-

olis: University of Minnesota Press, 2004); Graham Mooney, *Intrusive Interventions: Public Health, Domestic Space, and Infectious Disease Surveillance in England, 1840–1914* (Rochester, NY: University of Rochester Press, 2015).

17. Richardson, introduction, 2.

18. Richardson produced various poems and other literary productions, including a historical novel, *The Son of a Star* (1888). He published a series of histories of early medical men, which appeared in his later periodical, the *Asclepiad* (1888–95), which he wrote and produced singlehandedly. He was an active member of the Society of Antiquaries, becoming a fellow in 1877.

19. "The Late Sir Benjamin Ward Richardson: An Appreciation," *Public Health* 9 (1896–97): 122.

20. See Patrick Wallis, "Richardson, Sir Benjamin Ward," *Oxford Dictionary of National Biography*; Arthur Salusbury MacNalty, *Sir Benjamin Ward Richardson* (London: Harvey and Blythe, 1950); and Richardson's autobiography, *Vita Medica: Chapters of Medical Life and Work* (London: Longman, Green, and Co., 1897), 40, 158.

21. "Our Club," Royal College of Physicians, MS 520/26, p. 10. Ruth Richardson, "'Notorious Abominations': Architecture and the Public Health in 'The Builder' 1843–83," in Bynum et al., *Medical Periodicals*, 90–107. *The Builder* was launched in 1843, and George Godwin was to edit it from 1844 to 1883.

22. Richardson, *Vita Medica*, 228.

23. Benjamin Ward Richardson, *The Health of Nations: A Review of the Works of Edwin Chadwick, with a Biographical Dissertation,* 2 vols. (London: Longman, Green, and Co., 1887). Richardson notes in his introduction that his friendship with Chadwick began in 1853–54, "in the early days of the Epidemiological Society" (I, 1).

24. The Association of Public Sanitary Inspectors has transformed over the years into the current Chartered Institute of Environmental Health. Richardson was president from 1890 until his death in 1896. See http://www.cieh.org/about_us/history.html, accessed 1 November 2017.

25. Richardson, *Vita Medica*, 235.

26. Bynum and Wilson, "Periodical Knowledge," 41–43. They note, for example, that John Conolly, W. B. Carpenter, and E. A. Parkes all put in three- to five-year stints as editor of the *British and Foreign Medico-Chirurgical Review.*

27. Richardson, *Vita Medica*, 231.

28. The price of the weekly *Lancet* was six pence. For details of pricing of medical journals at this time, see Jean and Irvine Loudon, "Medicine, Politics and the Medical Periodical 1800–1850," in Bynum et al., *Medical Journals*, 59.

29. Richardson, *Vita Medica*, 231.

30. Ibid., 234.

31. This article and its sequel are signed simply "R." This is unusual in that most of the material is either unsigned or signed with accompanying detail as to professional position. It is possible that the "R" is Richardson himself, although much of the unattributed material is probably his, and some of the "original communications" are signed in full "Benjamin Ward Richardson" (17).

32. John Webster (1794–1876) MD, FRCP, FRS. Webster wrote a work on the cholera epidemic of 1832, and published regular reports on the state of health in London and on the conditions he encountered in lunatic asylums and prisons across Europe. See http://munksroll .rcplondon.ac.uk/Biography/Details/4674, and also https://www.wikitree.com/wiki/Webster -5142.

33. Rev. C. Girdleston, "On the Scientific Investigation of Sanitary Questions," *Journal of Public Health and Sanitary Review* 1 (1855): 29–31.

34. See Francis J. Allan, "The Early Days and Early Work of the Society, with Special Reference to the Registration of Disease," *Public Health* 13 (1900–1901): 76, where he notes, "There were many good men who were unsuccessful, including B. W. Richardson."

35. Robert Druitt, "Short Notes on Some of the Details of Sanitary Police," *Journal of Public Health and Sanitary Review* 1 (1855): 15–22. For details of Robert Druitt (1814–83), see http://munksroll.rcplondon.ac.uk/Biography/Details/1341.

36. [B. W. Richardson], "The Medical Police of London," *Journal of Public Health and Sanitary Review* 1 (1855): 325.

37. Allan, "Early Days," 74–75. See also Richardson's return to the issue in the first volume, under the general section "Hygienic Jurisprudence: Progress of Sanitary Legislation: Medical Officers of Health." Here he notes that his recommendations have been taken up, but fulminates against the narrow interpretation taken as to which grades of medical men would be suitable, thereby presumably excluding him (428–29).

38. Margaret Pelling, "'Progress, Difficulties, Suggestions and Reforms': 'Public Health' 1888–1974," *Public Health* 102 (1988): 212; "Society of Medical Officers of Health," *Lancet,* 19 May 1906, 1392–94.

39. "Richardson, "Medical Police," 325.

40. The problem has been, however, the sheer volume of material, which has made the reports rather daunting and therefore inaccessible. The recent digitization of the London reports by the Wellcome has helped to open them for scholarly use. See the website for "London's Pulse: Medical Officer of Health Reports 1848–1972," http://wellcomelibrary.org/moh/.

41. [Benjamin Ward Richardson], "The Registration of Disease in England," *Sanitary Review and Journal of Public Health* 3 (1857–58): 317–26, 322.

42. In his autobiography, Richardson gives further details of the exercise, noting that it encompassed diseases of the lower animals and the vegetable world, and that the number of participants, who were spread geographically from the Scilly Isles to the Shetlands, rose to sixty before he was forced to give up from sheer pressure of work. *Vita Medica,* 231–32.

43. Richardson, *Vita Medica,* 231–33.

44. Richardson, "Registration of Disease," 317.

45. Ibid., 320.

46. B. W. Richardson, "Water Supply in Relation to Health and Disease," *Journal of Public Health and Sanitary Review* 1 (1855): 130–40, 134.

47. See "Cholera Water Supply," 396; and "The Propagation of Cholera," which opens with the declaration, "The theory first advanced by Dr Snow regarding the mode of propagation of cholera is, with certain modifications, the value of which has been already discussed in the Journal, becoming very widely known and accepted" (402).

48. John Snow, "Cholera and the Water Supply in the South Districts of London, in 1854," *Journal of Public Health and Sanitary Review* 2 (1856): 239–57.

49. John Snow, "On the Comparative Mortality of Large Towns and Rural Districts, and the Causes by Which It Is Influenced," *Journal of Public Health and Sanitary Review, and Transactions of the Epidemiological Society of London* 1 (1855): 16–54.

50. John Snow, *On Chloroform and Other Anaesthetics, with a Memoir of the Author by Benjamin Ward Richardson* (London: John Churchill, 1858), xxii.

51. Ibid., xxi. The *Memoir* is the main biographical source for Snow, and it is Richardson's version of Snow's intervention in the vestry meeting which is replayed in all the popular accounts—creating, as Christopher Hamlin observed to me, the "myth of the Broad Street pump."

52. Richardson's famous talk on the ideal city of health, Hygeia, was given at the Social Science Congress in Brighton in 1875, when he was president of its Health section. It was printed in the NAPSS *Transactions*, and also published separately in 1876, with a dedication to Edwin Chadwick. The *Social Science Review* itself ranged widely, like its predecessor, including articles in its first volume titled "Forgotten Treasures: The Vision of Hystapses," 12–21, and "Art among the Buddhists," 124–35.

53. See, for example, "Capital Murder," I, 150–60; "Legislation and Prostitution," I, 26–30; and "The First Female Medical Practitioner in England," I, 476–77.

54. G. Mackenzie Bacon, "The Relation of Crime and Insanity: Illustrated by Recent Cases," I, 431–47; J. T. D., "The Cabmen of London," I, 410–16, 411.

55. As Lawrence Goldman argues, "To the Social Science Association, as to Victorians in general, the question of secondary education meant 'middle-class education.'" See his *Science, Reform, and Politics in Victorian Britain: The Social Science Association 1857–1886* (Cambridge: Cambridge University Press, 2002), 237.

56. "Middle-Class Schools," I, 327–35, 328. The article was in response to a recent piece by Matthew Arnold in *Macmillan's Magazine*.

57. The previous year, Richardson's friend, the doctor and sanitarian campaigner Edwin Lankester, had launched the *Journal of Social Science*, which ran between November 1865 and October 1866. For details on the founding of the *Transactions*, see Goldman, *Science, Reform, and Politics*, 87–89.

58. Charles R. Drysdale, "Arguments for and against Acclimatisation"; "Editorial: Improved Dwellings for the Poor"; J. De Blaine, "The London Printing Offices"; "An Appeal," *Sanitary Record and Social Observer, with Which Is Incorporated the Public Health* 1 (1869): 1, 14, 33, 42.

59. Ernest Hart edited the *British Medical Journal* from 1867 until his death in 1898, and the *Sanitary Record* from 1874 to 1887.

60. "L'Envoi," *Sanitary Record: A Journal of Public Health*, 4 July 1874, 8–9.

61. Ibid., 8.

62. Ibid.

63. See "Obituary of Fay Lankester," *British Medical Journal*, 21 June 1924, 1117; and Ernest Hart, "Objects of the National Health Society of London," *Public Health Papers and Reports* 19 (1893): 71–73. The society organized health lectures and training across England, and quickly acquired aristocratic patronage, becoming an uneasy mix of radical and conservative individu-

als and policies. It distributed numerous health pamphlets and instituted training courses, with examinations and certificates, in the more domestic aspects of sanitation and health.

64. Dorothea Beale, "Health and Education," and Mrs. Edmund Maurice, "Playgrounds for Poor Children," *Sanitary Record* 1 (1874): 6, 60. Mrs. Maurice was here drawing on the example of the Ladies' Sanitary Association, which had been organizing "park parties" for poor children.

65. "A Lady Analyst," *Sanitary Record*, 1 August 1874, 129.

66. George Johnson, "On Some Common Sources of Poisoning by Arsenic," *Sanitary Record*, 4 July 1874, 1–2. The article was spread over a series of issues, and had originally been delivered as a talk to the National Health Society on 18 June 1874.

67. "Putrid Cauliflowers," *Sanitary Record*, 23 November 1877, 335.

68. "The Goole Sewage Question," *Sanitary Record*, 26 September 1874, 228; "Tea Poisoning," *Sanitary Record*, 26 September 1874, 227.

69. E. M. King, "Dress Reform," *Sanitary Record*, 15 June 1881, 443–45. King was honorary secretary of the Rational Dress Society. "Vegetable Substances for Domestic Articles, Life Saving, and Sanitary Purposes," *Sanitary Record*, 15 June 1882, 528.

70. [Ernest Hart], "To Secretaries," *Sanitary Record*, 2 January 1875, 7.

71. Joseph Chamberlain, "On the Sanitary Condition of Large Towns," *Sanitary Record*, 23 January 1875, 51.

72. Francis George Heath, "A Plea from East London," *Sanitary Record*, 22 August 1874, 147.

73. As noted earlier, *Public Health* still carries the same title (with the new subtitle *Journal of the Society of Community Medicine*). The *Transactions* changed to the *Journal of the Sanitary Institute* in 1894, and is currently published as *Journal of the Royal Society for the Promotion of Health*.

74. [A. Wynter Blyth], preface to *Public Health, the Journal of the Society of Medical Officers of Health* 1 (1888): 1.

75. Pelling, "Progress, Difficulties," 213.

76. Ibid., 211.

77. Louis C. Parkes, *Jubilee Retrospect of the Royal Sanitary Institute, 1876–1926* (London: Sanitary Institute, 1926), 1–2.

78. "Congresses and Officers," *Transactions of the Sanitary Institute of Great Britain*, 1 (1880): xiv. The Duke of Northumberland was president of the Sanitary Institute itself, with vice presidents including William Farr, Richardson, and Chadwick. The council was chaired by Richardson.

79. "The Late Sir Benjamin Ward Richardson," 122.

80. Parkes, *Jubilee Retrospect*, 23.

81. De Chaumont had been a major contributor to the *Sanitary Record*, which had serialized his lectures on hygiene, from 31 October 1874.

82. F. De Chaumont, "On Certain Points with Reference to Drinking Water," and Henry C. Burdett, "The Unhealthiness of Public Institutions," *Transactions of the Sanitary Institute of Great Britain* 1 (1880): 64–70, 95–100.

83. B. W. Richardson, "Salutland: An Idea of a Healthy People," *Transactions of the Sanitary Institute of Great Britain* 1 (1880): 1–37.

84. "Annual Report: Congress and Exhibition," *Journal of the Sanitary Institute* (hereafter *JSI*) 19 (1898): 186–89.

85. By the 1880s, BAAS meetings were attracting more than three thousand participants. There has been a lot of work on the BAAS since Jack Morrell and Arnold Thackray's *Gentlemen of Science: Early Years of the British Association for the Advancement of Science* (Oxford, UK: Clarendon Press, 1982); and Louise Miskell has recently looked at the development of the scientific congress more generally, but she did not include the Sanitary Congress in her examples. See *Meeting Places: Scientific Congresses and Urban Identity in Victorian Britain* (Farnham, UK: Ashgate, 2013).

86. "Programme of Leeds Congress," *JSI* 18 (1897): 237–39.

87. "Leeds Exhibition, 1897: List of Awards," *JSI* 18 (1897): 369; "The Exhibition," *JSI* 19 (1898): 135; "Annual Report: Congress and Exhibition," *JSI* 19 (1898): 186. The reporting of the congress spread over two successive volumes of the *JSI*.

88. Robert Farquharson, "Inaugural Address," *JSI* 18 (1897): 257–76, 259. Farquharson (1836–1918) trained in medicine but abandoned it when he became of Laird of Finzean in 1878. He became a member of Parliament in 1880, and used his position to champion various public health causes. See "Obituary: The Right Hon. Robert Farquharson, P.C.," *British Medical Journal*, 15 June 1918, 684.

89. B. Womersley, "Presidential Address: Conference of Municipal Representatives," *JSI* 18 (1897): 327–34, 333.

90. A. Ravenhill, "Women as Hygiene Teachers," *JSI* 19 (1898): 126. "Resolutions Passed at the Congress Held at Leeds, 1897," *JSI* 19 (1898): 132–34.

91. [A.] Goodrich-Freer, "The Sanitary Aspects of Gardening as a Profession for Women," *JSI* 19 (1898): 129–30. On the Swanley College, see Donald L. Opitz, "'A Triumph of Brains over Brute': Women and Science at the Horticultural College, Swanley, 1890–1910," *Isis* 104 (2013): 30–62. Goodrich-Freer is now better known for her interest in spiritualism than for her support of female horticulture.

92. Goodrich-Freer, "Sanitary Aspects," 130.

93. Ibid.

94. Henry Crowther, FRMS, and R. Reynolds, FCS, "A Decade of Sunshine Observations at Leeds," *JSI* 18 (1897): 601–7.

95. "Discussion of Papers by H. Crowther and R. Reynolds and J. B. Cohen," *JSI* 18 (1897): 608–11, 609.

96. Ibid., 611.

97. J. B. Cohen, "Air Pollution in Towns," *JSI* 18 (1897): 607–8. Julius Berend Cohen (1859–1935) went on to become a professor of inorganic chemistry at the University of Leeds from 1904 to 1924, and is seen as one of the founders of the field. He was also a founder of the Leeds Smoke Abatement Society in 1890, and subsequently a leading figure in the national Coal Smoke Abatement Society when it was founded in 1898. For an excellent history of the campaigns against smoke pollution, see Peter Thorsheim, *Inventing Pollution: Coal, Smoke, and Culture in Britain since 1800* (Athens: Ohio University Press, 2006).

98. See, for example, the University of Leeds School of Earth and Environment website page on Professor Julius Cohen: http://www.see.leeds.ac.uk/research/essi/cohen-research -group/prof-julius-cohen/.

99. As Tom Crook has argued in *Governing Systems*, public health in Victorian England witnessed "centralization *and* localization; bureaucratization *and* democratization" (288).

100. See Katharine Anderson, *Predicting the Weather: Victorians and the Science of Meteorology* (Chicago: University of Chicago Press, 2005), 99–105.

101. G. J. Symons, "Presidential Address, Section III, Meteorology, Geology and Geography," *Transactions of the Sanitary Institute of Great Britain* 1 (1880): 173–89, 173.

102. Ibid., 178.

103. Christopher Hamlin, ed., *Sanitary Reform in the Provinces*, vol. 2 of Emerson, ed., *Sanitary Reform in Victorian Britain*, ix–x.

ACKNOWLEDGMENTS

This volume arises out of research conducted for "Constructing Scientific Communities: Citizen Science in the 19th and 21st Centuries," a large "Science in Culture" project, funded by the Arts and Humanities Research Council of the United Kingdom. For information on the project, and for further resources in science periodicals, see http://conscicom.web.ox.ac.uk/science-gossip. We would like to thank members of the project from our partner institutions: Julie Harvey, Paul Cooper, and John Tweddle at the Natural History Museum, London; Samuel Alberti and Thalia Knight at the Royal College of Surgeons of England; and Keith Moore at the Royal Society. We are also grateful to the members of our advisory board for their invaluable support and advice: Gillian Beer, Lorraine Daston, Alex Halliday, Julie Maxton, and Harriet Ritvo. In addition, we wish to thank Jim and Anne Secord and the other participants at the "Science in the Nineteenth-Century Periodical" workshop held at York University, Toronto, in April 2017. We are indebted to Karen Darling of the University of Chicago Press for her support and encouragement, and to two anonymous referees for their helpful suggestions for revision.

SELECT BIBLIOGRAPHY

This bibliography is intended as a guide to the principal scholarly literature on science periodicals in nineteenth-century Britain, but it is inevitably selective. Moreover, literature concerning other periods, periodicals, and territories has been largely excluded. For guides to the primary sources, see especially the lists by Bolton, Gascoigne, LeFanu, North, and Scudder. A useful searchable list of more than one thousand British science periodicals of the nineteenth century, based on those of Bolton and LeFanu, is provided at conscicom.web.ox.ac.uk.

Alberti, Samuel J. M. M. "Amateurs and Professionals in One County: Biology and Natural History in Late Victorian Yorkshire." *Journal of the History of Biology* 34 (2001): 115–47.

Albree, Joe, and Scott H. Brown. "'A Valuable Monument of Mathematical Genius': The *Ladies' Diary* (1704–1840)." *Historia Mathematica* 36 (2009): 10–47.

Allen, Bryce, Jian Qin, and F. W. Lancaster. "Persuasive Communities: A Longitudinal Analysis of References in the *Philosophical Transactions* of the Royal Society, 1665–1990." *Social Studies of Science* 24 (1994): 279–310.

Allen, David E. *Books and Naturalists.* London: Collins, 2010.

———. *The Naturalist in Britain: A Social History.* 2nd ed. Princeton, NJ: Princeton University Press, 1994 [1978].

———. "The Struggle for Specialist Journals: Natural History in the British Periodicals Market in the First Half of the Nineteenth Century." *Archives of Natural History* 23 (1996): 107–23.

Anderson, Patricia J. *The Printed Image and the Transformation of Popular Culture, 1790–1860.* Oxford, UK: Clarendon, 1991.

Anderson, Ronald. "The Referees' Assessment of Faraday's Electromagnetic Induction Paper of 1831." *Notes and Records of the Royal Society* 47 (1993): 243–56.

Atkinson, Dwight. "The Evolution of Medical Research Writing from 1735 to 1985: The Case of the *Edinburgh Medical Journal.*" *Applied Linguistics* 12 (1992): 337–74.

———. *Scientific Discourse in Sociohistorical Context: The "Philosophical Transactions of the Royal Society of London," 1675–1975.* London: Routledge, 1998.

Baldwin, Melinda. "'Keeping in the Race': Physics, Publication Speed and National Publishing Strategies in *Nature*, 1895–1939." *British Journal for the History of Science* 47 (2014): 257–79.

———. *Making "Nature": The History of a Scientific Journal.* Chicago: University of Chicago Press, 2015.

———. "The Shifting Ground of *Nature*: Establishing an Organ of Scientific Communication in Britain, 1869–1900." *History of Science* 50 (2012): 125–54.

———. "The Successors to the X Club? Late Victorian Naturalists and *Nature*, 1869–1900." In *Victorian Scientific Naturalism: Community, Identity, Continuity*, edited by Gowan Dawson and Bernard Lightman, 288–308. Chicago: University of Chicago Press, 2014.

———. "Tyndall and Stokes: Correspondence, Referee Reports, and the Physical Sciences in Victorian Britain." In *The Age of Scientific Naturalism: John Tyndall and His Contemporaries*, edited by Bernard Lightman and Michael Reidy, 171–86. London: Pickering and Chatto, 2014.

Barton, Ruth. "Just before *Nature*: The Purposes of Science and the Purposes of Popularization in Some English Popular Science Journals of the 1860s." *Annals of Science* 55 (1998): 1–33.

———. "Scientific Authority and Scientific Controversy in *Nature*: North Britain against the X Club." In *Culture and Science in the Nineteenth-Century Media*, edited by Louise Henson, Geoffrey Cantor, Gowan Dawson, Richard Noakes, Sally Shuttleworth, and Jonathan R. Topham, 223–35. Aldershot, UK: Ashgate, 2004.

Bartrip, P. W. J. "The *British Medical Journal*: A Retrospect." In *Medical Journals and Medical Knowledge: Historical Essays*, edited by W. F. Bynum, Stephen Lock, and Roy Porter, 126–45. London and New York: Routledge, 1992.

———. *Mirror of Medicine: A History of the "British Medical Journal."* Oxford, UK: British Medical Journal and Clarendon Press, 1990.

Bazerman, Charles. *Shaping Written Knowledge: The Genre and Activity of the Experimental Article in Science.* Madison: University of Wisconsin Press, 1988.

Belknap, Geoffrey. *From a Photograph: Authenticity, Science and the Periodical Press, 1870–1890.* London and New York: Bloomsbury Academic, 2016.

———. "Illustrating Natural History: Images, Periodicals, and the Making of Nineteenth-Century Scientific Communities." *British Journal for the History of Science* 51 (2018): 395–422.

Bickerton, David M. *Marc-Auguste and Charles Pictet, the Bibliothèque Britannique (1796–1815) and the Dissemination of British Literature of Science on the Continent.* Geneva: Slatkine Reprint, 1986.

Boig, Fletcher S., and Paul W. Howerton. "History and Development of Chemical Periodicals in the Field of Analytical Chemistry: 1877–1950." *Science* 115 (1952): 555–60.

———. "History and Development of Chemical Periodicals in the Field of Organic Chemistry: 1877–1949." *Science* 115 (1952): 25–31.

Bolton, Henry Carrington. *A Catalogue of Scientific and Technical Periodicals, 1665–1895, Together with Chronological Tables and a Library Check-List.* 2nd ed. Washington: Smithsonian Institution, 1897.

Bostetter, Mary. "The Journalism of Thomas Wakley." In *Innovators and Preachers: The Role of the Editor in Victorian England,* edited by Joel Wiener, 275–92. Westport, CT: Greenwood, 1985.

Bowler, Peter J. *Science for All: The Popularization of Science in Early Twentieth-Century Britain.* Chicago: University of Chicago Press, 2009.

Brake, Laurel, and Marysa Demoor, eds. *Dictionary of Nineteenth-Century Journalism in Great Britain and Ireland.* Ghent: Academia Press, and London: British Library, 2009.

Brock, W. H. "Brewster as a Scientific Journalist." In *Martyr of Science: Sir David Brewster, 1781–1868,* edited by A. D Morrison-Low and J. R. R. Christie, 37–42. Edinburgh: Royal Scottish Museum, 1984.

———. "British Science Periodicals and Culture, 1820–1850." *Victorian Periodicals Review* 21 (1988): 47–55.

———. "The *Chemical News,* 1859–1932." *Bulletin of the History of Chemistry* 12 (1992): 30–35.

———. "The Development of Commercial Science Journals in Victorian Britain." In *The Development of Science Publishing in Europe,* edited by A. J. Meadows, 95–122. Amsterdam: Elsevier, 1980.

———. "The Making of an Editor: The Case of William Crookes." In *Culture and Science in the Nineteenth-Century Media,* edited by Louise Henson, Geoffrey Cantor, Gowan Dawson, Richard Noakes, Sally Shuttleworth, and Jonathan R. Topham, 189–98. Aldershot, UK: Ashgate, 2004.

———. "Medicine and the Victorian Scientific Press." In *Medical Journals and Medical Knowledge: Historical Essays,* edited by W. F. Bynum, Stephen Lock, and Roy Porter, 70–89. London and New York: Routledge, 1992.

———. "Patronage and Publishing: Journals of Microscopy 1839–1989." *Journal of Microscopy* 155 (1989): 249–66.

———. "Science." In *Victorian Periodicals and Victorian Society,* edited by J. Don Vann and Rosemary T. VanArsdel, 81–96. Toronto: University of Toronto Press, 1994.

———. "Scientific Bibliographies and Bibliographers, and the History of the History of Science." In *Thornton and Tully's Scientific Books, Libraries, and Collectors: A Study of Bibliography and the Book Trade in Relation to the History of Science,* edited by Andrew Hunter, 298–332. 4th rev. ed. Aldershot, UK: Ashgate, 2000.

———. *William Crookes (1832–1919) and the Commercialization of Science.* Aldershot, UK: Ashgate, 2008.

Brock, W. H., and A. J. Meadows. *The Lamp of Learning: Taylor & Francis and the Development of Science Publishing.* 2nd ed. London: Taylor and Francis, 1998 [1984].

Broman, Thomas. "Periodical Literature." In *Books and the Sciences in History,* edited by

Marina Frasca-Spada and Nick Jardine, 225–38. Cambridge: Cambridge University Press, 2000.

Brooks, Michael. "*The Builder* in the 1840s: The Making of a Magazine, the Shaping of a Profession." *Victorian Periodicals Review* 14 (1981): 86–93.

Brown, Michael. "'Bats, Rats and Barristers': The *Lancet*, Libel and the Radical Stylistics of Early Nineteenth-Century English Medicine." *Social History* 39 (2014): 189–209.

Burnham, John C. "The Evolution of Editorial Peer Review." *Journal of the American Medical Association* 263 (1990): 1323–29.

Buttress, F. A. *Agricultural Periodicals of the British Isles, 1681–1900, and Their Location.* Cambridge: School of Agriculture, University of Cambridge, 1950.

Bynum, W. F., Stephen Lock, and Roy Porter, eds. *Medical Journals and Medical Knowledge: Historical Essays.* London and New York: Routledge, 1992.

Bynum, W. F., and Janice C. Wilson. "Periodical Knowledge: Medical Journals and Their Editors in Nineteenth-Century Britain." In *Medical Journals and Medical Knowledge: Historical Essays*, edited by W. F. Bynum, Stephen Lock, and Roy Porter, 6–28. London and New York: Routledge, 1992.

Cahan, David. "Institutions and Communities." In *From Natural Philosophy to the Sciences: Writing the History of Nineteenth-Century Science*, edited by David Cahan, 291–328. Chicago: University of Chicago Press, 2003.

Cantor, Geoffrey, Gowan Dawson, Graeme Gooday, Richard Noakes, Sally Shuttleworth, and Jonathan R. Topham. *Science in the Nineteenth-Century Periodical: Reading the Magazine of Nature.* Cambridge: Cambridge University Press, 2004.

Cantor, Geoffrey, and Sally Shuttleworth, eds. *Science Serialized: Representations of the Sciences in Nineteenth-Century Periodicals.* Cambridge, MA, and London: MIT Press, 2004.

Chernin, Eli. "The Early British and American Journals of Tropical Medicine and Hygiene: An Informal Survey." *Medical History* 36 (1992): 70–86.

Clarke, Imogen, and James Mussell. "Conservative Attitudes to Old-Established Organs: Oliver Lodge and the *Philosophical Magazine*." *Notes and Records of the Royal Society* 69 (2015): 321–36.

Connor, Jennifer J. "Folklore in Anglo-American Medical Journals, 1845–1897." *Canadian Folklore / Folklore Canadien* 7 (1985): 35–53.

Cook, Bernard A. "Agriculture." In *Victorian Periodicals and Victorian Society*, edited by J. Don Vann and Rosemary T. VanArsdel, 235–48. Toronto: University of Toronto Press, 1995.

Corsi, Pietro. "What Do You Mean by a Periodical? Forms and Functions." *Notes and Records of the Royal Society* 70 (2016): 325–41.

Costa, Shelley. "The *Ladies' Diary*: Gender, Mathematics and Civil Society in Early Eighteenth-Century England." *Osiris* 17 (2002): 49–73.

Crilly, Tony. "The *Cambridge Mathematical Journal* and Its Descendants: The Linchpin of a Research Community in the Early and Mid-Victorian Age." *Historia Mathematica* 31 (2004): 455–97.

Croarken, Mary. "Human Computers in Eighteenth- and Nineteenth-Century Britain." In

The Oxford Handbook of the History of Mathematics, edited by Eleanor Robson and Jacqueline Stedahll, 375–403. Oxford: Oxford University Press, 2009.

——. "Tabulating the Heavens: Computing the *Nautical Almanac* in 18th-Century England." *IEEE Annals of the History of Computing* 25 (2003): 48–61.

Csiszar, Alex. "Broken Pieces of Fact: The Scientific Periodical and the Politics of Search in Nineteenth-Century France and Britain." PhD diss., Harvard University, 2010.

——. "How Lives Became Lists and Scientific Papers Became Data: Cataloguing Authorship during the Nineteenth Century." *British Journal for the History of Science* 50 (2017): 23–60.

——. *The Scientific Journal: Authorship and the Politics of Knowledge in the Nineteenth Century*. Chicago: University of Chicago Press, 2018.

——. "Seriality and the Search for Order: Scientific Print and Its Problems during the Late Nineteenth Century." *History of Science* 48 (2010): 399–434.

——. "Troubled from the Beginning." *Nature* 532 (2016): 306–08.

Daly, Ann. "The *Dublin Medical Press* and Medical Authority in Ireland 1850–1890." PhD diss., National University of Ireland, Maynooth, 2008.

Dawson, Gowan. "Palaeontology in Parts: Richard Owen, William John Broderip, and the Serialization of Science in Early Victorian Britain." *Isis* 103 (2012): 637–67.

——. *Show Me the Bone: Reconstructing Prehistoric Monsters in Nineteenth-Century Britain and America*. Chicago: University of Chicago Press, 2016.

Dawson, Gowan, Chris Lintott, and Sally Shuttleworth. "Constructing Scientific Communities: Citizen Science in the Nineteenth and Twenty-First Centuries." *Journal of Victorian Culture* 20 (2015): 246–54.

Dear, Peter, ed. *The Literary Structure of Scientific Argument: Historical Studies*. Philadelphia: University of Pennsylvania Press, 1991.

Desmond, Adrian. *The Politics of Evolution: Morphology, Medicine, and Reform in Radical London*. Chicago: University of Chicago Press, 1989.

Desmond, Ray. *A Celebration of Flowers: Two Hundred Years of Curtis's Botanical Magazine*. Kew, UK: Royal Botanic Gardens, in association with Collingridge, 1987.

——. "Loudon and Nineteenth-Century Horticultural Journalism." In *John Claudius Loudon and the Early Nineteenth Century in Great Britain,* edited by Elisabeth B. MacDougall, 77–97. Washington: Dumbarton Oaks Trustees for Harvard University, 1980.

Despeaux, Sloan Evans. "Fit to Print? Referee Reports on Mathematics for the Nineteenth-Century Journals of the Royal Society of London." *Notes and Records of the Royal Society* 65 (2011): 233–52.

——. "International Mathematical Contributions to British Scientific Journals, 1800–1900." In *Mathematics Unbound: The Evolution of an International Mathematical Research Community*, edited by Karen Hunger Parshall and Adrian C. Rice, 61–88. Providence, RI: American Mathematical Society and London Mathematical Society, 2002.

——. "Launching Mathematical Research without a Formal Mandate: The Role of University-Affiliated Journals in Britain, 1837–1870." *Historia Mathematica* 34 (2007): 89–106.

——. "Mathematical Questions: A Convergence of Mathematical Practices in British Journals of the Eighteenth and Nineteenth Centuries." *Revue d'histoire des mathématiques* 20 (2014): 5–71.

———. "A Voice for Mathematics: Victorian Mathematical Journals and Societies." In *Mathematics in Victorian Britain*, edited by Raymond Flood, Adrian Rice, and Robin Wilson, 155–74. Oxford, UK: Oxford University Press, 2011.

Dewis, Sarah. *The Loudons and the Gardening Press: A Victorian Cultural Industry*. Farnham, UK, and Burlington, VT: Ashgate, 2014.

"The Economist," 1843–1943: A Centenary Volume. London: Oxford University Press, 1943.

Edwards, Marcia A. "The Library and Scientific Publications of the Zoological Society of London: Part II." In *The Zoological Society of London, 1826–1976 and Beyond*, edited by Solly Zuckermann, 253–67. London: Academic Press, 1976.

Ellegård, Alvar. *Darwin and the General Reader: The Reception of Darwin's Theory of Evolution in the British Periodical Press, 1859–72*. 2nd ed. Chicago: University of Chicago Press, 1990 [1958].

———. "The Readership of the Periodical Press in Mid-Victorian Britain." *Göteborgs Universitets Årsskrift* 63 (1957): 1–41.

English, Mary P. *Mordecai Cubitt Cooke: Victorian Naturalist, Mycologist, Teacher & Eccentric*. Bristol, UK: Biopress, 1987.

———. "Robert Hardwicke 1822–75: Publisher of Biological and Medical Books." *Annals of Natural History* 13 (1986): 25–37.

Fee, Elizabeth. "Science and the 'Woman Question,' 1860–1920: A Study of English Scientific Periodicals." PhD diss., Princeton University, 1978.

Fish, R. "The Library and Scientific Publications of the Zoological Society of London: Part I." In *The Zoological Society of London, 1826–1976 and Beyond*, edited by Solly Zuckermann, 233–52. London: Academic Press, 1976.

Frampton, Sally. "The Medical Press and Its Public." In *The Edinburgh History of the British and Irish Press*, vol. 2, *Expansion and Evolution, 1800–1900,* edited by David Finkelstein. Edinburgh: Edinburgh University Press, forthcoming 2020.

Frampton, Sally, and Jennifer Wallis. "Reading Medicine and Health in Periodicals." *Media History* 25 (2019): 1–5.

Fussell, G. E. "Early Farming Journals." *Economic History Review* 3 (1932): 417–22.

Fyfe, Aileen. "Journals, Learned Societies and Money: *Philosophical Transactions* ca. 1750–1900." *Notes and Records of the Royal Society* 69 (2015): 277–99.

———. "Journals and Periodicals." In *A Companion to the History of Science*, edited by Bernard Lightman, 387–99. Chichester, UK: Wiley Blackwell, 2016.

Fyfe, Aileen, Julie McDougall-Waters, and Noah Moxham. "350 Years of Scientific Periodicals." *Notes and Records of the Royal Society* 69 (2015): 227–39.

Fyfe, Aileen, and Noah Moxham. "Making Public ahead of Print: Meetings and Publications at the Royal Society, 1752–1892." *Notes and Records of the Royal Society* 70 (2016): 361–79.

Garrison, Fielding H. "The Medical and Scientific Periodicals of the 17th and 18th Centuries: With a Revised Catalogue and Checklist." *Bulletin of the Institute of the History of Medicine* 2 (1934): 285–343.

Gascoigne, R. M. *A Historical Catalogue of Scientific Periodicals, 1665–1900: With a Survey of their Development*. New York: Garland, 1985.

Goddard, Nicholas. "The Development and Influence of Agricultural Periodicals and News-papers, 1780–1880." *Agricultural History Review* 31 (1983): 116–31.

Gross, Alan G., Joseph E. Harmon, and Michael S. Reidy. *Communicating Science: The Scientific Article from the Seventeenth Century to the Present*. New York: Oxford University Press, 2002.

Harrison, Brian. "'A World of Which We Had No Conception': Liberalism and the English Temperance Press: 1830–1872." *Victorian Studies* 13 (1969): 125–58.

Hart, Julian Tudor. "The *British Medical Journal*, General Practitioners and the State, 1840–1990." In *Medical Journals and Medical Knowledge: Historical Essays*, edited by W. F. Bynum, Stephen Lock, and Roy Porter, 228–47. London and New York: Routledge, 1992.

Hemsley, W. Botting. *A New and Complete Index to the Botanical Magazine from Its Commencement in 1787 to the End of 1904, Including the First, Second, and Third Series; to Which Is Prefixed a History of the Magazine*. London: Lovell Reeve & Co., 1906.

Henson, Louise, Geoffrey Cantor, Gowan Dawson, Richard Noakes, Sally Shuttleworth, and Jonathan R. Topham, eds. *Culture and Science in the Nineteenth-Century Media*. Aldershot, UK: Ashgate, 2004.

Hinton, D. A. "Popular Science in England, 1830–1870." PhD diss., University of Bath, 1979.

Hopwood, Nick, Simon Schaffer, and James A. Secord. "Seriality and Scientific Objects in the Nineteenth Century." *History of Science* 48 (2010): 251–85.

Hostettler, John. *Thomas Wakley: An Improbable Radical*. Chichester, UK: Barry Rose Law Publishers, 1993.

Jackson, Derek, and Brian Launder. "Osborne Reynolds and the Publication of his Papers on Turbulent Flow." *Annual Review of Fluid Mechanics* 39 (2007): 19–35.

Katzen, May F. "The Changing Appearance of Research Journals in Science and Technology: An Analysis and a Case Study." In *The Development of Science Publishing in Europe*, edited by A. J. Meadows, 177–214. Amsterdam: Elsevier, 1980.

Kjærgaard, Peter C. "'Within the Bounds of Science': Redirecting Controversies to *Nature*." In *Culture and Science in the Nineteenth-Century Media*, edited by Louise Henson, Geoffrey Cantor, Gowan Dawson, Richard Noakes, Sally Shuttleworth, and Jonathan R. Topham, 211–21. Aldershot, UK: Ashgate, 2004.

Knight, David. *Natural Science Books in English, 1600–1900*. Reprint. London: Portman Books, 1989 [1972].

———. "Science and Culture in Mid-Victorian Britain: The Reviews and William Crookes' 'Quarterly Journal of Science.'" *Nuncius* 11 (1996): 43–54.

Kronick, David A. *"Devant la Deluge" and Other Essays on Early Modern Scientific Communication*. Lanham, MD: Scarecrow Press, 2004.

———. "The Fielding H. Garrison List of Medical and Scientific Periodicals of the 17th and 18th Centuries; Addenda et Corrigenda." *Bulletin of the History of Medicine* 32 (1958): 456–74.

———. *A History of Scientific and Technical Periodicals: The Origin and Development of the*

Scientific and Technological Press, 1665–1790. 2nd ed. Metuchen, NJ: Scarecrow Press, 1976 [1962].

———. "Medical 'Publishing Societies' in Eighteenth-Century Britain." *Bulletin of the Medical Library Association* 82 (1994): 277–82.

———. *Scientific and Technical Periodicals of the Seventeenth and Eighteenth Centuries: A Guide*. Metuchen, NJ: Scarecrow Press, 1991.

Lancashire, Julie Ann. "The Popularisation of Science in General Science Periodicals in Britain, 1890–1939." PhD diss., University of Kent at Canterbury, 1988.

Lee, Charles E. "John Herepath and the Birth of Railway Journals." *Journal of the Railway and Canal Historical Society* 15 (1969): 1–7.

LeFanu, William R. "British Periodicals of Medicine: A Chronological List. Part 1: 1684–1899." *Bulletin of the Institute of the History of Medicine* 5 (1937): 735–61, 827–55.

———. *British Periodicals of Medicine: A Chronological List, 1640–1899*, revised ed., edited by Jean Loudon. Oxford, UK: Wellcome Unit for the History of Medicine, 1984.

Lightman, Bernard. "*Knowledge* Confronts *Nature*: Richard Proctor and Popular Science Periodicals." In *Culture and Science in the Nineteenth-Century Media*, edited by Louise Henson, Geoffrey Cantor, Gowan Dawson, Richard Noakes, Sally Shuttleworth, and Jonathan R. Topham, 199–210. Aldershot, UK: Ashgate, 2004.

———. "The Mid-Victorian Period and the *Astronomical Register* (1863–1886): 'A Medium of Communication for Amateurs and Others.'" *Public Understanding of Science* 27 (2018): 629–36.

———. "Popularizers, Participation and the Transformations of Nineteenth-Century Publishing: From the 1860s to the 1880s." *Notes and Records of the Royal Society* 70 (2016): 343–59.

———. *Victorian Popularizers of Science: Designing Nature for New Audiences*. Chicago: University of Chicago Press, 2007.

Lilley, S. "'Nicholson's Journal,' 1797–1813." *Annals of Science* 6 (1948–50): 78–101.

Loudon, Jean, and Irvine Loudon. "Medicine, Politics, and the Medical Periodical, 1800–50." In *Medical Journals and Medical Knowledge: Historical Essays*, edited by W. F. Bynum, Stephen Lock, and Roy Porter, 49–69. London and New York: Routledge, 1992.

MacLeod, Roy, et al. "Centenary Supplement." *Nature* 224 (1969): 417–76.

Madden, Lionel. *The Nineteenth-Century Periodical Press in Britain: A Bibliography of Modern Studies 1901–1971*. New York and London: Garland, 1976.

Manten, A. A. "The Development of European Scientific Journal Publishing before 1850." In *The Development of Science Publishing in Europe*, edited by A. J. Meadows, 1–22. Amsterdam: Elsevier, 1980.

Manzer, Bruce M. *The Abstract Journal, 1792–1920: Origin, Development and Diffusion*. Metuchen, NJ: Scarecrow Press, 1977.

Meadows, A. J. "Access to the Results of Scientific Research: Developments in Victorian Britain." In *The Development of Science Publishing in Europe*, edited by A. J. Meadows, 43–62. Amsterdam: Elsevier, 1980.

———. *Communication in Science*. London: Butterworths, 1974.

———. *Science and Controversy: A Biography of Sir Norman Lockyer, Founder of "Nature."* London: Macmillan, 1972.

Meadows, A. J., ed. *The Development of Science Publishing in Europe.* Amsterdam: Elsevier, 1980.

Meinel, Christoph. "Structural Changes in International Scientific Communication: The Case of Chemistry." In *Atti del V Convegno Nazionale di Storia e Fondamenti della Chimica*, 47–61. Perugia: Accademia Nazionale delle Scienze, 1993.

Mold, Alex, and Virginia Berridge. "Using Digitised Medical Journals in a Cross European Project on Addiction History." *Media History* 25 (2019): 85–99.

Morton, Leslie T. "The Growth of Medical Periodical Literature." In *Thornton's Medical Books, Libraries, and Collectors*, edited by Alain Besson, 221–38. 3rd rev. ed. Aldershot, UK: Gower, 1990.

Moss, David J., and Chris Hosgood. "The Financial and Trade Press." In *Victorian Periodicals and Victorian Society,* edited by J. Don Vann and Rosemary T. VanArsdel, 199–218. Toronto: University of Toronto Press, 1994.

Moulds, Alison. "The 'Medical-Women Question' and the Multivocality of the Victorian Medical Press, 1869–1900." *Media History* 25 (2019): 6–22.

Moxham, Noah, and Aileen Fyfe. "The Royal Society and the Prehistory of Peer Review, 1665–1965." *Historical Journal* 61 (2018): 863–89.

Mussell, James. "Arthur Cowper Ranyward, *Knowledge* and the Reproduction of Astronomical Photographs in the Later Nineteenth-Century Periodical Press." *British Journal for the History of Science* 42 (2009): 345–80.

———. "Bug-Hunting Editors: Competing Interpretations of Nature in Late Nineteenth-Century Natural History Periodicals." In *(Re)creating Science in Nineteenth-Century Britain: An Interdisciplinary Approach*, edited by Amanda Mordavsky Caleb, 81–96. Newcastle, UK: Cambridge Scholars Publishing, 2007.

———. *The Nineteenth-Century Press in the Digital Age.* Basingstoke, UK: Palgrave Macmillan, 2012.

———. "Science and Journalism" and "Science Popularization." In *Dictionary of Nineteenth-Century Journalism in Great Britain and Ireland*, edited by Laurel Brake and Marysa Demoor, 559–60 and 560–61. Ghent: Academia Press, and London: British Library, 2009.

———. *Science, Time, and Space in the Late Nineteenth-Century Periodical Press: Movable Types.* Aldershot, UK: Ashgate, 2007.

———. "Science and the Timeliness of Reproduced Photographs in the Late Nineteenth-Century Periodical Press." In *The Lure of Illustration*, edited by Laurel Brake and Marysa Demoor, 203–19. Basingstoke, UK: Palgrave, 2009.

———. " 'This Is Ours and for Us:' The *Mechanic's Magazine* and Low Scientific Culture in Regency London." In *Repositioning Victorian Sciences*, edited by David Clifford, Elisabeth Wadge, Alex Warwick, and Martin Willis, 107–18. London: Anthem Press, 2006.

Niessen, Olwen C. "Temperance." In *Victorian Periodicals and Victorian Society,* edited by J. Don Vann and Rosemary T. VanArsdel, 251–77. Toronto: University of Toronto Press, 1994.

Neylon, Cameron. "Communities Need Journals." *Notes and Records of the Royal Society* 70 (2016): 383–85.

North, John S., ed. *The Waterloo Directory of English Newspapers and Periodicals, 1800–1900*, series 1 and 2. 20 vols. Waterloo, ON: North Waterloo Academic Press, 1997–.

———. *The Waterloo Directory of Irish Newspapers and Periodicals, 1800–1900*. Waterloo, ON: North Waterloo Academic Press, 1986.

———. *The Waterloo Directory of Scottish Newspapers and Periodicals, 1800–1900*. 2 vols. Waterloo, ON: North Waterloo Academic Press, 1989.

Palmer, John E. C. "Railway Periodicals of the Nineteenth Century Published in the British Isles: A Bibliographical Guide." Postgraduate diploma diss., University of London, 1959.

Palmer, John E. C., and Harold W. Paar. "Transport." In *Victorian Periodicals and Victorian Society*, edited by J. Don Vann and Rosemary T. VanArsdel, 179–98. Toronto: University of Toronto Press, 1994.

Peterson, M. Jeanne. "Specialist Journals and Professional Rivalries in Victorian Medicine." *Victorian Periodicals Review* 12 (1979): 25–32.

———. "Medicine." In *Victorian Periodicals and Victorian Society*, edited by J. Don Vann and Rosemary T. VanArsdel, 22–44. Toronto: University of Toronto Press, 1994.

Pieffer, Jeanne, Maria Conforti, and Patrizia Delpiano. Introduction to special issue "Scholarly Journals in Early Modern Europe: Communication and the Construction of Knowledge." *Archives internationales d'histoire des sciences* 63 (2013): 5–24.

Pladek, Brittany. "'A Variety of Tastes': The *Lancet* in the Early-Nineteenth-Century Periodical Press." *Bulletin of the History of Medicine* 85 (2011): 560–86.

Porter, Roy. "The Rise of Medical Journalism in Britain to 1800." In *Medical Journals and Medical Knowledge: Historical Essays*, edited by W. F. Bynum, Stephen Lock, and Roy Porter, 6–28. London and New York: Routledge, 1992.

Potts, Jason, John Hartley, Lucy Montgomery, Cameron Neylon, and Ellie Rennie. "A Journal Is a Club: A New Economic Model for Scholarly Publishing." *Prometheus* 35 (2017): 75–92.

Price, Derek J. de Solla. *Little Science, Big Science*. New York and London: Columbia University Press, 1963.

———. "Networks of Scientific Papers." *Science* 149 (1965): 510–15.

Rayward, W. Boyd. "The Search for Subject Access to the Catalogue of Scientific Papers, 1800–1900." In *The Variety of Librarianship: Essays in Honour of John Wallace Metcalfe*, edited by W. Boyd Rayward, 146–70. Sydney: Library Association of Australia, 1976.

Richardson, Ruth. "'Notorious Abominations': Architecture and the Public Health in *The Builder*, 1843–83." In *Medical Journals and Medical Knowledge: Historical Essays*, edited by W. F. Bynum, Stephen Lock, and Roy Porter, 90–107. London and New York: Routledge, 1992.

Richardson, Ruth, ed. *Vintage Papers from the "Lancet."* Edinburgh: Elsevier, 2006.

Richardson, Ruth, and Robert Thorne. "Architecture." In *Victorian Periodicals and Victorian Society*, edited by J. Don Vann and Rosemary T. VanArsdel, 45–61. Toronto: University of Toronto Press, 1994.

———. *The "Builder:" Illustrations Index*. London: Builder Group and Hutton, 1994.

Rogal, Samuel J. "A Checklist of Medical Journals Published in England during the Seventeenth, Eighteenth, and Nineteenth Centuries." *British Studies Monitor* 9 (1980): 3–25.

Roos, David A. "The Aims and Intentions of *Nature*." In *Victorian Science and Victorian Values*, edited by James Paradis and Thomas Postlewait, 159–80. New York: New York Academy of Sciences, 1981.

Rowlette, Robert J. *The Medical Press and Circular, 1839–1939: A Hundred Years in the Life of a Medical Journal*. London: Medical Press and Circular, 1939.

Rudwick, Martin J. S. "The Emergence of a Visual Language for Geological Science 1760–1840." *History of Science* 14 (1976): 149–95.

———. "Historical Origins of the Geological Society's *Journal*." In *Milestones in Geology: Reviews to Celebrate 150 Volumes of the "Journal of the Geological Society,"* edited by M. J. Le Bas, 3–6. London: Geological Society, 1995.

Ruth, Jennifer. "'Gross Humbug' or 'The Language of Truth'? The Case of the *Zoist*." *Victorian Periodicals Review* 32 (1999): 299–323.

Scudder, Samuel H. *Catalogue of Scientific Serials of All Countries Including the Transactions of Learned Societies in the Natural, Physical and Mathematical Sciences, 1633–1876*. Reprint. New York: Kraus Reprint Corp., 1965 [1879].

Secord, James A. "Knowledge in Transit." *Isis* 95 (2004): 654–72.

———. "Science, Technology, and Mathematics." In *The History of the Book in Britain*, vol. 6, *1830–1914*, edited by David McKitterick, 443–74. Cambridge: Cambridge University Press, 2009.

———. *Victorian Sensation: The Extraordinary Publication, Reception, and Secret Authorship of "Vestiges of the Natural History of Creation."* Chicago: University of Chicago Press, 2000.

———. *Visions of Science: Books and Readers at the Dawn of the Victorian Age*. Oxford, UK: Oxford University Press, 2014.

Shaw, Jean. "Patterns of Journal Publication in Scientific Natural History from 1800 to 1939." In *The Development of Science Publishing in Europe*, edited by A. J. Meadows, 149–76. Amsterdam: Elsevier, 1980.

Sheets-Pyenson, Susan. "Darwin's Data: His Reading of Natural History Journals, 1837–1842." *Journal of the History of Biology* 14 (1981): 231–48.

———. "From the North to Red Lion Court: The Creation and Early Years of the *Annals of Natural History*." *Archives of Natural History* 10 (1981): 221–49.

———. "A Measure of Success: The Publication of Natural History Journals in Early Victorian Britain." *Publishing History* 9 (1981): 21–36.

———. "Popular Science Periodicals in Paris and London: The Emergence of a Low Scientific Culture, 1820–1875." *Annals of Science* 42 (1985): 549–72.

Shepherd, Michael. "Psychiatric Journals and the Evolution of Psychological Medicine." In *Medical Journals and Medical Knowledge: Historical Essays*, edited by W. F. Bynum, Stephen Lock, and Roy Porter, 188–206. London and New York: Routledge, 1992.

Shuttleworth, Sally, and Berris Charnley. "Science Periodicals in the Nineteenth and Twenty-First Centuries." *Notes and Records of the Royal Society* 70 (2016): 297–304.

Shuttleworth, Sally, and Berris Charnley, eds. "Science Periodicals in the Nineteenth and

Twenty-First Centuries." Special issue. *Notes and Records of the Royal Society* 70 (2016): 297–404.

Smith, S. D. "Coffee, Microscopy, and the *Lancet's* Analytical Sanitary Commission." *Social History of Medicine* 14 (2001): 171–97.

Sprigge, Squire. *The Life and Times of Thomas Wakley.* London: Longman, Green & Co., 1897.

[Sprigge, Squire, ed.] "The Centenary of the *Lancet*." Special issue. *Lancet* 202 (1923): 685–764.

Strange, Philip. "Two Electrical Periodicals: The *Electrician* and the *Electrical Review*, 1880–1890." *IEE Proceedings* 132A (1985): 574–81.

Taunton, Matthew. "Mining Press." In *Dictionary of Nineteenth-Century Journalism in Great Britain and Ireland*, edited by Laurel Brake and Marysa Demoor, 412. Ghent: Academia Press, and London: British Library, 2009.

Thornton, John L. and R. I. J. Tully. "The Growth of Scientific Periodical Literature." In *Scientific Books, Libraries and Collectors: A Study of Bibliography and the Book Trade in Relation to the History of Science*, 277–93. 3rd rev. ed. London: Library Association, 1971.

Topham, Jonathan R. "Anthologizing the Book of Nature: The Circulation of Knowledge and the Origins of the Scientific Journal in Late Georgian Britain." In *The Circulation of Knowledge between Britain, India and China: The Early-Modern World to the Twentieth Century*, edited by Bernard Lightman, Gordon McOuat, and Larry Stewart, 119–52. Leiden and Boston: Brill, 2013.

———. "John Limbird, Thomas Byerley, and the Production of Cheap Periodicals in the 1820s." *Book History* 8 (2005): 75–106.

———. "Periodicals." In "'An Infinite Variety of Arguments': The *Bridgewater Treatises* and British Natural Theology in the 1830s," 553–615. PhD diss., University of Lancaster, 1993.

———. "Science, Mathematics, and Medicine." In *The History of Oxford University Press: Volume II: 1780 to 1896*, edited by Simon Eliot, 513–57. Oxford, UK: Oxford University Press, 2013.

———. "The Scientific, the Literary and the Popular: Commerce and the Reimagining of the Scientific Journal in Britain, 1813–1825." *Notes and Records of the Royal Society* 70 (2016): 305–24.

———. "Scientific and Medical Books, 1780–1830." In *The Cambridge History of the Book in Britain*, Volume 5, *1695–1830*, edited by Michael Turner and Michael Suarez, 827–33. Cambridge: Cambridge University Press, 2009.

———. "Scientific Publishing and the Reading of Science in Nineteenth-Century Britain: A Historiographical Survey and Guide to Sources." *Studies in History and Philosophy of Science* 31A (2000): 559–612.

———. "Technicians of Print and the Making of Natural Knowledge." *Studies in History and Philosophy of Science* 35 (2004): 391–400.

Tucker, Albert. "Military." In *Victorian Periodicals and Victorian Society*, edited by J. Don Vann and Rosemary T. VanArsdel, 62–80. Toronto: University of Toronto Press, 1994.

Uffelman, Larry K. *The Nineteenth-Century Periodical Press in Britain: A Bibliography of Modern Studies, 1972–1987.* Edwardsville: Southern Illinois University, 1992.

Vadillo, Ana Parejo. "Transport Press," In *Dictionary of Nineteenth-Century Journalism in Great Britain and Ireland*, edited by Laurel Brake and Marysa Demoor, 639–40. Ghent: Academia Press, and London: British Library, 2009.

Vann, J. Don, and Rosemary T. VanArsdel, eds. *Victorian Periodicals: A Guide to Research*. 2 vols. New York: Modern Language Association of America, 1978–89.

———. *Victorian Periodicals and Victorian Society*. Toronto: University of Toronto Press, 1994.

Wale, Matthew. "The Sympathy of a Crowd: Periodicals and the Practices of Natural History in Nineteenth-Century Britain." PhD diss., University of Leicester, 2018.

Ward, William S. *British Periodicals and Newspapers, 1789–1832: A Bibliography of Secondary Sources*. Lexington: University Press of Kentucky, 1973.

Watts, Iain. "'We Want No Authors': William Nicholson and the Contested Role of the Scientific Journal in Britain, 1797–1813." *British Journal for the History of Science* 47 (2014): 397–419.

Wolff, Michael, John S. North, and Dorothy Deering, *The Waterloo Directory of Victorian Periodicals, 1824–1900, Phase I*. Waterloo, ON: University of Waterloo, [1970].

Wolmar, Christian. "Railway Press." In *Dictionary of Nineteenth-Century Journalism in Great Britain and Ireland*, edited by Laurel Brake and Marysa Demoor, 527. Ghent: Academia Press, and London: British Library, 2009.

Womack, Elizabeth Coggin. "Window Gardening and the Regulation of the Home in Victorian Periodicals." *Victorian Periodicals Review* 51 (2018): 269–88.

Wood, Karen. "Making and Circulating Knowledge through Sir William Hamilton's *Campi Phlegraei*." *British Journal for the History of Science* 39 (2006): 67–96.

Young, Robert M. "Natural Theology, Victorian Periodicals, and the Fragmentation of the Common Context." In *Darwin's Metaphor: Nature's Place in Victorian Culture*, 126–63. Cambridge: Cambridge University Press, 1985.

Zuckermann, Harriet, and Robert K. Merton. "Patterns of Evaluation in Science: Institutionalization, Structure and Functions of the Referee System." *Minerva* 9 (1971): 66–100.

CONTRIBUTORS

Geoffrey Belknap, National Science and Media Museum, Bradford, West Yorkshire, BD1 1NQ, United Kingdom

Alex Csiszar, Department of the History of Science, Harvard University, 1 Oxford Street, Cambridge, MA 02139, USA

Gowan Dawson, School of Arts, University of Leicester, University Road, Leicester, LE1 7RH, United Kingdom

Sally Frampton, The Oxford Research Centre for the Humanities (TORCH), Radcliffe Observatory Quarter, University of Oxford, Woodstock Road, Oxford, OX2 6GG, United Kingdom

Graeme Gooday, School of Philosophy, Religion, and History of Science, University of Leeds, Woodhouse Lane, Leeds, LS2 9JT, United Kingdom

Bernard Lightman, Humanities Department, York University, 4700 Keele Street, Toronto, Ontario, M3J 1P3, Canada

Sally Shuttleworth, St. Anne's College, Oxford, OX2 6HS, United Kingdom

Jon Topham, School of Philosophy, Religion, and History of Science, University of Leeds, Leeds, LS2 9JT, United Kingdom

Matthew Wale, School of Arts, University of Leicester, University Road, Leicester, LE1 7RH, United Kingdom

INDEX